Transgovernance

Louis Meuleman

Editor

Transgovernance

Advancing Sustainability Governance

 Springer

Editor
Louis Meuleman
PublicStrategy
Brussels, Belgium

University of Massachussets
Boston, USA

Free University
Amsterdam, The Netherlands

This book is a result of the TransGov project of the Institute for Advanced Sustainability Studies (IASS). In November 2011 a first TransGov report was presented: 'Transgovernance: The Quest for Governance of Sustainable Development', authored by Roeland J. in 't Veld. This report is downloadable as an open source publication at www.iass-potsdam.de.

ISBN 978-3-642-28008-5   ISBN 978-3-642-28009-2 (eBook)
DOI 10.1007/978-3-642-28009-2
Springer Heidelberg New York Dordrecht London

Library of Congress Control Number: 2012943839

Printed on acid-free paper

Springer is part of Springer Science+Business Media (www.springer.com)

Foreword

Klaus Töpfer

We live, as we always have, in turbulent times. Social systems are reflexive in nature and can and will change pathways step by step, be it over time or instantly and abruptly. These changes may occur because of learning processes in a society or in the political culture or may result from changing power structures. These processes may be smooth and incremental or disruptive and powerful.

Two main changes have altered this already challenging social fabric of the twentieth century. First, we live in the era of the anthropocene (Paul Crutzen). This means: Humankind has become a quasi-planetary force, as the first symposium of Nobel laureates organised in Potsdam, Germany, stated in its declaration in 2007. This immensely productive first symposium has also provided the seed for the Institute for Advanced Sustainability Studies (IASS).

Second, changes at a planetary level take place increasingly often, ever faster than before and with increasing amplitude. Indeed, it is becoming more difficult to differentiate between changes and disasters caused by natural forces and man-made catastrophes, as the reasons are overlapping. The terminology of a 'great acceleration' is no longer reserved for special moments in human history. This situation seems to be a companion of our times.

The main question we are confronted with is as follows: Will we be able to respond to these challenges effectively? Even more fundamentally: Are we able to understand the driving forces, and are we in a situation to reduce the complexity of these planetary interdependencies as a precondition for concrete and targeted policy making? Do we make sense of what we witness in the reality of life or is it just happening to us? Will it be possible to transform the wealth of knowledge available into actions and will we be able to take full advantage of the breath of engaged citizens? Do we have to complement the acceleration of changes we are facing nowadays with an acceleration of ideas and solutions as well as ever larger systematic and holistic changes? Or do we have the chance to reduce the complexity of change in order to realise a piecemeal engineering procedure with the chance to react to new insights and knowledge and to new or changing values in an open democratic society?

At the same time: What can we learn from the puzzling fact that some twenty years of sustainability governance – the overarching frame for our considerations – have not led us very far, to say it politely? Should we consider slowing down our actions and reactions and taking time to reflect in order to move forward more effectively? Are we aware of the recommendation formulated by the Spanish philosopher Balthasar Gracian in the sixteenth century: *'The most difficult part in running is to stand still'*. Are we running too fast whilst not sufficiently questioning the direction in which we are running?

The foundation of the IASS and its cluster 'Global Contract for Sustainability' exists to address questions like these. As a consequence, the TransGov project was started in the summer of 2010 as the first fully fledged research project of IASS. Its aim is ambitious and manifold: First, to bring together new and existing ideas on governance for sustainable development and to develop new, that is, 'advanced' insights from them. Second, to provide a platform for exchange for scientists, including four research fellows forming the forerunners at the IASS literally, and practitioners. Third, in addressing 'Science for Sustainable Transformations: Towards Effective Governance', TransGov has laid out a conceptional matrix for further projects at the IASS in order to find effective ways for science-society collaborations. This book presents the work done by the TransGov project team. It complements the project synthesis report, written by Roeland in 't Veld, which was launched in November 2011.

We are facing enormous environmental, social and economic challenges as well as opportunities at all levels. These are often not identified early enough, not analysed deeply enough or not systematically integrated into actions. The problems are interconnected, but the levels at which solutions may occur are also linked. People in modern societies are increasingly concerned that they are living in a *'Nebenfolgengesellschaft'*. The fact that science and technology are constantly cultivating deeper insights into the construction patterns of nature and life means that there are far-reaching consequences both in time and space which are not adequately considered. For instance, there is a suspicion that the economic increase measured via the GNP is mainly due to overcoming the previous negative consequences of the growth.

In the year of the UN Conference on Sustainable Development in Rio de Janeiro in 2012, it remains essential to address a huge implementation gap with regard to agreed-upon goals and targets. One approach is to define a new set of goals and targets, which fit the purpose and are better than those implemented 20 years ago. These considerations lead to the proposal to work out in Rio+20 additional Millennium *Sustainable* Development Goals (MSDGs), correcting the failure to concentrate at the UN Millennium Assembly on millennium goals more or less globalising the 'Western way of development' to the developing countries as well. The integration of the sustainable component in the MSDGs would put forward rights and obligations both for developed and developing countries to a culturally diversified 'development'.

Another way of addressing the gap between knowledge and action, or between words on paper – constituting numerous declarations and Calls for Action – and practice, is to define new approaches such as green economy or, even more important and challenging, a green society. The challenge is to design new institutional arrangements for governing sustainable development, changing technology and behaviour, and asking for efficiency as intensively as for sufficiency. These are just two of the areas which this book addresses through its individual chapters.

One of the main building blocks of the TransGov project has been the concept of 'knowledge democracy' (in 't Veld 2010), which addresses these changes and new dimensions, providing, for example, a better understanding as to why different traditional ways of developing solutions are frequently not suited to the problem for which they were created. Complementing this with Ulrich Beck's approach regarding Second Modernity reveals that classic institutions and approaches will not just disappear but will coexist with new forms. How to handle such transformation processes within the conditions of open democratic societies concerns me a great deal.

Sense-making mechanisms and chains in economy and in modern technologies as well as scientific findings are increasingly global in nature. At the same time – and this is something I have been following for years from a distance and from 'within' – a renaissance of 'culture' or 'traditions' can be observed. As a consequence, rigidity in thinking and acting, for example favouring one-dimensional concepts instead of accepting if not appreciating diversity, will certainly not succeed in bringing us closer to sustainable societies. This is one of the main messages this book explores from different angles.

We must change course significantly and transform practices across different sectors of society, as clearly stated by the 2011 report of the German Advisory Council on Global Change (WBGU). With this said, the question still remains: How do we think and initiate such transformations? TransGov makes the case that many transformation processes will have to occur, more often than not, simultaneously, partially overlapping, at different places at the same time and exercised by people who are multiply engaged in different forums, roles and levels. It also helps us to understand why 'intraventions' in many cases work better than interventions. Hence, 'the global' does not take centre stage at TransGov in order to tackle large-scale problems successfully. For example, the emergence of new and powerful citizens' initiatives comes to mind. The 'Stuttgart 21' case in Germany kept us busy thinking throughout the implementation of the project. Participation of the general public, as integrated in modern regional planning and building legislation, is no longer able to stabilise the peace-making function of legally based processes. Processes leading to a transformation of the German energy system, the so-called 'Energiewende', after Japan's nuclear disaster of Fukushima at the beginning of 2011, are another case in point. This has resulted in a call for a 'Gemeinschaftswerk', a common effort. In times of knowledge democracies, it is less of an issue whether or not citizens are allowed to participate and to raise concerns. Their active engagement, namely intraventions, in domains until now covered by governmental actors becomes a necessary condition for effective governance towards sustainable development.

If co-evolution of science and practice is meant to be not just another fancy term which refers to thinking about the science-practice interface, is it the only way to put forth successfully knowledge-based solutions towards sustainability? Answering this question positively is an easy task. To transform science and practice accordingly – that is, production of useful knowledge here and knowledge-based decision-making there – is not easy at all. However, since providing a platform or interface for science and societal interaction is the mission of the IASS, it was a logical consequence to put TransGov first, in order to reflect on such challenges in more conceptional terms in the first place.

Finally, TransGov is without any doubt the beginning rather than the end of our work on governance for sustainable development. Follow-on activities on governance research at IASS are implemented by focussing on concrete issues. For example, IASS is expanding its work on soils – almost a 'forgotten' resource despite its paramount importance – and will set up a knowledge-based monitoring process for the *'Energiewende'*. Insights from TransGov will help to design these research activities, inform knowledge exchange platforms therein and put forward recommendations concerning the 'how to' of these challenges. The cultural dimension will continue to play a major role in our work. In doing so, culture and governance will alter their roles as 'dependent' and 'independent' variables, respectively, if one wishes to phrase it this way. Topics such as short-lived climate forcers present straightforward governance challenges if one addresses their drivers and possible response options. In addition, it goes without saying that any critical assessment of climate engineering has at its core a governance challenge as well.

Sustainable development as decided upon at the UN Conference on Environment and Development in Rio de Janeiro in 1992 is more than an ecological concept. At this very time, we are confronted with a financial architecture which is far from sustainable and which is even threatening to destroy the sustainable fundament for social stability and environmental responsibility. The massive financial turbulences we are witnessing are irrefutable evidence of the fact that modern societies are living under the dictatorship of short-termism, externalising social and environmental costs due to the prices we are currently paying for goods and services. The financial disaster is nothing less than the oath of disclosure of this short-termism. It is, therefore, a must that we also think 'out of the box' with regard to reshaping the financial architecture in a way which ensures it meets the conditions for sustainability. Finally, theories of sustainable development, historical analysis and regional comparisons and reflections on transdisciplinarity more generally will continue as cross-cutting themes of the IASS and its clusters. TransGov and its findings will help shaping these research agendas. Hence, I hope that the IASS with this research project is able to present a modest but at the same time bold contribution to the discussion on how to improve governance for sustainable development – for the planet as well as for people and their places.

Prof. Dr. Dr. h. c. mult. Klaus Töpfer
Klaus Töpfer is founding director of the Institute for Advanced Sustainability Studies in Potsdam, Germany

Acknowledgements

This collection of 'think pieces' on advancing the governance of sustainable development is the result of a remarkable cross-fertilisation between scholars from different disciplines, practitioners from different levels of government and business and civil society representatives.

The first to express my gratitude to is Klaus Töpfer, who created with the Institute for Advanced Sustainability Studies a challenging arena in which new ideas emerge easily. He not only provided us with the workplace and creative space we needed, but has also been actively participating in our discussions from the very beginning. In the TransGov team, Roeland in 't Veld played a crucial role. He is the author of the first product of TransGov, the report 'Transgovernance: The Quest for Governance of Sustainable Development', and he has stimulated us with provocative questions. Also Günther Bachmann, as member of the steering group, never stopped asking uneasy questions, which stimulated out of the box thinking. The four TransGov research fellows, Stefan Jungcurt, Jamel Napolitano, Alexander Perez-Carmona and Falk Schmidt, who have each contributed a chapter in this volume, have been a pleasure and joy to work with. Although I needed to manage the project from a distance, the communication has always been easy, open and constructive. I am grateful to the staff of IASS who were essential for the project and have always been very supportive. It has also been a pleasure to work with IFOK with regard to the organisation of events and other communication issues.

I would also like to thank the participants in the challenging workshops we have organised, as well as the high-level practitioners we have interviewed, for their willingness to share their insights and for their precious time. Without Madelon Eelderink, I would not have finished the editing of this volume so fast and so thoroughly. Madelon, you have been an enormous support! Last but not least, I am grateful to all those around the TransGov team, our partners and other close friends and relatives, for having endured that all of us made the project a first priority. I suppose I am not the only one who has recently received the urgent request: "No more (book) projects!" I promise to try, Inge. . ..

Louis Meuleman
Project director TransGov

Contents

Introduction: Transgovernance

Louis Meuleman

Transgovernance: Advancing Sustainability Governance analyses the question what recent and ongoing changes in the relations between politics, science and media – together characterised as the emergence of a knowledge democracy – may imply for governance of sustainable development, on global and other levels of societal decision-making, and the other way around: How can the discussion on sustainable development contribute to a knowledge democracy?

This volume is one of the results of the IASS project 'Science for Sustainable TRANSformations: Towards Effective GOVernance' (TransGov). This was the first project in the IASS cluster 'Global Contract for Sustainability'. The overall objective was to present a new context for the sustainability governance debate, because the mainstream debate in many respects lacks diversity, variety, and reflexivity. Reframing classical questions in order to find new options for societal decision-making and identifying starting points and strategies towards effective governance of transitions to sustainability are important. Our analysis of contemporary attempts to understanding transitions towards sustainable societies took the starting point that we live in tense and turbulent times, in which simple answers to complex questions not only fail in understanding the challenges but also hinder solutions and are sometimes even counterproductive. Our aim, therefore, was not to reduce complexity in our analyses but to find ways to address and appreciate it.

As a matter of fact, many of the current arrangements for collective decision-making and action on global and other levels are not leading towards sustainable societies. Therefore, these arrangements should be open to change and may need to be rethought in a considerably new way to begin with. The failure of mainstream sustainability governance arrangements can be explained as a consequence of misconceptions, such as the belief that only centralised and legally-binding regulatory arrangements are the best option, that hegemonic thinking is preferable or at least has to be accepted over pluralist and tolerant attitudes towards other values, that there is no realistic alternative to mainstream thinking about economic growth, that science can and should always be 'objective' and indisputable, that participation of civil society and business is only a fashion or an add-on and that institutions (as the rules of the game) should be formal by nature and should lead to the creation of formal organisations. New governance features have to deal with these underlying notions that are deeply buried into prevailing governance thinking. However, as our reference to 'Second Modernity' (Ulrich Beck) makes clear, this will not be an issue of just replacing 'old' arrangements with 'new' ones, but to find ways and means to govern the relationship between 'old' and 'new' governance arrangements in a fruitful rather than conflictive manner.

This volume contains contributions from the TransGov research team members. The eight 'think pieces' take a range of different angles, such as international relations, governance and metagovernance theory, cultural diversity, economics and knowledge management. We aimed at offering insights regarding institutions and transformation processes and to dig a bit deeper in the paradigms behind contemporary sustainability governance. The chapters focus on different subjects, taking various perspectives, borrowed from many social science theories. In doing so, we tried not to hassle to new insights for the sake of translating them into solutions. However, we have put forward some new ideas and finally provide some useful recommendations as well.

For example, on one hand, international scholarly and political discussions on International Environmental Governance (IEG) and the Institutional Framework for Sustainable Development (IFSD) could profit from such a linkage of governance theory to conceptional thinking from, for example, sociology, cultural anthropology, psychology and various schools of economics. Therefore, the Rio+20 process is often used as a reference point for our considerations without dominating or restricting the analysis. On the other hand, those engaged in cultural studies may perceive this project as an entry point into sustainability studies, as TransGov pays much attention to this concern.

We tried hard to practise what we preach. If we believe that variety is important in sustainability governance and in thinking about it, variety should also be reflected in this book. Thus, while a general conceptional framework guided our thinking throughout the research process – put forward by the synthesis report written by Roeland in 't Veld, of which the summary and recommendations are presented in Chap. 8 of this volume – each chapter has its own merit and can be read separately. The chapters are presented in two parts. The first contains general reflections on the challenges of sustainability governance. The second illustrates the current discussions regarding a number of very topical themes.

Knowledge Democracy

Part 1 opens with a reflection by Roeland in 't Veld (Chap. 1) on the problems sustainable development faces in relation to the tensions he has framed as 'knowledge democracy'. The argument begins with the observation that the concept of sustainable development is all over the place, maybe because it is very broad and vague. The vagueness of the concept has a Janus face. It has been called a unifying concept because its vagueness breeds a consensus that might be utilised later on. It is an asset if it triggers action. On the other hand, if sustainable development is everything, maybe it is nothing. Although the concept may be vague, it has overwhelming appeal on political agendas, programmes and dialogues. The precautionary principle is the nucleus of a powerful moral imperative. The multidimensional nature of the concept, covering ecological, economic and social aspects of change, relates to our needs for integration. Sustainable development as a concept bears a persuasive character. Actors of all kinds may contribute to it – citizens, enterprises, NGOs, governments, etc.

Thinking about the governance of sustainable development leads us to the recognition of a multi-level, multi-scale, multi-disciplinary character of the *problematique*. Moreover, development refers to change, to transitions and transformations. Governance of sustainable development, therefore, has to cope with complex dynamics. The concept of knowledge democracy sheds new light on the emerging relationships between politics, media and science. It shows how the emergence of participatory democracy besides representative democracy, the revolutionary rise of social media besides corporate media, the emergence of transdisciplinary trajectories besides classical disciplinary science lead to explosions of complex interactions. The chapter discusses the variety of possible future variants of knowledge democracies, quiet and turbulent ones, in relation to the quest for sustainable development. The main conclusion is that strategies for sustainability may vary with the types of knowledge democracies around.

Cultural Diversity

Chapter 2 by Louis Meuleman concentrates on the crucial role of cultural diversity in sustainability governance. The cultural dimension is often considered an obstacle to sustainability, but there are good arguments to reconsider this. It is argued that many sustainability policies – at least when put into practice – deny complexity and uncertainty, favour centralised negotiations and institutions, view governments as exclusive decision-makers and imply hegemony of Western economic and political principles. Part of the mainstream language of sustainability governance is centralist and refers to monolithic concepts (*the* economy, *the* climate, *the* Earth System. . .) rather than embracing diversity and complexity. These concepts are – as of now – dominating the discourse.

This chapter aims to shed light on the problematic relations between cultural diversity, sustainable development and (meta)governance. These three concepts have a normative character, which is a good predictor of trouble as soon as they interact. It argues that the implementation deficit of sustainable development can be traced back to three problems: a neglect of the opportunities cultural diversity offers, an implicit preference for central steering, a dominance of top-down political solutions, such as the idea of a global carbon tax or a global Kyoto regime, and an underestimation of the 'wickedness' of many sustainability challenges. It is concluded that unless sustainable development became more inclined to work with diversity concepts, reflexive and dynamic, it will most probably fail.

This requires institutions, instruments, processes and actor involvement based on compatibility of values and traditions, rather than on commonality or integration, and on situationally effective combinations of ideas from hierarchical, network and market governance. The consequence of this is that we need an approach beyond traditional forms of governance, towards a culturally sensitive metagovernance for sustainable development; beyond disciplinary scientific research, towards more transdisciplinarity; beyond borders formed by states and other institutions, towards trans-border approaches; beyond conventional means to measuring progress, towards new and more interactive measuring methods; beyond linear forms of innovation, towards open innovation; beyond cultural integration or assimilation, towards looking for compatibility. In other words, governance for sustainable transformations requires what we have framed in this volume as *transgovernance*. The chapter ends with recommendations on how to apply culturally informed metagovernance embedded in the broader approach of transgovernance.

Growth: A Discussion of the Margins of Economic and Ecological Thought

In the third chapter, Alexander Perez-Carmona reviews the long debate on economic growth, by which he touches one of the most disputed dimensions of sustainable development. The aim of this chapter is to analyse the economic growth debate hitherto and to review and compare two alternatives to it: the stationary-state economy of Herman Daly and de-growth of Serge Latouche. The growth debate emerged out of the convergence of several ecological and political factors in the late 1960s. The position of economists became divided on the issue, with the majority maintaining the growth commitment. It was, however, the study 'Limits to Growth' that really projected the debate beyond academia. The debate remained strongly polarised until the Brundtland report was published, settling the issue politically. The report, a product inevitably of compromise, neglected many important issues, for example the phenomenon of social-engineered wants already well-documented at that time. Furthermore, it ended up recommending what otherwise would have been pursued, such as improvements in energy-matter efficiency, while ignoring scale effects (Jevons paradox) already known too.

International Politics and Cooperation

In Chap. 4, Jamel Napolitano focuses on sustainability governance and international politics and cooperation. This chapter argues for a lecture on the notion of development as strongly linked to the uneven distribution of material and non-material sources of power among groups. It thus analyses the rise of a public environmentalist awareness as a challenge to the capitalist pattern of production and consumption. Finally, the chapter shed some light on the process of mainstreaming these claims by subsuming them within the Western model of societal transformation, under the new, catchy label of sustainable development.

Pressing for institutional solutions to environmental depletion has meant to further spread the sustainability goal worldwide. On the other hand, it has also implied a kind of betrayal of the truly transformative instances of many social movements and local communities, which were seeking for a revolutionary, rather than reformative, path to societal change. After having set the stage in Part 1, the second part of the book presents selected issues with a high political and scientific relevance.

Planetary Boundaries

Falk Schmidt critically reviews one of the new discourses of sustainability science, namely the challenge to govern planetary boundaries. Schmidt argues that it seems intuitive to identify boundaries of an earth system which is increasingly threatened by human activities. Being aware of and hence studying boundaries may be necessary for effective governance of sustainable development. However, can the planetary boundaries function as useful 'warning signs' in this respect? The answer presented in his chapter is: *yes, but.* Schmidt argues that these boundaries cannot be described exclusively by scientific knowledge claims. They have to be identified by science-society or transdisciplinary deliberations. He provides two recommendations for sustainability governance: to better institutionalise integrative transdisciplinary assessment processes along the lines of the interconnected nature of the planetary boundaries and to foster cross-sectoral linkages in order to institutionalise more integrative and yet context-sensitive governance arrangements. These insights are briefly confronted in his chapter, with options for institutional reform in the context of the Rio+20 process.

Governance of Emergencies

In Chap. 6, Günther Bachmann presents his thoughts on the notion of global environmental emergencies in the context of knowledge-based action towards sustainable development. Bachmann argues that responding to emergency situations is about immediate decisions and action. If carried out incorrectly or badly performed, it not only fails in substance but is likely to destroy and delegitimise any further attempts to transform constraints and contingencies which have caused the emergency situation in the first place. This is why emergency response should play a role in governance concepts. Bachmann refers to examples of hazardous substances, impacts of climate change and nuclear accidents, all of them producing nonconventional risks that need transgovernance features beyond national borders.

Bachmann suggests that neither the recent debates on international environmental governance nor those focusing on the multilateral governance framework for sustainable development emphasise sufficiently the issue of emergency response. More often than not, dealing with emergency control is regarded as a strictly national task. This chapter argues that this is inadequate, because the character of emergencies is changing. Whereas conventional emergencies are mostly local, it is clear that limited and calculable nuclear accidents and the adverse effects of climate change demonstrate that the modern generation of emergencies has the potential to surpass geographic limits and national borders and to be long term. Therefore, this chapter argues that emergency control policies may play an important role in clustering change processes and transition efforts, at least under certain conditions and whilst framed by the concept of transgovernance.

Boundary Work

Finally, in Chap. 7, Stefan Jungcurt concentrates on boundary work between science and society. He investigates how a systemic approach to the analysis of interactions between knowledge production and decision-making on sustainable development could be shaped.

The concept of boundary work has been put forward as an analytical approach towards the study of interactions between science and policy. While the concept has been useful as a case-study approach, there are several weaknesses and constraints when using the concept in a more systemic analysis of the interactions between knowledge production and sustainable development decision-making at the international level, for example its inability to capture the diversity of institutions involved in such boundary work and a lack of conceptualisation of the impacts of the specific conditions of intergovernmental decision-making, such as rules for representation and the mode of negotiation. This chapter suggests complementing the concept of boundary work with a configuration approach based on a two-dimensional conceptualisation of the boundary space in international decision-making that allows the positioning of institutions with regard to their degree of politicisation and their position in terms of national and regional representation. Such an approach, which is in line with what transgovernance requires, could be a useful guide in the further conceptualisation and application of the boundary concept.

The Quest for Governance of Sustainable Development

The TransGov project not only resulted in this academic volume with reviewed chapters but also produced a separate monograph authored by Roeland in 't Veld, summarising and enriching the main lines of discussion within the project, focusing on practical suggestions for decision-makers in governments and other relevant social actors. This report, in line with our view on science as a transdisciplinary exercise, is open source and, therefore, freely available at www.iass-potsdam.de. Of course, the open source mechanism also applies to this volume, which is available at www.springer.de. In order to show the links between both publications, we found it useful to include the summary and recommendations of the report *'Transgovernance: The Quest for (Global) Governance of Sustainable Development'* in this volume (Chap. 8).

Part 1
Reflections on Sustainability Governance

Chapter 1
Sustainable Development Within Knowledge Democracies: An Emerging Governance Problem

Roeland Jaap in 't Veld

Abstract Sustainable development is all over the place. The concept is broad and vague. The vagueness of the concept has a Janus face. It has been called a unifying concept because its vagueness breeds a consensus that might be utilised later on. Vagueness is an asset if it triggers action. On the other hand, if sustainable development is everything, maybe it is nothing... Although – or maybe because – the concept is vague, it has overwhelming appeal on political agendas, programmes and dialogues. The precautionary principle is the nucleus of a powerful moral imperative. The multidimensional nature of the concept, covering ecological, economic and social aspects of change relates to our needs for integration. Sustainable development as a concept bears a persuasive character. Actors of all kinds may contribute to it, citizens, enterprises, NGOs, governments et cetera.

Thinking about the governance of sustainable development leads us to the recognition of a multi-level, multi-scale, multi-disciplinary character of the *problematique*. Moreover, the term development refers to change, to transitions and transformations. Governance of sustainable development therefore has to cope with complex dynamics. This chapter deals with the specific consequences of sustainability governance inside knowledge democracies. The concept of knowledge democracy sheds new light on the emerging relationships between politics, media and science. It shows how the emergence of participatory democracy besides representative democracy, the revolutionary rise of social media besides corporate media, the emergence of transdisciplinary trajectories besides classical disciplinary science lead to explosions of complex interactions. We will digress upon the variety of possible future variants of knowledge democracies, quiet and turbulent ones, in relation to the quest for sustainable development. Our main conclusion will be that

R.J. in 't Veld (✉)
Waterbieskreek 40 2353 JH Leiderdorp, Netherlands
e-mail: roelintveld@hotmail.com

L. Meuleman (ed.), *Transgovernance*,
DOI 10.1007/978-3-642-28009-2_1, © The Author(s) 2013

strategies for sustainability may vary with the types of knowledge democracies around.

1.1 Introduction

Since the introduction of the concept of knowledge democracy with the meaning of enabling a new focus on the relationships between knowledge production and dissemination (in 't Veld 2010a), the functioning of the media and the evolution of our democratic institutions and processes, we have seen remarkable proof of the vitality of the concept. The concept obliges us to realise that the institutional frameworks of today's societies may appear to be deficient as far as the undercurrents, trends and other developments demand change. Reconsidering the events in 2011 in the Maghreb, the Middle East and some other regions, the crucial role of social media besides phenomena of participatory democracy demand our attention.

Democracy is without any doubt the most successful governance concept for societies during the two last centuries. It is a strong brand, even used by rulers who do not meet any substantial democratic criterion. Representation gradually became the predominant mechanism by which the population at large, through elections, provides a body with a general authorisation to take decisions in all public domains for a certain period of time. Representative parliamentary democracy became the icon of advanced nation-states.

The recent decline of representative parliamentary democracy has been called upon by many authors. On the micro-level the earlier consistent individual position of an ideologically-based consistent value pattern has disappeared. The values are present but the glue of a focal ideological principle is not any longer at stock. Fragmentation of values has led to individualisation, to uniqueness but thereby also to the impossibility of being represented in a general manner by a single actor such as a member of parliament. More fundamentally media-politics destroy the original meaning of representation. On the meso-level the development of political parties to marketeers in the political realm destroys their capacity for designing consistent broad political strategies. Like willow trees they move with the winds of the supposed voters' preferences. And on the macro-level media-politics dominate. Volatility therefore will probably increase.

The debate on the future of democracy has not yet led to major innovations in advanced national societies in Europe, contrary to sweeping innovation elsewhere. Established political actors try to tackle populism with trusted resources: a combination of anti-populist rhetoric and adoption of the populist agenda. Some of the media have responded by attempting to become 'more populist than populists themselves', almost always at the expense of analytical depth. In other parts of the world the longing for democracy leads to sweeping movements.

The development in different parts of the world partially points in a variety of directions: city government in parts of South America is characterised by remarkable citizens' participation in many cases, while in Asia the rule of law is introduced without classical democracy in influential nations. The recent developments in the Middle East still await thorough evaluation.

Meanwhile, the worldwide web as well as the evolution of social media provides for a drastic change in the rules of the game. A better educated public has wide access to information, and selects it more and more by itself, instead of relying on media filters as produced by classical media. Moreover citizens themselves have become media. They may produce, in some cases soon world-famous, YouTube videos at home or down town. Even more, social media have to the surprise of many shown to be of decisive importance in drastic changes of government and governance in several North-African states in 2011.

The relationships between corporate, top-down media and politics may change considerably as a consequence of the rise of social media because politicians may utilise social media in order to create direct communication with voters, so their dependence on the top-down media diminishes. The corporate media are not any longer the necessary, only intermediaries between politicians and voters. Nevertheless, people get tired of social media already too, because the latter produce also much pulp, and the costs of finding trustworthy information are high; confusion and ambiguity are all over the place. The crucial combination of a network society and media-politics provides new problems and tensions. The political agenda is increasingly filled with so-called wicked problems, characterised by the absence of consensus both on the relevant values and the necessary knowledge and information. Uncertainty and complexity prevail.

Today's societies are characterised by an increasing intensity and speed of reflexive mechanisms. Reflexive mechanisms in a more or less lenient political environment cause overwhelming volatility of bodies of knowledge related to social systems. As all available knowledge is utilised to facilitate reflexive processes, the result of such processes might establish new relationships that undermine the existing knowledge. Social reality has thus become unpredictable in principle.

Voß and Kemp in their introductory chapter to Reflexive Governance to Sustainable Development (2006) deal with reflexivity and distinguish first- and second-order reflexivity. First-order reflexivity

> refers to how modernity deals with its own implications and side effects, the mechanism by which modern societies grow in cycles of producing problems and solutions to these problems that produce new problems. The reality of modern society is thus a result of self-confrontation. (Voß and Kemp 2006: 6).

Second-order reflexivity concerns 'the cognitive reconstruction of this cycle'. It 'entails the application of modern rational analysis not only to the self-induced problems but also to its own working, conditions and effects'. It may be clear that we mainly deal with second-order reflexivity in the terminology of Voß and Kemp.

The relationships between science and politics demand new designs in an environment of media-politics, wicked problems and reflexivity. The classical theory on boundary work as published by Jasanoff and others in order to master the existing gaps between science and politics is nowadays widely accepted among experts. The underlying insight is that scientific knowledge by its very structure never directly relates to action, because it is fragmented, partial, conditional and immunised. This observation is valid for both mono- and multi- disciplinary knowledge. Thus, translation activities are always necessary in order to utilise scientific knowledge for policy purposes.

The literature on transdisciplinary research is dominated by process-directed normative studies. It appears to me that the core concept of transdisciplinarity is to be defined as the trajectory in a multi-actor environment from both sources: from a political agenda and existing expertise, to a robust, plausible perspective for action.

In the third part of the chapter we reflect upon the specific consequences of the mixing of governance of sustainability and knowledge democracies. The final part of this chapter is devoted to observations on quiet and turbulent democracies as very different typologies of potential evolutionary patterns of knowledge democracy.

1.2 Sustainable Development

We consider our world through the veils of fundamental normative perspectives that shape our beliefs, our inspiration and our actions. One of the many disputes between Plato and Aristoteles concerned the question whether mankind is either part of nature or has a subject-object relation to nature. The anthropocentric character of the concept of nature became gradually stronger in the Western world. The Christian religion defined the duty of men towards nature as steward-ship, *Verwalterstelle*, but did seldom practice it. The era of *Aufklärung*, Enlightenment has delivered the perspective of humankind as the master of the universe, with the perspective of a world governed by reason and by science. But the shadow of Faust was always near. More recently the metaphor of the exhaustion of the earth, caused by human irresponsibility, has come to the forefront in disputes. Economic growth then may be sinful. Perez-Carmona treats this issue more fundamentally in Chap. 3 of this volume. On the other side of the spectre, commentators consider technological innovation as the great liberator of the human race, because it will eradicate poverty, hunger and many other shortcomings.

Statistics indicate that we on the average live longer and in better health than ever before, but the pursuit of happiness relates to more than statistics. Our values on distributive justice urge us to pay attention to differences. Many of the normative perspectives on the environment are formulated in terms of threats that demand immediate action. While increasing wealth appears to reduce the willingness to accept risks of wealthy people, these threats are shaped as extreme risks.

It has been generally accepted nowadays that mankind is able to bring about irreversible change that partially diminishes the options of future generations. The normative insight derived from this principle is formulated as the precautionary principle. This principle leads to the norm that we should abstain from action that reduces the valuable future options for choice. Moreover the concept of sustainability now concerns the three major dimensions of human societies, the economic, social and ecological dimension, collected as the three P's people, planet, profit. Van Londen and De Ruijter (2011: 10) define the concept of sustainable development as the reconciliation of three imperatives: (a) the ecological imperative, to remain within planetary bio-physical carrying capacity; (b) the economic imperative, to ensure an adequate material standard of living; and (c) the social imperative, to provide social structures – including systems of governance – that effectively propagate and sustain the values that people want to live by, in order to maximise human welfare.

The reconciliatory character of sustainable development raises specific questions as to the judgment on changes that lead to improvement in two dimensions but to deterioration in the third. Until now we lack a satisfactory interdimensional measuring rod in order to judge upon this type of changes. This deficiency is serious because as a consequence we are unable to provide convincing criteria to judge upon policy options in a comparative manner.

Many different dialogues about sustainable development take place simultaneously: cities, states, enterprises and families discuss sustainable development in their own specific environment. They use common words, but in various rationalities. Sustainable development is a container notion. The use of the singular form fits in holistic viewpoints. The supporters of these viewpoints speak about *the* climate, *the* earth, *the* emissions, *the* planetary boundaries (Meuleman 2010b). All of these are at stake, and disasters threaten. Such constructs enable us subsequently to deal with a *global* challenge that should be met in a well-coordinated manner. Thus, the normative construction of the *problematique* leads to a specific line of argumentation on governance. The supporters of this view may be found in international organisations that make continuous efforts to produce consensus on international binding agreements, in order to prevent disasters. Basic metaphors like the exhaustion of the earth, and planetary boundaries, then are very useful.

However, people do not experience *the* climate but a climate in the neighbourhood. They pursue a good life according to their own values and in many cases try to find a satisfactory relationship to the surrounding nature. Their visible world is not abstract or systemic but specific and concrete. Likewise, until a few years ago, climatologists distinguished many different climates. Entrepreneurs make attempts to design and apply more sustainable technologies. They act in a specific environment too, not in an abstract universe. So Perceptions are not only context-bound but also acceptable ways of dealing with problematic issues. Thus, major discrepancies may exist here between the systemic world on one hand and the daily life world on the other.

The Western world has developed environmental policies during the last half century. In the international realm younger nation-states, often former colonies, more recently also become aware of the disagreeable side effects of economic growth. They want to counterbalance these effects in their own manner. In the diplomatic arena they however are confronted continuously with urgent calls to participate in bargaining processes on treaties with the former colonial powers. These partners now urge for dramatic reductions of emissions and the like. Quota for a certain future year are symbols of urgency. The young nation that is coping with the need for reduction of backwardness in technologies and is just starting to think about clean technologies will not feel inspired by the short term limits set by others. It will experience those as unnatural.

Moreover, the language of international traditional diplomacy would not necessarily be accepted by all relevant actors because some could interpret this language as an expression of hegemony by former colonial powers. Cultural diversity should be recognised both as a component of sustainability and as a complicating factor, that prohibits progress in reaching consensus on collective action. Meuleman devotes a chapter in this book to these questions (Chap. 2).

A society needs a certain cohesion, that is produced as a moral order, based on consensus on some fundamental values and norms. Therefore, culture within a society is also the sharing of some common substantial and relational values. A society consists of configurations. A configuration possesses a specific culture but as observed earlier, this leads to outside walls, and tensions arise. In particular the tensions between emerging identities on one side, accompanied necessarily by outer walls, and the need for cohesion and collective action on the other will never disappear. Shaping governance thus is walking a high wire.

We should argue that biodiversity and cultural diversity both are components of sustainability. We may mourn about the loss of a language somewhere on this globe as about the loss of a species. But our general attitude towards cultural diversity in daily practice is far more critical than towards biodiversity. We do not believe that each culture is intrinsically good. On the contrary, some cultures are horrifying to many. As sustainability also implies the economic and social dimension, we realise that 'diversity always is a bedfellow of inequality' (Van Londen and De Ruijter 2011: 14). Inequality might be a threat to sustainable development, so our attitude towards cultural diversity is ambiguous.

According to the concept of second modernity (Beck 1992) it is probable that from the tense relations between emerging opposites variety increases. Striving at sustainable development urges us to take these tensions fully into account when dealing with governance.

Because sustainable development is a long range trajectory, with considerable uncertainty and lack of forecasting options, the notion of resilience is crucial: like Noah we can act sensibly without any certainty on future events by answering the question how to avoid a disaster, *in casu* by building an Ark. Nowadays for instance it is uncertain which theory on climate change is the right one, but once the theory that allies climate change to carbon emissions is there, the justification of measures to reduce emissions can be based on the resilience norm: in order to avoid disasters we have to take into account the feasible theoretical viewpoints irrespective of our beliefs.

Some supporters of strict environmental policies consider the sustainability concept as a watered-down notion. Like T.S. Eliot (where is the wisdom we lost in knowledge, where is the knowledge we lost in information?) they ask themselves: where is the attention for the environment we lost in sustainability?

We should realise in accordance with the view of Grunwald (2004), Grin (2006) and others that the plurality of notions of sustainable development and their normative origins and connotations lead to the necessity of considering the recommendable knowledge-producing and policy-making processes as reflexive. In Grunwalds terminology:

> The normative character of the imperative of sustainability, its inseparable connection with deep-rooted societal structures and values, the long-term nature of many relevant developments, as well as the often necessary inclusion of societal groups and actors, result in specific demands on scientific problem-solving contributions. Research for sustainable development is a particularly marked type of post-normal science (Funtowicz and Ravetz 1993: 151)

Therefore we will argue that dealing with reflexivity and transdisciplinarity are necessary once we strive at sustainable development.

1.3   Knowledge Democracy

1.3.1   The Overwhelming Success of Democracy

In 2011 again sweeping moves may be observed, and loud outcries may be heard demanding more democracy in different parts of the world. As we argued earlier (in 't Veld 2010a), democracy is the most successful governance concept for societies as well as a strong brand. Even the most cruel dictatorships call themselves democracies.

Democracy according to Abraham Lincoln is a very broad concept: 'government of the people, by the people and for the people'. Some centuries later Schumpeter (1943) however defines it in a minimal manner:

> [....] the democratic method is that institutional arrangement for arriving at political decisions in which individuals acquire the power to decide by means of a competitive struggle for the people's vote.

From the Greek philosopher Plato onwards, (who inherited some insights from the vedas) the continuous debates on the relative merits of democracy versus aristocracy, of consensual versus majoritarian typologies of democracy, of unicentric versus pluricentric concepts of democracy enrich our thinking.

In the course of the last two centuries, a group of related types of representative constitutional democracy became the predominant format of the nation-state. It enjoyed unheard popularity, and still does, all over the globe. All Western and most Southern political leaders preach democracy as an all-healing recipe. Representation gradually became the predominant mechanism by which the population at large, through elections, provides a body with a general authorisation to take decisions in all public domains for a certain period of time.

specific negative influence on political representation as media-politics develop. All these trends appear to cause the gradual disappearance of checks and balances, among which adequate protection against arbitrary or random political action. We will digress upon these options later. Another group of far more optimistic experts indicates that ICT enables new types of democracy that could prove to deliver adequate countervailing powers against the just listed threats.

The debate on the future of democracy in advanced European States has not yet led to major innovations. Established political actors try to tackle populism with trusted resources: a combination of anti-populist rhetoric and adoption of the populist agenda.

We are aware that the development in other parts of the world partially point in another direction: city government in South America is characterised by remarkable citizens' participation in many cases, while in Asia the rule of law is introduced without classical democracy in important nations.

However, recent changes add to the complexity of the relations mentioned so far.

1.3.3   Wide Access to Information for Everyone

As Fig. 1.2 shows, we envisage a world now in which representative democracy is supplemented with, not replaced by participatory democracy, in which social media are added to classical corporate top-down media, and in which disciplinary science is increasingly accompanied by transdisciplinary trajectories. The evolutionary patterns in each corner of the triangle are not without tensions: the inner institutions feel threatened by the younger, outer ones. Each of the corners in the triangle is prone to profound change, indicated in the second-order relationships:

- The bottom-up media do not only supplement the classical media, but also compete with them.
- Participatory democracy is complementary to representative democracy but is also considered as a threat to the latter.
- Transdisciplinary design or research is not only a bridge between classical science and the real world but also produces deviant knowledge and insights, in some cases hostile to the disciplinary viewpoints.

The evolution of the worldwide web and the mobile phone, as well as the evolution of social media provide for a drastic change in the rules of the game. Acts of harassment on weblogs become political facts; virtual allegations become unchecked urban myths and pressure groups design increasingly easier ways to find endorsement on the internet. US president Obama's campaign was trendsetting for the latter.

Internet, better education and other societal changes have made knowledge accessible to many more people than in the past. This leads to an abundance of knowledge and information that needs to be interpreted. It also leads to different types of knowledge: not only scientific knowledge appears to be relevant, but also citizens' knowledge. This is a huge challenge for policy-makers, for scientists and

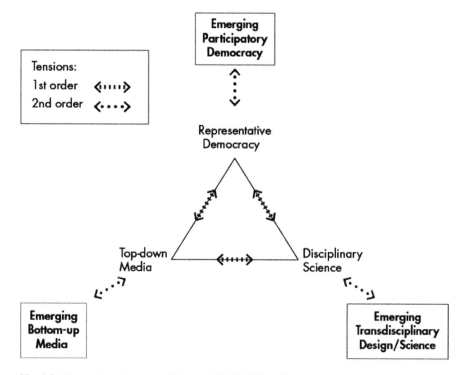

Fig. 1.2   Two orders of tensions (After in 't Veld 2010a: 11)

for the media. Politics is not just about how knowledge can be selected for political decisions, but also about how democratic decision-making processes should change in order to incorporate the different types of knowledge adequately.

A majority of the population now utilises social media. Castells (2009) speaks about 'mass self communication'. Moreover, citizens themselves have become media: any citizen may produce a YouTube video that becomes famous in a few days: icons in political turmoil with great political momentum may be created by amateurs, as the recent events in Iran in 2010 already showed us. The Maghreb and Middle East uprisings in 2011 were influenced decisively – according to many observers – by social media. The classical media suffer from the new ones: not only in a commercial sense, but also because of the influence of the new media. We call the new media the bottom-up media in order to distinguish them from the classical media, the top-down media. This distinction does not imply that the top is more powerful than the bottom. An increasing series of empirical counterproofs is available.

Many of the new media do not know an editing function: nobody accepts the obligation to select the rubbish from the trustworthy materials. This results in very high costs for the recipient of the information in order to make the aforementioned selection. The developments in and with the media are confusing. Our capacity to observe appears deficient. Information and knowledge of very different origins are available within a second but it is hard to judge upon quality. As usual in second modernity the top-down media do not disappear altogether but develop innovative strategies, accepting internet options and modes of cooperation with social media. The social media are in the process of discovering their own deficiencies, and in some cases organise a revival of editorial functions.

The wicked character of many problems on the political agenda sheds a fascinating light on the complexities caused by the interaction of top-down and bottom-up media. Inclusion and exclusion get new dimensions: while the Dutch authorities promoted a campaign of vaccination in order to protect young girls against future cervical cancer in the official media, the target group itself communicated on MSN Messenger, including series of very negative rumours. A woman in a flower shop started a website that got more hits for some time than the aggregate number of hits for all websites of Dutch ministries. This website produced very negative information on vaccination in general, and sketched considerable risks. As a consequence a large part of the target group refused vaccination. Like ships in the night, the different streams of information passed each other. Thus important real life consequences came forward from this multiplicity of information channels and content.

As mentioned above, we can distinguish 'top-down media' and 'bottom-up media'. Both contribute to the agenda setting of politics. The top-down media operate in structural interdependency with politics. The expression 'media-politics' is devoted to this interdependency. The bottom-up media are to a considerable degree independent from both the top-down media and politics. Participation in decision preparation and -making may be invited by public authorities, but uninvited participation takes place too, in particular with support of bottom-up media. We are not in the position yet to draw consolidated conclusions on this development: it is fluid, it is fast, and it is reflexive itself so also unpredictable.

1.3.4   From Knowledge Economy to Knowledge Democracy

During the last decade, an influential debate was conducted on the 'knowledge-based economy'. This concept even became the main policy objective of the European Union, the Lisbon Strategy. However, there are signs that the strength of the argument for the knowledge-based economy is weakening rapidly. The current worldwide economic crisis leads to new, very challenging questions. These questions refer mainly to the institutional frameworks of today's societies. It is therefore time for a transition to a new concept that concentrates on institutional and functional innovation. As the industrial economy has been combined with mass democracy through universal suffrage and later by the rise of mass media, one might suggest that the logical successor of knowledge economy is a new type of governance, to be called 'knowledge democracy'.

Which challenges and threats will we be facing? How will the respectable parliamentary and new direct forms of democracy mix, and which roles will knowledge play in the transition towards a knowledge democracy? The crucial combination of a network society and media-politics provides new problems and tensions. Earlier we concentrated upon the roles of knowledge and information in today's democracies. We further developed the concept of knowledge democracy in order to analyse whether we might be able to deal with these problems and tensions. Now we want to discover what new tensions are arising once we practice knowledge democracy.

Today policy-making in many instances is evidence- or knowledge- based, providing both legitimacy and effectiveness, according to the supporters. Effectiveness is assured as the knowledge concerns true statements on the relationships between political interventions and their societal effects, so is their claim worded. Legitimacy according to them is furthered when the policies are based upon the 'objective' truth. It is not difficult to undermine this belief.

Scientific research is a specific form of research, aimed at the creation or accumulation of scientific knowledge. Classical scientific research is performed within disciplines, specialised branches of science with specific theories and methodologies. This monodisciplinary knowledge is formalised in a particular way methodologically: it is for example subject to peer review. It is often put into a rule-based form, such as: 'A implies B' in a particular set of circumstances, whenever these circumstances occur. Such an assertion is known as a hypothesis. 'The more a child participates in sports, the less likely the child is to turn to drugs', is a statement which could originate from empirical research and which probably holds true for white families in European cities from 1990 to the present time. But not for rural areas in Colombia. And why should this statement hold true for the future? Scientific knowledge is therefore by definition both fragmented and conditional. Its scientific value is dependent on the correct application of the agreed methodology. Scientific knowledge lays claim to validity and is a protection against criticism. What we are talking about here is what is called 'normal research'.

It is difficult to integrate different areas of scientific knowledge because scientific knowledge is by its very nature fragmented. And its conditional character means that in order to apply the knowledge in real-world situations, it is necessary to verify whether the conditions set have been complied with. In terms of the future, this question can never be definitively answered. This means that every application of social scientific knowledge for the purpose of policy bears an element of risk.

If a policy-maker – in the course of preparing policy proposals – wishes to apply an assertion which is based on a rule, such as 'for every X, under condition Y: A implies B', she first has to verify:

- 'Is the X that I am talking about the same X as in the assumption?'
- 'Are the conditions which I am faced with the same as the Y in the assumption?'
- 'Is there really an A in my situation?'
- 'Will the implication still apply at the time when the policy is implemented?'

In particular the last question is a nasty one because the consciousness of reflexivity urges us to wonder whether the drug dealers might have reflected upon the research results too, and might have ensured for themselves a position in the boards of the sports clubs.

This implies that applying scientific knowledge in policy does not always and should not follow the accepted route of meeting the methodological requirements which applied when the knowledge in question was developed. The application of scientific knowledge in a political and governmental context is an exercise in uncertainty, partly based on suppositions and it also requires competences other than scientific ones, such as social intelligence and well-developed social intuition. It appears necessary to link scientific knowledge to other types of insights without detracting from its relevance and usefulness. Combining knowledge from different scientific disciplines and mixing it with other insights is an opportunity to try to maintain the relevance and usefulness of such knowledge in the relevant application. Multi-, inter- and transdisciplinary developments in research are in full swing. Anyone who realises this, cannot fail to be impressed by the speculative nature of many elements of the methods used. The precision of a great deal of scientific knowledge very soon gets lost in these methods. Robust concepts are often unrefined.

As Silvio Funtowicz has explained over and over again, this image of evidence based policies based upon 'sound' knowledge is not adequate according to the advanced science model. We will elaborate upon this later.

Let us now state that knowledge on social systems by definition is volatile as a consequence of the reflexivity we will discuss below. The predominant position of wicked problems on political agenda's as indicated earlier is the main cause that linear problem solution strategies cannot be used. Wicked problems cannot be solved, they can be managed. In many cases interactive processes are part of effective management. Elements of participatory democracy as well as transdisciplinarity may be involved, to be dealt with later on.

1.4 Reflexivity

Today's societies are characterised by an increasing intensity and speed of reflexive mechanisms. I define reflexive mechanisms as events and arrangements that bring about a redefinition of the action perspectives, the focal strategies of the groups and people involved, as a consequence of mindful or thoughtful considerations concerning the frames, identities, underlying structures of themselves as well as other relevant stakeholders. Defined in this manner, reflexivity has to do with a particular kind of learning potential. Reflexive systems have the ability to re-orientate themselves and adapt accordingly based on available self-knowledge.

Reflexive mechanisms in a more or less lenient political environment cause overwhelming volatility of bodies of knowledge related to social systems. As all available knowledge is utilised to facilitate reflexive processes, the result of such processes might establish new relationships that undermine the existing knowledge. Social reality has then become unpredictable in principle. The efficacy of reflexive mechanisms is furthered by institutional arrangements that enable individual liberty and tolerance.

In a tyrannical environment reflexive learning may take place, but it is not spontaneously transformed into a change in behaviour because that change probably is illegal, and severely punished. Insofar as tyranny is negatively correlated with democracy, a democratic environment will prove to be more apt for reflexivity. Extreme profiles in courage however do show behavioural consequences of reflexive learning in tyrannical environments (for example Havel, Mandela).

It is necessary to develop this notion of reflexive learning further because it is of utmost importance for the design of an advanced way of thinking on policy-making: we should realise that a social theory of any kind may never be used to create policy measures without an additional research effort on the specific issue. Such an effort should include the question whether it is probable or plausible that the theory is already undermined by reflexive reactions in or around the target group of the measure. This latter effort will never deliver results with an absolute truth claim. Uncertainty is overwhelmingly present there too. The policy dialogue will then be characterised by different layers of uncertainty, and so by a discussion on the impact of the different layers of uncertainty too.

Evidence-based policy-making as a normative concept probably bears some relevance when it concerns the application of a physical, chemical or biological scientific theory. But it becomes a hazardous pretention if the decision support comes from a theory in the social sciences for the reasons just explained. In particular the claims of economics in important fields as education and health are sometimes preposterous. More modesty would fit once the complexity jump that results from reflexive systems is internalised by the expert. Thus, the fashionable approach towards evidence-based policies in social domains should be moderated in a more modest and thoughtful framework.

Knowledge democracy could become an emerging concept with political, ideological and persuasive meaning. The analogy with the concept of knowledge economy is clear: the latter brought political attention for the economic meaning of research and development, a focus on the quality of education and political support for larger public budgets for the domains under consideration. The human capital theory – although deficient from a scientific point of view – became the predominating policy paradigm in educational policies.

The concept of knowledge economy has developed as a rather vague persuasive notion concerning the relationships between advanced research and education on one hand and economic prosperity on the other. The 'container'-character of the concept has not prohibited favourable effects. It has proven to cause a more conscious approach to the relationships between knowledge production and dissemination on one hand and economic innovation on the other. Education has been recognised fully as a crucial factor in the pursuit of economic progress.

The concept is meant to enable a new focus on the relationships between knowledge production and dissemination, the functioning of the media and our democratic institutions. The emerging concept of knowledge democracy moreover obliges us to realise that the institutional frameworks of today's societies may appear to be deficient insofar as the above mentioned undercurrents, trends and other developments demand change. We explored the directions for institutional change during the conference.

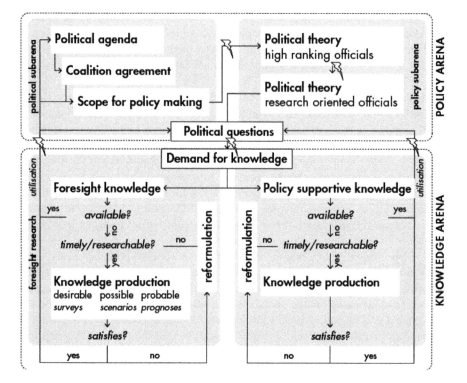

Fig. 1.3 Bottlenecks between the realm of politics, policy-making and useful research (After in 't Veld [Ed.] 2000/2009)

In the perspective of new relationships between politics, media and science also classical problems demand new solutions: the concept of knowledge democracy concerns a *problematique* that relates to the intensification of knowledge in politics. I developed a heuristic scheme in order to think more accurately about the bottlenecks that threaten optimal trajectories between the realm of politics, policy-making and useful research (Fig. 1.3). The thunderbolts show possible bottlenecks in the processes of articulation of the demand for knowledge, as well as the utilisation of knowledge, for instance:

- The actual political agenda may not correspond with the existing policy theories that are either laid down in existing policies, legal systems budgeting rules et cetera and/or are embraced by the top civil servants.
- The translation of policy questions in knowledge demand may prove to be extremely difficult, for instance because the policy objectives bear a symbolic character, or because the policy questions are wicked in nature, lacking underlying consensus on values.
- Inconvenient truth, newly produced knowledge that attacks the existing policy theories, will probably not be applied in policy-making.

- Research will produce knowledge in the future but the need is urgent, and the political agenda is slightly volatile so there is a general problem of timeliness. In order to recognise the time lags just described on one hand and the legitimate demand for useful new knowledge on the other we should attempt to design the policy agenda in the near future instead of only the present one, but that is a dangerous activity.

The aforementioned bottlenecks can be reformulated as problems that demand a solution or at least improvements.

The media are far from neutral or passive. The illusion that they are a neutral mirror of reality belongs to a forgotten past. We have already shed light on the relationships between politics and media. Media create realities, they also produce knowledge, and moreover report on citizens' knowledge. They are the reporters on scientific findings but also competitors of scientists. The same goes for the relationships between media and citizens. This increasing complexity demands efforts in order to gain insight. Other important questions are for instance:

- How do media deal with scientific knowledge, and in particular how do they select the new knowledge to be reported on from the vast supply of new knowledge?
- How can scientific knowledge and citizens' science both be utilised in processes within politics?
- How can conflicts between both types of knowledge be solved?
- How do supervisors and regulators deal with citizens' science?

A number of questions concerning the functioning of the democratic institutions themselves as far as application of knowledge is concerned are very relevant:

- How do parliaments deal with different types of knowledge?
- How do parliaments not only use but also produce knowledge?
- Is parliamentary research to be trusted since parliamentary research committees never lose their power orientation?
- How do parliaments deal with their dependence on information from ministries?
- Which challenges and threats will we be facing? How will parliamentary and new direct forms of democracy mix, and which roles will knowledge play in the transition towards a durable and sustainable knowledge democracy?

In the framework of a knowledge democracy this scheme becomes far more complicated: the policy-knowledge interaction is not any longer restricted to the official political institutions but spreads inevitably over society as a whole: citizen's groups and initiatives develop viewpoints over any major issue. Moreover citizens utilise social media independent from authorities either in order to mobilise support for ideas, or to attack existing policy theories. Science is involved in fierce competition, in continuous marketing efforts in order to gain support for viewpoints, based upon research, aiming at the acquisition of public resources for further research. Advocacy coalitions between the proponents of a certain policy theory, the scientific representatives of related scientific theoretical viewpoints, and sympathetic NGOs and citizen's initiatives are borne, live and disassemble later on.

1.5   Transdisciplinarity

Much valuable scientific work has been performed on the relationships between science and politics, in order to answer the last question partially. Jasanoff and others have argued that it would be wise to design an independent boundary function in order to foster the quality of the translation. The classical theory on boundary work in order to master the existing gaps between science and politics is nowadays widely accepted among experts. The underlying insight is that scientific knowledge by its very structure never directly relates to action, because it is fragmented, partial, conditional and immunised. This observation is valid for both mono- and multi- disciplinary knowledge. Thus translation activities are always necessary in order to utilise scientific knowledge for policy purposes. Pohl, Scholz, Nowotny, Regeer and Bunders, and many others have explored this vast domain and developed the concept of transdisciplinarity in a number of variations.

The literature on transdisciplinary research is dominated by process-directed normative studies. Many authors suggest that transdisiciplinary research is just a specific category of scientific research, characterised by the acceptance of some normative bases for scientific reasoning. Here another viewpoint is defended: it appears clear that the core concept of transdisciplinarity is to be defined as the trajectory in a multi-actor environment, a trajectory that leads from two sources: a political agenda and existing scientific expertise, to a robust, plausible perspective for action. This trajectory bears the character of a communicative and argumentative process. Funtowicz's later models contain both solutions and caveats on this thorny road.

The terminology of the main authors is still more hesitant and still bears the word 'research' in the title. It appears fair, however, to acknowledge that the core activity of transdisciplinarity is design, more than research. Researchers of course may contribute to design. Figure 1.4 illustrates the twofold tense relationships between the corners of the triangle. The original, inner institutional framework was fit for the application of the fruits of disciplinary science, in order to solve rather simple policy problems within the framework of representative democracy. Society was ordered clearly in terms of ideological patterns and classical top-down media fulfilled their roles. The first-order relationships show this picture. The second order relationships describe the evolution of each corner. As a consequence of that evolution we are confronted with tensions, threats and opportunities around the outer corners of the triangle that are indicated in third-order relationships. As we may observe the outer points of the extended triangle also strengthen and stimulate each other. Transdisciplinarity nears participatory democracy, and social media play crucial roles in large scale communication processes. So the tensions relate mainly to the inside-outside relations in the triangle while the stimuli relate to the

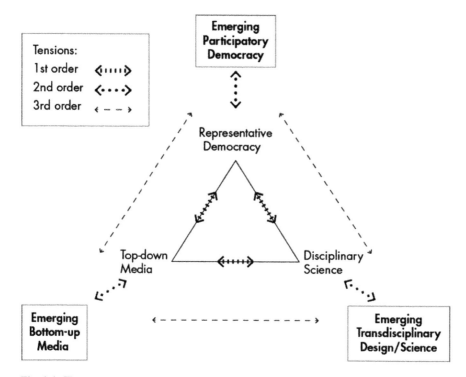

Fig. 1.4 The emergence of the knowledge democracy concept

outer point of the corners. Hardly any empirical research is available here yet. Figure 1.5 shows some of the relations between each inner and each outer corner. This type of relations also has far reaching consequences for the governance of sustainable development in knowledge democracies. These fourth order relations might prove to be very diversified: for instance, bottom-up media might be utilised by representative democracy but also cause conflicts as shown in the case study on vaccination mentioned above. Citizen's initiatives might internalise fruits of disciplinary science, but also application problems might be caused by it. Top-down media might orga-nise transdisciplinary trajectories, but they could prove to be boomerangs for those media themselves, et cetera.

In any society, a wide diversity of actors possesses relevant knowledge concerning important societal problems. In a knowledge democracy both dominant and non-dominant actors could and maybe should have equal access and ability to put this knowledge forward in the process of solving societal problems. We did already explain why disciplinary knowledge on its own is not fit to solve broader societal problems.

During the past centuries the specialisation tendency dominated in science, destroying the practical meaning of the *uomo universale*, and leading to more and more disciplines and sub-disciplines. Sometimes innovation was brought about by new combinations of those, called multi-disciplinary or interdisciplinary cooperation or even mergers. According to the earlier terminology, transdisciplinary

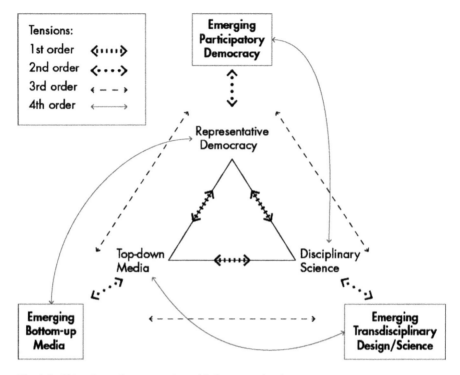

Fig. 1.5 Old and new forms co-exist and influence each other

research developed during the 1980s and early 1990s of the past century. Multidisciplinary and interdisciplinary research than can be placed in a continuum between monodisciplinary research and transdisciplinary research. Klein (2001: 7) at the start of this century defined transdisciplinarity as:

> A new form of learning and problem-solving involving co-operation between different parts of society and science in order to meet complex challenges of society. Transdisciplinary research starts from tangible, real-world problems. Solutions are devised in collaboration with multiple stakeholders.

So she already states that cooperation and mutual learning are key notions in transdisciplinary trajectories.

It is doubtful whether it is fair to describe transdisciplinarity as research. The end product of the cooperation is an action perspective, not a truth claim. Not validity but plausibility, social robustness and support are the decisive criteria. From the perspective of knowledge democracy, we can distinguish two important dimensions in transdisciplinary approaches:

- The degree of knowledge input of lay groups that is included in a specific transdisciplinary project and
- The degree in which non-dominant actors are explicitly involved in the decision-making of the development process of policies or research agendas.

This results in two different styles of transdisciplinary approaches. We discuss the similarities and differences of these different styles and approaches. We conclude this paragraph with a discussion on transdisciplinary research styles in relation to forms of democracy – on the one hand basic and representative democracy and on the other hand deliberative democracy.

Transdisciplinary efforts are embedded in local scientific, cultural and political practices that are differentiated in varied cultures and governance styles. Based on the wide diversity of transdisciplinary efforts we can ask the following questions: what similarities and differences of these programmes are relevant from the perspective of knowledge democracy? Which specific characteristics need to be analysed if we want to understand how transdisciplinary efforts can contribute to the process of knowledge democratisation? An initial look reveals a difference in time scales. We have examples of transdisciplinary research processes that take only a few months (for example, some consultation exercises), while there are also programmes that take over 10 years, and all options in between. The methods and tools used also appear to be quite diverse. Regarding involvement of non-scientific actors for example, they range from interviews to group sessions in all kinds of designs (focus groups, expert meeting, dialogues, citizen juries et cetera).

Notwithstanding these differences, we observe following Bunders et al. 2010, that in scholarly literature the core of transdisciplinary research is most often presented as a shared set of principles. Principles differ from theories, methods, tools and conditions because they refer to the attitudes of the researcher-participant; the researcher is said to perform genuine transdisciplinary research as long as he or she acknowledges and acts in accordance with the intention of these principles. These principles relate to process demands like joint problem definition, orientation towards robust action perspectives, et cetera. As such, a set of principles describes the intentions that guide the researcher in choices he or she has to make for the design of the project or programme, which is the choice of methods, tools and the sequence of these. In other words, 'the approach' is the manner in which the issue at stake is approached. This is in line with the wide-spread convention of labelling specific realisations of transdisciplinary research as 'approaches'.

If one concentrates on the essentials of transdisciplinarity as communication and argumentation, the demands for specific attitudes and even principles concerning the other participants besides researchers are as crucial. The policy-makers will tend to accept those scientific viewpoints that are closely related to the predominant policy theory if present. They however should develop a certain willingness to open up for other scientific insights because the aim of the exercise could be to end up with resilient proposals, having answered the question how to avoid disasters. This demands a sophisticated degree of reflexivity on their part.

Once all participants are touched by the need for mutual adapting, learning and the common goal of a resilient design, the transdisciplinary process could really be successful in the sense of supporting sustainable development. Considering the existing literature one might observe that these conditions are seldom fulfilled.

1.6 Governance of Sustainable Development in Knowledge Democracies

Knowledge democracies are examples of second modernity: they develop in evolutionary patterns characterised by tense relationships between opposite institutions: participatory democracy besides representative democracy, social media besides corporate media, transdisciplinarity besides disciplinary science, and not instead of! The outer corners of the evolving triangle seem to reinforce each other: social media enable participatory democracy, while some categories of transdisciplinarity demand participatory democracy to a certain degree also.

Sustainable development is also profoundly related to second modernity: fragmegration and *glocalisation* illustrate tense relations that characterise the dynamics. It is a fundamental transition or transformation. It is a multi-scale, multi-level, multi-aspect *problematique*. Transition demands restructuration in the landscape-regime-niches environment. Sustainable development knows a number of wicked problems. Uncertainty and complexity prevail besides lack of consensus both on values and on knowledge. Wicked problems cannot be solved by hierarchical order, but can be managed in a multi-actor environment. Finally, sustainable development is a long term *problematique* that demands long term decisions. This type of decisions – dependent on the structure of the problem – either demands an attitude of persistence or of resilience.

Because of reflexivity exogenous steering impulses are not effective in the long run unless the values that determined the steering actions are internalised by the social system under consideration. So exogenous *inter*ventions in general are deficient. We should instead start to think about *intra*ventions as principles of governance. These again point in the direction of participatory democracy, but now considered as a condition for effectiveness. The great governance institutions 'hierarchy', 'market' and 'network' will be amalgamated in a slightly different way in knowledge democracies that aim at transitions: the different actors should move in a manner that can be described as congruency.

Governance of sustainable development should not overconcentrate on global binding environmental agreements as the major tool for progress. The transaction costs of these agreements are often very high, and their effectiveness is often deficient. Second modernity points to regional treaties besides global ones, voluntary agreements besides binding ones, local programmes besides national ones, city developments besides nation-state ones. Moreover, we could design all kinds of private-public arrangements that could stimulate both technological evolution in a favourable direction and unify forces towards societal evolution in a sustainable direction.

1.7   Future Perspectives

In this final paragraph I formulate my insights concerning the predominant tensions and challenges that have to be envisaged: concentrated in the question whether democracies – and more in particular knowledge democracies – can participate favourably to the governance of sustainable development. It is already hard to imagine how the evolutionary tendencies in politics, media and science that all lead to more multiplicity, uncertainty and lack of traditional legitimacy and authority, will have to be coped with simultaneously, and sustainable development is one of these extremely complex and vague issues.

We have proposed to replace the concept of knowledge economy by that of knowledge democracy as a focal item of global agendas. The purpose is to illustrate the necessity to respond to the actual evolutionary patterns of advanced societies. These patterns are interwoven technological and social complex transitions in the triangle politics-science-media. Of course the concept has a persuasive nature. We have fabricated the triangle politics-media-science in order to illuminate the connections and tensions between them. The analysis by Turnhout (2010) on the character of the concept knowledge democracy, leading to the conclusion that it is potentially both utopian and totalitarian should be properly interpreted as an early warning signal. Applications of institutional and procedural requirements in knowledge democracies, such as participatory decision-making processes, should continuously be tested in the contingent environments of empirical reality. The danger of totalitarian and technocratic misadventures is always present, but accidents can be avoided if one is prepared to take a careful look into the value patterns of all concerned actors. This danger is reinforced once more as we realize that sustainable development itself is also persuasive, that it easily might be utopian too. And if we would accept some of the suggestions that due to planetary boundaries and other threats, the command to lead to sustainable development could also bear a totalitarian character itself. Therefore there is ample reason for a lot of attention on arrangements that could fight hasty hypes and other uttering of ultra- persuasive politics. Moreover the present dangers once more underline the necessity of diversified approaches and plurality of methods.

Public authorities within systems of representative democracy are facing legitimacy and effectiveness problems. Representation in its historical shape has eroded because of structural changes in value patterns, and because of the educational level of the population. Legitimacy and effectiveness of governing and steering in a classical manner are fundamentally undermined.

Politicians are far from stupid. They have designed lots of strategies in order to cope with the recently emerged complications. The phenomenon of the spin doctor with the unique assignment to bend available knowledge and information in a favourable direction, and if necessary to provide useful information – invented or not – was temporarily successful until the increasing revulsion of spin doctors enforced them to go under cover. Politicians themselves participate massively in social media. More refined practices have developed in order to influence the so called independent audits and evaluations of public policies and programmes by selecting experts who supposedly would give a positive judgement.

Another category of the strategies of these public authorities in order to regain legitimacy is the introduction of citizen participation. Often it remains completely unclear whether this participation should contribute to either the collection of support or to the process of enriching the content of the decision. This is important because the preferable shape of the processes will depend upon the objectives of the participation.

When we think about participatory democracy we usually refer to notions like civil society, stakeholders-citizens, interested parties, et cetera. In the context of deliberation or participation around a certain issue some public authority usually decides who the desirable partners are. This type of 'guided participation' is often tolerated if the boundaries of an invited group are experienced as 'logical'. However the framing of the problem is decisive for the acceptance of the 'logic'.

Media play crucial roles in any democracy. We have elaborated upon the tensions and other interactions between top-down and bottom-up media earlier in this chapter, and stressed the point that much is still unknown. In September 2010 for instance, the Chair of Dutch parliament suggested the members of parliament to abstain from the use of Twitter during parliamentary debate, because the different streams of information – the official debate in parliament and the Twitter stream-would be 'unmanageable'. From the viewpoint of checks and balances, and taking into account the fact that we live in a world where frequently too much rather than too little information is available, the key role of the media requires a certain degree of self-reflection regarding the presentation of scientific and other policy-relevant knowledge. The question stays, if both top-down and bottom-up media are able to fulfill such a requirement. As Stephan Jungurt explains in this volume, we should refine our viewpoints with respect to bridging gaps between science and policy in the context of international decision-making on sustainable development.

The process of formulating research agendas becomes increasingly important in a knowledge democracy. It cannot any longer be left to scientists alone. Broad participation is desirable. For assessing the need and usefulness of the generation of knowledge by policy oriented research programmes, more reflection in advance is needed. Knowledge democracy therefore appears to demand at least twofold complex participation processes: the transdisciplinary character is necessary to transform scientific insights to robust, plausible action perspectives, and the contribution of stakeholders and citizens is necessary to assure that the decision to be taken will be accepted and effective. Moreover in many cases the specific knowledge of stakeholders and citizens is also necessary to enrich the content of the decisions to be taken sufficiently. All participants have legitimate interests of very different kinds that have to be accommodated. The multi-purpose setup of the processes will vary with the different relative intensities of the objectives: the amalgamation of values, knowledge and interests, the enrichment of content and the gathering of support.

The classical political game will have to change profoundly, and this may be the most important motive for the fierce resistance from many politicians against reform in a participatory direction. Loss of power is the main fear. To accept a role as process architect instead of the position as the final decision-maker is risky because many fear that the voters may not support the architects, but will favour the politicians who present themselves as leaders in substantial solutions.

The quest for acceptable mixed systems of representative and participatory democracy will appear on many agendas in the years to come, and is a focal research question in the knowledge democracy research programme. Democracies have basic characteristics that other regimes do not know: the dynamics are determined by periodical elections that may lead to power shifts. Each politician inside a democracy is profoundly aware of and sensitive for this. The supposed preferences of voters are the guides of action. Many authors have argued that as a consequence of this, democracies are biased towards short term orientations. If this interpretation would be right, democracies are not fit to govern long term problems that demand action in contrast with short term viewpoints. This would cause serious bottle necks with respect to the precautionary principle. But the above mentioned interpretation is inaccurate: exactly because of the indicated dependencies democracies will be very well suited to produce decisions in accordance with the precautionary principle once the formation of citizens' preferences is dominated by the same principle. Once again by this consideration the importance of value dynamics stimulated by value oriented learning processes of populations at large is underlined. Although the most urgent recommendations concern the processes aspects of decision-making, transdisciplinarity and participatory democracy, one may also wonder if structures should change and institutions should be reformed. In general we would argue that institutional redundancy is often recommendable because it will enlarge the resilience of a governance system.

The most apparent characteristic of most democracies – after honouring the will of the people through elections and participatory democracy – is the presence of checks and balances. The rule of law already moderates the power of the executive branch of government. The *trias politica* is the most powerful concept in order to moderate the absoluteness of power, but it is supplemented by numerous other arrangements that serve the same purpose. However it is exactly the recent history of emerging knowledge democracies that puts the checks and balances at risk: this history is full of new populist political parties, currents and undercurrents that flourish in an atmosphere where traditional authority of institutions, professionals and scientists is under attack and fading away. Classical media served the purpose of reporting on the exercise of power, thereby contributing to checks and balances. The perverting power of media tycoons shifted this contribution to the exercise of power by the media themselves and destroyed checks and balances. The social media may contribute to control of power, but it is to early to standardise the conditions under which this favourable function could develop.

In order to produce an adequate scheme for analysis the presumption is formulated that nation-states can be divided in two opposite evolutionary types. This of course is simplification. In reality we may observe in one and the same nation-state spurs of various even contradictory developments. Observing both the available literature and the emerging practice of knowledge democracy in a number of in particular European nation-states I was struck by the differences in the

observable tensions between science, politics and media. We therefore design a
distinction between quiet and turbulent democracies. After having defined them we
will analyse the consequences for the governance of sustainable development.

In the quiet democracies the main characteristics to be observed are:

- In important domains there may be conflicts on the preferable substance or
  content of policies, as based on value differences and variations, but the knowledge base for those policies is generally not contested; therefore problems do not
  bear a wicked character. Moreover complicated two-level conflicts, relating both
  to the substance of policies and the credibility of the different knowledge
  sources, remain absent or at least an exception.
- The mutual dependence of politics and media is not very strongly developed.
  Politicians have realised that the locus for political debate should be parliament,
  and therefore oppose actively to the transfer of political dialogue to mass media
  orchestrated by journalists; media-politics are not predominant.
- Different types of knowledge – such as scientific knowledge, local knowledge
  and/or citizens' knowledge – are integrated in participatory processes for policy
  preparation, aiming at socially robust and plausible perspectives for action;
  boundary actors and institutions play important roles.
- The societal attention for the maintenance of adequate checks and balances is
  considerable; not only the respect for the classical *trias politica* is cherished, but
  also the awareness on the desirability of free basic research and education – free
  in the meaning of: not influenced by either politics or media – is intense.

In the turbulent democracies we find the following phenomena:

- Many political problems are perceived as wicked: neither on the value aspect nor
  on the knowledge or information side consensus exists. Many two-level conflicts
  complicate the political realm. In political environments with a strong meta-
  value, that leads to a high degree of tolerance and mutual respect: this situation
  will lead to the development of transdisciplinary trajectories with considerable
  participation. Populist politics on the contrary will aim at the decrease of this
  type of complexity by establishing a clear, simple and predominating view both
  on values and substance.
- The mutual dependence of politics and media is clearly visible: hypes prevail,
  the political agenda is mainly determined by media utterances, scandals and
  abuses give rise to political action. In extreme instances (for example Italy
  around 2010) the reigning political coalition also rules an important proportion
  of the top-down media. Publics frequently manifest themselves in relation to
  specific hypes.
- Where media-politics dominate, the space for broad citizens' participation in
  policy preparation appears to be limited because politicians and media wish to
  establish a collective monopoly on information-gathering and dissemination.
  Therefore, the stronger the mutual dependence of politics and top-down media
  manifests itself, the more possibilities for unhampered – in the sense of not
  orchestrated by mass media – influential argumentation and communication

seem to be limited. But on the other hand we observed earlier that the existing technologies enable groups of citizens by internet application as YouTube, MSN Messenger, e-mail and Twitter to create their own mass media, to produce their own expressions of interests and views in a manner that cannot be controlled by commercialised or professionalised media.

- Populist politics disrespect checks and balances: the perceived necessity of transparency of authority demands hierarchy in the political realm; populist politicians will continuously criticise any disagreeable action of uncontrolled professionals, and will try to minimise their influence and to maximise their dependence. Moreover the internal structure of the public sector will be stream-lined according to hierarchical principles: as a consequence of which the discre-tion of agencies and other semi-autonomous bodies, but also of inspectorates and supervisors will be diminished. It should be mentioned that the response of established political parties to the successful populists is often a pattern of imitation: the agenda's shift towards the populist issues and vie points.
- In the presence of populist success the attack at checks and balances is often formulated as defence of democracy: independent public decision-making power, for instance by judges, is described as essentially undemocratic. This sometimes leads to a plea to gain political control over the judiciary.

The foregoing static comparison neglects of course the important and necessary analysis of dynamic developments. Castells in particular words his forecasts in terms of accumulative developments, such as the fatal transition of media-politics to populism, or worse. Our observations on the increasing importance of reflexive mechanisms however hamper us to formulate any deterministic forecasts, laws or regularities as to societal developments. Scenarios, simulations and explorations could serve as catalysers to enlarge our sensitivity for potential developments, but the fundamental character of the existing uncertainty and complexity prohibit us to consider them as building stones for direct action. The indirect use could be that we try to design action perspectives that are robust, for example, do not have disastrous consequences in either of the feasible scenarios. It may be clear that the possibilities for such designs are more feasible in quiet than in turbulent democracies. In addition, the increasing complexity of societal problems should not lead to the prohibition of controversial research; to the contrary: such a pluralist approach of research may open new strategies for problems still unforeseen. In case of doubt as to the scientific integrity of knowledge for policy, it is useful to organise discussions on the desirable research agendas, aiming at wide bandwidths of the opinions, and to seek a common knowledge base, as described by many authors in this book. As a matter of course also oppositional parties in parliaments should be included in these processes. The effectiveness of these institutional arrangements may differ in different domains, so careful choices should be made.

Looking at sustainable development as a major issue in all knowledge democracies I feel comfortable in the observation that the opportunities for consis-tent long term policies towards sustainability are more favourable in quiet than in turbulent democracies. The clashes between insights produced by transdisciplinary adventures on one hand and the political priorities on the other will be bitter in turbulent democracies. Recently in these environments boundary functions have disappeared, as ministers themselves claim to be competent to fulfil these functions themselves. On the national level the degree of participation is rather waining than expanding. This appears to create an unbalance in the relations between science and politics, but the scientific world often has remained completely silent. Parliament attempts to decrease its dependence on information from ministries by strengthening its own research activities, but so far the results are of varying quality, to put it mildly.

As we have been able to observe, the relationships between science and media in turbulent environments also lead to scandals and turmoil. The IPPC clashes have weakened the political positions of pro- sustainability actors. Thus the internal conflicts in science are aggravated and magnified by media simplifications that on their turn influence political positions on sustainability issues. On the other hand, in many democracies top civil servants are sincerely involved in efforts to strengthen the knowledge intensity of policy preparation. But their position is weakened in turbulent democracies too, because politicians tend to argue that civil servant do not need discretionary space. This secret war is hardly visible on the surface of the political realm.

Disturbing reflexive phenomena complicate the picture further: ministries design strategic research agendas, but actual research activities sometimes move in another direction. The number of public affairs officers and controllers at ministries increases at the cost of cognitive experts. The cleansing operations – often under the label of 'lean and mean' – in order to reduce the number of relatively independent advisory bodies in the public domain as well as the increasing hierarchy of the political realm support the hypothesis that the evolutionary pattern of turbulent democracies could be characterised as the gradual decrease of that type of checks and balances that may be defined as shock dampers. The extreme phenomena of populist politics to be observed may be summarised in the expression 'fact-free politics'. This expression means that political opinions are formulated irrespective of available information and knowledge so instead of knowledge the driving force for action is conviction, passion or will or a command from elsewhere. Of course the erosion of scientific authority has facilitated this phenomenon, because politicians with fact-free proposals can successfully defend themselves by pointing at the internal dissensus between scientists, or the earlier mistakes made by scientists, planning offices, and the like. It is even possible that the options for fact free politics are influenced positively by the awareness of the characteristic of reflexivity of social systems. If forecasting is impossible, why then rely on science that produced causal relationships with only temporary validities? If evaluation produces meaning, the empirical evidence to be produced later on the results of fact free politics will reveal its deficiencies. But much time will be lost then.

The international world in which endeavours to further sustainable development are taking place is still more varied than the national context of knowledge democracies, because not only quiet and turbulent democracies are present there, but also regimes that could not be described properly as democracies at all. Strange but understandable alliances can be observed: if one would take the degree of authoritarian exercise of power as a measuring rod for regimes, one might observe that the most authoritarian turbulent democracies and the moderated non-democratic regimes find each other quite easily.

As we find ourselves more and more in environments of turbulent democracies, it is important to formulate conditions under which the pursuit of sustainable development is still feasible. How to fight hype orientations, short term oriented populism, fact-free politics?

Earlier in this chapter we have shown that transdisplinarity and participatory democracy are prime methodologies within knowledge democracies to produce those intraventions that reveal the basic values of a society. To protect these opportunities appears to be the first obligation of responsible actors within turbulent democracies. Tensions might become intense, and relationships tight because it is the core belief of the populist that consultations are superfluous because he essentially *is* the people. Of course, reflexivity is also a source of hope and optimism concerning future change.

Acknowledgements I would like to thank the TransGov team and John Grin for their useful comments on earlier versions of this Chapter.

References

Beck U (1992) Risk society. Towards a new modernity. Sage, London

Biesta G (2007) Towards the knowledge democracy? Knowledge production and the civic role of the university. Stud Philos Educ 26(5):467–479

Buchanan J, Congleton RD (1998) Politics by principle, not interest. Cambridge University Press, Cambridge

Bulkeley H, Mol APJ (2003) Participation and environmental governance: consensus, ambivalence and debate. Environ Val 12:143–154

Bunders JFG, Leyderdorff L (1987) The causes and consequences of collaborations between scientists and non-scientific groups. In: Blume S, Bunders JFG, Leyderdorff L, Whitley R (eds) The social direction of the public sciences, sociology of the sciences yearbook, vol XI. D. Reidel, Dordrecht, pp 331–347

Bunders JFG, Regeer B (2009) Knowledge co-creation: interaction between science and society. RMNO, The Hague

Bunders JFG et al (2010) How can transdisciplinary research contribute to knowledge democracy? In: In 't Veld RJ (ed) Knowledge democracy. Consequences for science, politics, and media. Heidelberg, Springer, pp 125–150

Castells M (1996) The rise of the network society, the information age: economy, society and culture, vol I. Blackwell, Cambridge/Oxford

Castells M (2009) Communication power. Oxford University Press, Oxford

Caswill C, Shove E (2000) Introducing interactive social science. Sci Public Policy 27(3):154–157

Coleman S (1999) Can the new media invigorate democracy? Polit Q 70(2):16–22

Dahrendorf R (2002) Über Grenzen. Lebenserinnerungen. C.H. Beck, München

De Bruijn H (2006) One fight, one team: the 9/11 commission report on intelligence, fragmentation and information. Public Adm 84(2):267–287

De Bruijn JA (2007) Managing performance in the public sector, 2nd edn. Routledge, London/ New York/Melbourne

De Bruijn JA, Ten Heuvelhof EF (1999) Scientific expertise in complex decision-making processes. Sci Public Policy 26(3):179–184

De Bruijn HJ, In 't Veld RJ, Ten Heuvelhof EF (2010) Process management: why project management fails in complex decision making. Springer, Heidelberg

De Zeeuw A, In 't Veld R, Van Soest D, Meuleman L, Hoogewoning P (2008) Social cost-benefit analyses for environmental policy-making. RMNO, The Hague

Defila R, Di Guilio A (1999) Evaluating trandisciplinary research. Panorama 1:1–28

Dworkin R (2002) Sovereign virtue: the theory and practice of equality. Harvard University Press, Cambridge

Frey BS, Jegen R (2001) Motivation crowding theory: a survey of empirical evidence. J Econ Surv 15:589–611

Funtowicz S, Ravetz J (1991) A new scientific methodology for global environmental issues. In: Constanza R (ed) Ecological economics: the science and management of sustainability. Columbia University Press, New York, pp 137–152

Funtowicz S, Ravetz J (1992) Three types of risk assessment and the emergence of post-normal science. In: Krimsky S, Golding D (eds) Social theories of risk. Praeger, Westport, pp 211–232

Funtowicz S, Ravetz J (1993) Science for the post-normal age. Futures 25:739–755

Gaber I (2007) Too much of a good thing: the "problem" of political communications in mass media democracy. J Public Aff 7:219–234

Gallopín GC, Funtowicz S, O'Connor M, Ravetz J (2001) Science for the twenty-first century: from social contract to the scientific core. Int J Soc Sci 168(2):219–229

Gaventa J (1991) Toward a knowledge democracy: viewpoints on participatory research in North America. In: Fals-Borda O, Rahman MA (eds) Action and knowledge: breaking the monopoly with participatory action-research. Apex Press, New York, pp 121–133

Gaynor D (1996) Democracy in the age of information: a reconception of the public sphere. http:// bit.ly/oHQWYr. Accessed 4 Nov 2009

Geels FW (2002) Technological transitions as evolutionary reconfiguration processes: a multi-level perspective and a case-study. Res Policy 31:1257–1274

Gee D (2008) Costs of inaction (or delayed action) to reduce exposures to hazardous agents: some lessons from history. Paper to the SCBA conference, The Hague, 17 Jan 2008

Gee D (2009) Evaluating and communicating scientific evidence on environment and on health. Presentation to EEAC, EEA, Copenhagen, 12 June 2009

Gerstl-Pepin C (2007) Introduction to the special issue on media, democracy, and the politics of education. Peabody J Educ 82(2):1–9

Gibbons M, Limoges C, Nowotny H, Schwartzman S, Scott P et al (1994) The new production of knowledge: the dynamics of science and research in contemporary societies. Sage, London/ Thousand Oaks/New Delhi

Global Report on Human Settlements (2011) Cities and climate change. UN Habitat, London

Grin J (2006) Reflexive modernization as a governance issue or: designing and shaping re-structuration. In: Voß JP, Bauknecht D, Kemp R (eds) Reflexive governance for sustainable development. Edward Elgar, Cheltenham, pp 57–81

Grin J, Rotmans J, Schot J (2010) Transitions to sustainable development. Routledge, New York

Grunwald A (2004) Strategic knowledge for sustainable development: the need for reflexivity and learning at the interface between science and society. Int J Foresight Innov Policy 1(1–2): 150–167

Hajer M (2003) Policy without polity: policy analysis and the institutional void. Policy Sci 36(2):175–195

Hajer M (2005) Setting the stage, a dramaturgy of policy deliberation. Adm Soc 36:624–647

Hajer M, Wagenaar H (2003) Deliberative policy analysis: understanding governance in the network society. Cambridge University Press, Cambridge, UK

Hoogeveen H, Verkooijen P (2010) Transforming sustainable development diplomacy: lessons learned from global forest governance. Ph.D. thesis, Wageningen

Hoppe R (2008) Scientific advice and public policy: expert advisers' and policy-makers' discourses on boundary work. Poiesis Praxis: Int J Technol Assess Ethics Sci 6(3–4):235–263

Hoppe R (2010) Lost in translation? A boundary work perspective on making climate change governable. In: Driessen P, Leroy P, Van Viersen W (eds) From climate change to social change: perspectives on science-policy interactions. Earthscan, London

Hoppe R (2011) The governance of problems: puzzling, powering, and participation. Policy Press, Bristol

In 't Veld RJ (ed) (2000/2009) Willingly and knowingly. The roles of knowledge about nature and the environment in policy processes. RMNO, The Hague

In 't Veld RJ (2001/2008) The rehabilitation of Cassandra. A methodological discourse on future research for environmental and spatial policy. WRR/RMNO/NRLO, The Hague

In 't Veld RJ (2009) Towards knowledge democracy. Consequences for science, politics and the media. Paper for the international conference towards knowledge democracy, 25–27 Aug Leiden

In 't Veld RJ (ed) (2010a) Towards knowledge democracy. Consequences for science, politics and the media. Springer, Heidelberg

In 't Veld RJ (2010b) Kennisdemocratie. SDU, The Hague

In 't Veld RJ, Verhey AJM (2000/2009) Willingly and knowingly: about the relationship between values, knowledge production and use of knowledge in environmental policy. In: In 't Veld RJ (ed) Willingly and knowingly: the roles of knowledge about nature and environment in policy processes. RMNO, The Hague, pp 105–145

In 't Veld RJ, Maassen van den Brink H, Morin P, Van Rij V, Van der Veen H et al (2007) Horizon scan report 2007. Towards a future oriented policy and knowledge agenda. COS, The Hague

Jasanoff S (1990) The fifth branch: advisers as policy makers. Harvard University Press, Cambridge, MA

Jasanoff S (2003) Technologies of humility: citizen participation in governing science. Minerva 41:223–244

Jasanoff S (ed) (2004) States of knowledge: the co-production of science and social order. Routledge, London/New York

Jasanoff S (2005) Designs on nature. Science and democracy in Europe and the United States. Princeton University Press, Princeton

Jasanoff S, Martello ML (eds) (2004) Earthly politics. Local and global in environmental governance. MIT Press, Cambridge, MA/London

Kickert WJM, Koppenjan JFM, Klijn EH (1997) Managing complex networks: strategies for the public sector. Sage, London

Klein JT, Grossenbacher-Mansuy W et al (2001) Transdisciplinarity: Joint problem solving among science, technology, and society. An effective way for managing complexity. Basel, Birkhauser

Lindblom CE, Cohen DK (1979) Usable knowledge: social science and social problem solving. Yale University Press, New Haven

Meuleman L (2008) Public management and the metagovernance of hierarchies, networks and markets. Springer, Heidelberg

Meuleman L (2011) Metagoverning governance styles: broadening the public manager's action perspective. In: Torfing J, Triantafillou P (eds) Interactive policy making, metagovernance and democracy. ECPR Press, Colchester, pp 95–110

Meuleman L (2012) Cultural diversity and sustainability metagovernance. In: Meuleman L (ed) Transgovernance: advancing sustainability governance. Springer, Heidelberg, pp 63–121

Meuleman L, In 't Veld RJ (2009) Sustainable development and the governance of long-term decisions. RMNO/EEAC, The Hague

Meuleman L (2010a) The cultural dimension of Metagovernance: why Governance doctrines may fail. Public Org Rev. doi: 10.1007/s11115-009-0088-5. Online first: 12 Aug 2009

Meuleman L (2010b) Metagovernance of climate policies: moving towards more variation. Paper presented at the Unitar/Yale conference strengthening institutions to address climate change and advance a green economy, Yale University, New Haven, 17–19 Sep 2010

Napolitano J (2012) Development, sustainability and international politics. In: Meuleman L (ed) Transgovernance: advancing sustainability governance. Springer, Heidelberg, pp 223–283

Nowotny H, Scott P, Gibbons M (2002) Re-thinking science: knowledge and the public in an age of uncertainty. Polity Press, Cambridge, UK

Nussbaum MC (2006) Frontiers of justice. Belknap Press, Cambridge

Perez Carmona A (2012) Growth: a discussion of the margins of economic and ecological thought. In: Meuleman L (ed) Transgovernance: advancing sustainability Governance. Springer, Heidelberg, pp 121–223

Petschow U, Rosenau J, Von Weizsaecker EU (2005) Governance and sustainability. Greenleaf Publishing, Sheffield

Pohl C, Hirsch Hadorn G (2007) Principles for designing transciplinary research, proposed by the Swiss academies of arts and sciences. Oekom, München

Pohl C, Hirsch Hadorn G (2008) Methodological challenges of transdisciplinary research. Nat Sci Soc 16(1):111–121

Pollitt C, Bouckaert G (2000) Public management reform. A comparative analysis. Oxford University Press, Oxford

Regeer B, Mager S, Van Oorsouw Y (2011) Licence to grow. VU University Press, Amsterdam

Scholz RW (2011) Environmental literacy in science and society: from knowledge to decision. Cambridge University Press, Cambridge, UK

Scholz RW, Stauffacher M (2007) Managing transition in clusters: area development negotiations as a tool for sustaining traditional industries in a Swiss prealpine region. Environ Plan A 39:2518–2539

Schumpeter J (1943) Capitalism, socialism and democracy. Allen and Unwin, London

Schwarz M, Elffers J (2011) Sustainism is the new modernism. DAP, New York

Selin H, Najam A (eds) (2011) Beyond Rio + 20: governance for a green economy. Boston University, Boston

Sen AK (1999) Development as freedom. Alfred A. Knopf, New York

Surowiecki J (2004) The wisdom of crowds: why the many are smarter than the few and how collective wisdom shapes business, economics, society and nations. Anchor Books, London/ New York

Teisman GR, Van Buuren MW, Gerrits L (2009) Managing complex governance systems. Routledge, New York

Turnhout E (2010) Heads in the clouds: knowledge democracy as a utopian dream. In: In 't Veld RJ (ed) Knowledge democracy, consequences for science, politics, and media. Springer, Heidelberg, pp 25–37

Van Londen S, De Ruijter A (2011) Sustainable diversity. In: Janssens M et al (eds) The sustainability of cultural diversity. Edward Elgar, Cheltenham, pp 3–31

Van Twist MJ, Termeer CJ (1991) Introduction to configuration approach: a process theory for societal steering. In: In 't Veld RJ, Termeer CJAM, Schaap L, Van Twist MJW (eds) Autopoiesis and configuration theory: new approaches to societal steering. Kluwer, Dordrecht, pp 19–30

Voß J, Bauknecht D, Kemp R (2006) Reflexive governance for sustainable development. Edward Elgar, Cheltenham

Walker B, Holling CS, Carpenter SR, Kinzig A (2004) Resilience, adaptability and transformability in social-ecological systems. Ecol Soc 9(2):5

Walter A, Helgenberger S, Wiek A, Scholz RW (2007) Measuring societal effects of transdisciplinary research projects: design and application of an evaluation method. Eval Program Plan 30:325–338

Weber L (1979) L'Analyse économique des dépenses publiques. Presses Universitaires de France, Paris

Webler T, Tuler S (2000) Fairness and competence in citizen participation: theoretical reflections from a case study. Adm Soc 32(5):566–595

Weick KE, Sutcliffe KM (2001) Managing the unexpected: assuring high performance in an age of complexity. Jossey-Bass, San Francisco

Weinberg A (1972) Science and trans-science. Minerva 33:209–222

Weingart P (1999) Scientific expertise and political accountability: paradoxes of science in politics. Sci Public Policy 26:151–161

Wynne B (1991) Knowledges in context. Sci Technol Hum Value 16:111–121

Wynne B (1996) May the sheep safely graze? A reflexive view of the expert-lay knowledge divide. In: Lash S, Szerszynski B, Wynne B (eds) Risk, environment and modernity, towards a new ecology. Sage, London/Thousand Oaks/New Delhi

Wynne B (2006) Public engagement as a means of restoring public trust in science – hitting the notes, but missing the music? Community Genet 9:211–220

Wynne B (2007) Risky delusions: misunderstanding science and misperforming publics in the GE crops issue. In: Taylor I, Barrett K (eds) Genetic engineering: decision making under uncertainty. University of British Columbia Press, Vancouver

Yearley S (2000) Making systematic sense of public discontents with expert knowledge: two analytical approaches and a case study. Public Underst Sci 9:105–122

Chapter 2
Cultural Diversity and Sustainability Metagovernance

Louis Meuleman

Abstract In the 20 years since the United Nations summit on sustainable development in Rio de Janeiro in 1992, the world has become more diverse, turbulent, fast
and multi-polar. Tensions between old and new forms of politics, science and media,
representing the emergence of what has been framed as the knowledge democracy,
have brought about new challenges for sustainability governance. However, the
existing governance frameworks seem to deny this social complexity and uncertainty. They also favour centralised negotiations and institutions, view governments
as exclusive decision makers, and imply hegemony of Western economic, political
and cultural principles. This is also reflected in the language of sustainability
governance: it is centralist and is referring to monolithic concepts (*the* economy,
the climate, *the* Earth System) rather than embracing diversity and complexity.

This chapter sheds light on the problematic relations between cultural diversity,
sustainable development and governance. These three concepts share a normative
character, which is always a good predictor of trouble if interaction takes place.

It is argued that the implementation deficit of sustainable development can
be traced back to three problems: a neglect of the opportunities which cultural
diversity offers, an implicit preference for central top-down political solutions, and
an underestimation of the 'wickedness' of many sustainability challenges. It is
concluded that sustainability governance should be more culturally sensitive, reflexive
and dynamic. This requires institutions, instruments, processes, and actor involvement
based on compatibility of values and traditions rather than on commonality or
integration. It also calls for situationally effective combinations of ideas from hierarchical, network and market governance. This implies an approach beyond traditional
forms of governance, towards a culturally sensitive metagovernance for sustainable
development, beyond disciplinary scientific research, beyond states and other existing

L. Meuleman (✉)
PublicStrategy, Avenue du Haut-Pont 11, Brussels 1060, Belgium
e-mail: louismeuleman@hotmail.com

L. Meuleman (ed.), *Transgovernance*,
DOI 10.1007/978-3-642-28009-2_2, © The Author(s) 2013

institutional borders, beyond existing ways to measure progress, beyond linear forms of innovation, and beyond cultural integration or assimilation, towards looking for compatibility. Governance for sustainable transformations requires what we have framed in this volume as *transgovernance*.

2.1 Introduction

2.1.1 Sustainability Governance from Rio to Rio

The Rio-Summit of 1992 marked the beginning of a new era. In the aftermath of the bipolar world, new actors, new challenges as well as new potential solutions emerged onto the scene. However, the mainstream view concerning this increasing diversity – both in scholarly circles and by policy makers – viewed it as being at odds with effective responses to global challenges related to environment and development. They also felt that it supported common (inter-) governmental approaches. The main merit of the Rio-Summit was that it sketched out a new set of challenges and opportunities running counter to this mainstream perception. It was concluded that issues such as cultural diversity, diversity of actors, diversity of institutional mechanisms, and response actions cutting-across well-established sectoral boundaries exemplified the very notion of sustainable development and hence could not be alienated any longer from an agenda for action.

Twenty years later, we live in a different world; it has become hot, crowded, spiky, turbulent and multi-polar. Governments on all levels and international organisations are struggling to implement sustainability strategies, and at the same time find it difficult to embrace the notion of diversity in their attempts to put the objectives of the Rio Declaration into action. This poses the question of whether it is possible to change the common perception of diversity from a potential hindrance to a genuine part of the solution. Would this imply that cultural diversity should be translated into political and institutional diversity? If this is the case, it might threaten vested interests. In addition, how would this relate to the broadly shared conviction that universal aims are also needed, such as human rights? The paradoxical challenge is that sustainability requires shared objectives, which should be achieved by diverse actions pursued through a multitude of governance arrangements at different levels and with different actor constellations, while recognising the varying needs of different countries and communities within them.

The question is whether the growing recognition of the need for adaptivity to different situations could help in bridging the gap between shared objectives and diverse action. The majority of the different situations referred to above pertain to the climate change debate, but also emerge from more general lessons learned in many issue areas spelled out by Agenda 21. As the focus of sustainable development has shifted increasingly towards implementation, compared to the days of Rio 1992, we may raise the question, 20 years later, of whether too strong a focus on common actions, on legally-binding and global agreements alone, has also contributed to the existing lack of implementation.

Another contextual change compared to 20 years ago is that the Internet have made communication and exchange of ideas extremely fast. Social media have partly taken over from classical media. The world is much more knowledge-based than two decades ago. However, at the same time classical natural and social sciences have lost part of their 'natural' authority. Like the media, the field of knowledge production has become more diffused and more participatory. Additionally, in the sciences there are tensions between classical, disciplinary science and *transd*isciplinary knowledge development in which practical and lay knowledge is taken on board. Last but not least, political systems are moving towards more participative forms in many parts of the world. This does not mean that there is a clear convergence towards one type of democracy, but that pluralism is also increasing in this domain. The turbulence and tensions within, and between old and new forms of politics, media and science has been framed as the emerging 'knowledge democracy' (in 't Veld 2010b).

Other conditions that co-determine which governance designs for sustainable development could work well in certain situations are a nation or region's history, and the existing institutional frameworks. The latter are 'frozen' expressions of policy theories from, in some cases, decades ago.

It is with reference to this context that this chapter analyses how cultural diversity might contribute to, rather than hinder, sustainability governance. The key question it addresses is: How can cultural diversity contribute to sustainable development (meta)governance, and how can it be prevented from being a hindrance? Before we embark on this analysis, a short discussion on the ambiguity of the term sustainable development is presented.

2.1.2 Sustainable Development: A Value-Laden Concept

Sustainable development, as defined by the 1987 World Commission on Environment and Development (Brundtland Commission), is:

> development that meets the needs of the present without compromising the ability of future generations to meet their own needs.

The focus on the freedom of choice of future generations in this widely-used definition makes sustainability a modern social-ecological version of Kant's categorical imperative: 'everybody should act in a way that the maxim of this behaviour could become a maxim applicable to all' (Spangenberg 2005: 31). Since the definition of the Brundtland Commission, the normative idea of linking sustainability and development has evolved. During the Rio World Summit on Sustainable Development in 1992, the Brundtland norm of intergenerational justice was elaborated with the aim of bringing about a balance between social, ecological and economic systems, using the terms people, planet and profit or prosperity. The term development can be seen neutrally; like in biological systems, development can be constructive or destructive. With development of societies, economic development (growth) can be the intention, but on occasions the motivation is also with regards to the selection of parameters combined in the Human Development Index. In the context of sustainable development, development usually points at a process character: sustainable development is a societal learning process aiming at developing more sustainable societies. However, it is also based on the concept of progress that may not be shared by all who are ipso facto interested in sustainable development. Moreover, the term development can be seen as culturally related to processes of colonialism, capitalism (including neo-liberalism) and resource exploitation (Oswald Spring 2009).

2.1.2.1   The Cultural Dimension

Like all normative political concepts, sustainable development means something different in different cultural and politico-administrative contexts, for example in Western welfare states, in emerging democracies, and in non-democracies. The concept is used differently in BRIC[1] countries than in many African countries. Policy makers in Vietnam or Bangladesh use definitions which differ from those employed by their counterparts in Germany or Paraguay. Moreover, sustainable development triggers different discussions at the global UN headquarters than it does locally, for example between villagers and professionals implementing an irrigation strategy in a Nepali village. Even though this insight seems trivial, it is surprising to witness the struggles of the international community to draft culturally and, moreover, context-sensitive policies for sustainable development. A concept like the 'Green Economy' adds a case in point with regards to how difficult conceptual discussions can become if it is not based to a certain degree on a shared understanding.

The fact that sustainable development is a normative concept with a Western cultural flavour implies that it may conflict with non-Western cultures. Indeed, this has happened in the past and remains a frequent occurrence. However, the fact that there is, apart from the 1987 Brundtland definition, no global agreement on the exact meaning of sustainable development, also presents an advantage: the concept is in principle adaptable to different cultures. In China, for example, since around 2002, the Communist Party has pursued a 'harmonious society' in a 'harmonious world', with a development model which is similar to the Brundtland definition (Ferro 2009).

2.1.2.2   Top-Down Governance

Environmental governance and sustainability governance are currently dominated by a top-down practice of steering, at least on the global level, although many

---

[1] Brazil, Russia, India, and China.

politicians orally suggest a preference for cooperation and participation. Examples of the top-down approach can be found in climate policy. Climate change has politically and scientifically become framed as a global (upcoming) disaster, for which centralist and legally binding agreements are usually presented as the sole solution. This has its merits because it enables a bird's eye, global perspective and is an expression of political urgency, but also has downsides. The impacts of climate change vary enormously in different geographical areas, and some argue that the centralist frame has also centralised the research budgets. The result is that there is a lack of money for research regarding climate change mitigation and adaptation on specific regional situations, and that it has led to a focus on globalised knowledge which 'erases geographical and cultural difference and in which scale collapses to the global' (Hulme 2010). Hulme argues that in a world which 'possesses a multiplicity of climates and a multiplicity of cultures, values and ways of life', such globalised knowledge is de-contextualised top-down and detached from meaning-making. Examples mentioned are global climate models, global planetary conditions to define sustainability, global indices of human vulnerability to climate change, the Stern Review with its singe metric of globalised monetary value, and the 2° climate change target. Barnett and Campbell (2010) show how such an attitude leads to consider Pacific islands as uniform objects: they are always pictured as vulnerable, powerless and ignorant.

With regard to sustainable development, it is illustrated that much of the energy during the intergovernmental discussions for the preparation of the UNCSD 'Rio' 2012 conference has concentrated on the roles and institutional form of a global sustainability organisation.

2.1.2.3 'Wicked' Sustainability Challenges

This hierarchical bias contradicts the complexity and what political scientists call 'wickedness' of the challenges of sustainable development. The notion of 'wicked problems' (Rittel and Webber 1973), which refers to a situation where there is neither consensus on values nor on knowledge (Fig. 2.1) is crucial for understanding sustainability: 'wicked problems' are a permanent sources of conflict. Some of the typical characteristics of wicked problems are (Rittel and Webber 1973: 162–166):

- Every wicked problem is essentially unique.
- There is no definitive description of a wicked problem.
- Solutions to wicked problems are not true-or-false but good-or-bad.
- Every implemented solution to a wicked problem is a 'one-shot operation' which leaves traces: it changes the problem.
- There are no criteria which enable one to prove that all solutions to a wicked problem have been identified and considered.

In addition, the result of tackling such problems is often path-dependent, and the problems are characterised by lock-in effects with regard to physical (long lead

Values Knowledge	Consensus	Disagreement
Consensus	Technical	Political
Disagreement	Scientific	**WICKED**

Fig. 2.1 Typology of problems

time, bounded by the use of a specific technique or infrastructure) and social (mentality, life styles). The sustainable development agenda is filled with wicked problems. Examples are the future of energy production (how can we become independent of fossil fuels and of nuclear energy?), infrastructure projects (how can we improve railway systems without destroying historical cities and natural sites?), biofuels (how can we increase the use of biofuels without decreasing the land surface available for food production?) and climate change (how can we achieve a global agreement on carbon-neutral economies while acknowledging the right of developing nations to increase their prosperity?). Wicked problems are a product of the increasing complexity and uncertainty of the physical world as well as our societies, and of our cognitive capabilities and values to cope with these issues.

The point is that governance based on hierarchical or market mechanisms often fails when it is applied to wicked problems (Meuleman 2008: 348). A hierarchical view assumes that there are clearly defined problems and that there can be a clear line of command in the problem-solving process. Market governance assumes that the 'invisible hand' of (internal or external to organisations) markets solves problems when the 'right' incentives and instruments are in place.

Wicked problems are value-laden, as are the terms 'governance' and 'sustainable development', and they are also characterised by disagreement on the level of values. Therefore, values and traditions, and hence the cultural dimension, must be included in sustainability governance. Wicked problems escape the logics of hierarchies and markets. Network governance accepts chaos and unpredictability, and also assumes that value conflicts are part of the game and should be dealt with. Therefore, dealing with wicked sustainability problems seems to require at least a substantial network governance dimension in the total approach. The usefulness of additional legal constructions and market-type incentives depends on the context. In this chapter I will focus on the cultural dimension of sustainability governance, but it is also necessary to relate this to the two other themes, the centralist bias and the neglect of wickedness of sustainability problems. After this short introduction, the next step is to discuss the relation between cultural diversity and sustainable development (Sect. 2.2). This will be linked to the governance debate (Sect. 2.3).

Following this, the role of cultural diversity will be discussed in relation to arguments in favour of uniformity (Sect. 2.4). A framework for a positive contribution of cultural diversity to sustainability governance will be developed, based on insights into the governance of governance, or *meta*governance (Sect. 2.5). Sect. 2.6 puts this in the broader context of transgovernance, and conclusions are drawn in Sect. 2.7.

2.2 The Cultural Dimension of Sustainable Development

If the normative dimension of sustainable development is relevant, as we have seen in the first section, then we should discuss the cultural dimension of sustainability. This section first defines culture, and then introduces ideas about the relation between cultural diversity and glocalisation, sustainability governance, and biodiversity. Finally, commonly used arguments are presented for considering cultural diversity as a hindrance to sustainable diversity.

2.2.1 Cultures

Culture can be defined as the values, attitudes, beliefs, orientations, and underlying assumptions prevalent among (a group of) people in a society. Cultures are dynamic patterns of assumptions in a given group. They can also be seen as systems of symbolic communication (Lévi Strauss 1958). In this general definition, the role of human agency and of power should also be included. The latter is significant when tackling the universal character of cultural values.

Behaviour is not part of a culture, but is driven by culture. However, the relation between values and behaviour is part of the discourse on cultural diversity. For example, if we value altruism, and at the same time behave in an egotistic way, we create a tension; 'living your values' therefore may be a relevant expression in the sustainability debate.

The concept of culture changes with the development of our societies. In the beginning of the twenty-first century:

> Culture increasingly stands for ambivalent, ambiguous and paradoxical frames of reference and action. It is increasingly difficult to distinguish between them in a world of shifting alliances and configurations, a world without hegemony, a world where no agency, group or person can still define reality for others, a world rife with turbulence, instability and complexity. In such a world, culture does not succeed in providing clear recipes for action. (Van Londen and De Ruijter 2011: 7)

Although the idea that hegemony no longer plays a crucial role is contestable, the central argument in this quotation is important: cultures are dynamical. If we think that cultures should have operational value for sustainability governance, then the approach that 'culture is an instrument, a vehicle in order to organise diversity (in interests, views, et cetera)' (De Ruijter 1995: 219) may be quite useful.

2.2.2   Cultural Diversity and Identity: The Paradox
of Glocalisation

The cultural dimension of sustainable development can be illustrated with the emergence of what is called globalisation, a phenomenon that has changed the world economically and politically in a dramatic way. Capital looking for new markets and for cheap resources has changed the game. Western (economic) values have dominated the world economy for some time, but with increasing speed, non-Western economies are taking over, or at least co-determining the shape of globalisation.

During the 1990s, the global consumption culture which is responsible for many environmental problems was boosted by the emergence of neo-liberal regimes and their pro-market policies in many Asian, African and Latin American developing countries (Haque 1999: 204). High-consumption lifestyles threaten both the natural environment and the maintenance of the cultural dignity of many societies. National cultural priorities are being sacrificed in favour of global competitive trends. It can be argued that focusing on market principles marginalises the long-term values of cultural and biological diversity (Appadurai 2002: 18, 19). More-over, the globalisation of Northern consumption culture is leading to levels of resource use which are unsustainable, and which may lead to more violent conflict and massive ecological as well as humanitarian degradation (VanDeveer 2011: 45).

Economic globalisation and the ICT revolution have made the world more 'flat', which gives cultures which absorb foreign ideas and meld those with their own traditions an advantage (Friedman 2006: 410) although it is at the same time also 'spiky': differences between e.g. wealth have never been as large as these years (Florida 2005). This has changed the homogeneity of cultures: there are not many nations anymore which are geographically congruent with culturally solidary societies like Japan or Norway (Von Barloewen and Zouari 2010).

The pressure of globalisation has provoked counter-reactions in the form of nationalism, regionalism, localism and renewed ethnicity (Verweel and De Ruijter 2003: 5). Indeed, glocalisation and localisation are two faces of the same trend (Hall 1991). This paradox has been framed as *glocalisation* (Robertson 1995). Globalisation may have made cultures increasingly ambivalent, ambiguous and paradoxical, but the counter reaction – localisation – is equally important for effective governance. It is important that people work from their own values, values in which they believe and which make sense to them, because:

> A sense of identity provides the feeling of security from which one can encounter other cultures with an open mind (Hofstede and Hofstede. 2005: 365)

Therefore, it seems that globalisation on the one hand endangers cultural diversity, but on the other hand stimulates people to discover the rich diversity of cultures with its potential for innovation. *Glocalisation* is an example of the type of fruitful paradoxes that Beck et al. (2003) suggest will become more abundant in the current 'second modernity'. The latter is a concept which explains characteristics of contemporary societies like plurality, ambivalence, ambiguity and contradiction, and claims that a meta-change is taking place from 'first' modernity which is based upon the nation state, socially and in terms of possession of hierarchical knowledge, towards a second form of modernity.

2.2.3 Cultural Diversity and Sustainability Governance

If cultural diversity has 'survived' globalisation, and is linked with people's identity, it should be a powerful asset in the sustainability debate. Nurse (2006: 45) attempts to make cultural diversity the fourth dimension of sustainable development, besides the environmental, social and economic dimensions, because:

> ... sustainable development is only achievable if there is harmony and alignment between the objectives of cultural diversity and that of social equity, environmental responsibility and economic viability.

Taking cultural diversity into account in sustainability governance is important for many reasons, including the following[2]:

- Different cultures are effective when it comes to living in different environments. One model does not work everywhere.
- Different cultures carry different types of wisdom – we need access to all the wisdom we can get.
- Multiple cultures mean multiple options for humanity. We need all the options we can get.
- Communities and societies are structured around identity, which includes a sense of place or home. Without attention to this, people lose their connection to place and are not interested in doing things to protect places over the long term.
- Culture links the larger goals of survival to specific moral visions, and thus makes it attractive (and essential) for people.
- If we succeed in eliminating cultural diversity, it is open to being replaced by other world views such as consumerism or fascism or whatever-ism is being promoted by the strongest, wealthiest or least ethical self-interested party.

There is a huge contrast between these arguments and the little attention that the cultural dimension of sustainability has received in social sciences. Although it is broadly accepted that values, traditions and history co-determine how decisions on public issues are made in different localities, regions and nations (Kickert 2003), the nature of the relations between cultures and governance has largely been neglected. Additionally, in other disciplines this interdisciplinary theme has low priority. For example, although geography investigates man's relationship to his environment, references to game theory in geographic literature were almost non-existent in the 1960s (Gould 1969). Anthropologists and ecologists were the

---

[2] Personal communication Deborah Rogers.

first to investigate how cultures and the physical environment influence each other and what this meant for the feasibility of types of common decision-making. The first link between anthropology and biology, with its ecosystem concept as a useful unit of analysis, dates from 1963 (Moran 1990: 11). The sociologist Hofstede (1980) was an early investigator of the impacts of cultures on decision-making (in business environments) and of the differences between national cultures. In political science, Thompson et al. (1990: 1, 5) (see also Sect. 2.3.2) were forerunners. Their seminal book 'Cultural theory' links cultures or 'ways of life' with the concept of governance.

It seems therefore that if social sciences are to produce meaningful knowledge for sustainability governance, interdisciplinary, or moreover transdisciplinary approaches are crucial. This could start within the social sciences, where anthropology and political science are often organised within one and the same faculty, without significant cross-fertilisation. New scientific approaches also require a new vocabulary (Van Londen and De Ruijter 2011: 23) as well as better cooperation between natural and social sciences (Bennet 1990: 454). In a recent attempt to analyse the relations between cultural diversity and sustainable development, this research topic is framed as *sustainable diversity*. It is defined as:

> ... the ability to structure and manage diversity in such a way that this diversity results in or promotes (ecological and social) sustainability, implying stable and acceptable relationships within and between (groups of) people involving the maintenance of biological diversity, improving material standards of living overall, and equal (or at least fair) access to scarce resources of all kinds as (paid) labour, health, housing, education, income or whatever. This definition (...) sketches the paradox of sustainable diversity: the realisation of equal rights and opportunities under conditions of diversity. (Van Londen and De Ruijter 2011: 17)

In this definition, the concept of equality (in e.g. equal rights) plays a central role. There is an entire literature on defining the concepts of equality and equity. Bronfenbrenner (1973), for example, argues that equality is in essence, an (objective) matter of fact. He distinguishes it from equity, which he frames as a matter of ethical and therefore subjective judgment.

It can therefore be concluded that research on the cultural dimension of sustainable development is lagging behind, although cultures and sustainability are in principle mutually embracing concepts. The consequences are far-reaching: it implies that political decisions regarding sustainability on all governmental levels are ill-informed with knowledge about values, traditions and practices, and therefore also ill-informed about the possibility of implementation. One could ask why, if science does not put the issue on the agenda, practitioners – decision makers on sustainable development – are not pushing for it. The next section attempts to answer this question.

2.2.4   Cultural Diversity as a Hindrance

A key reason why cultural diversity does not appear on the agenda of most environmental and sustainable development policy debates could be that cultural arguments can be easily presented as an obstacle to sustainability, and as an excuse for inaction. Existing *un*sustainable practices are sometimes based on essential values and traditions. National governments argue that for ownership in implementation processes, national circumstances and capacities must be taken into account, but it is often the nations lagging behind in implementation who use this argument. Existing unsustainable practices can be based on high-tech or high-profit approaches, but also on non-factual (scientifically tested) indigenous knowledge. An example of the first is the economic growth paradigm, which, as long as it is based on using more resources, is physically impossible in the long run. Another argument is that diversity can counteract equality. It can be seen as a 'bedfellow of inequality' (Van Londen and De Ruijter 2011: 14); it can be argued that economic globalisation has not created more equality, but has made the experienced social deficit even larger: 'Poor people can now catch a glimpse of the 'rich life''.

Some argue that cultural differences can lead to misunderstanding and disagreement, and therefore form one of the risks which must be dealt with when sustainability partnerships are established and maintained; overcoming these differences takes time and effort (Van Huijstee et al. 2007: 84). Finally, the combination of bad communication as well as existing traditions and values can lead to unintended use or even destruction of sustainable technologies. Local, off-grid renewable energy technologies like solar home systems, biogas cook stoves and small hydropower units are sometimes implemented without an understanding of local cultures, which can lead to unsatisfactory results (Sovacool 2011). In addition, in some countries, discussing cultural diversity is a societal taboo. The USA, a nation which is these days less culturally diverse than it was 200 years ago (Parillo 1994), provides a strong example:

> Awareness of their subjective culture is particularly difficult for Americans since they often interpret cultural factors as characteristics of individual personality. This view of internalised cultural patterns, disregarding their social origins, is a characteristic of American culture. It is not a universal point of view. (Stewart and Bennet 1991)

This attitude hinders reflection on the merits and risks of living in culturally pluralist societies, and reflects a hegemonic attitude towards other cultures. Because values are what you believe in, it is only logical that people consider their own culture 'better' than the cultures of other people. However, cultural hegemonism creates tensions between cultures. With regards to the implementation of sustainable development strategies which are inherently normative, not dealing with cultural differences is a good recipe for further stagnation. In Sect. 2.3.4 alternatives to such a hegemonic attitude are discussed, of which pluralism is the most important.

2.2.5 Cultural Diversity and Biodiversity

An important contribution to raising attention for the cultural dimension of sustainability originates from the analogy with biological diversity. Analogies are among the policy maker's best friends: they suggest a clear logic and, moreover, causality when the inconvenient message would be one of complexity. Hence, it is no surprise that many have suggested an analogy between cultural and biological diversity. The UN Millennium Declaration (2000), for example, considers 'respect for nature' as one of the fundamental values for humanity. Therefore, it concludes that respect for biological diversity also implies respect for human diversity. The declaration takes one step further in arguing that cultural diversity contributes to sustainability because it links universal development goals to plausible and specific moral visions; biodiversity provides an enabling environment for it. The Millennium Declaration builds on the conviction that humankind is part of nature. This conviction is deeply embedded in the East, for example in Chinese and Japanese cultures. The idea that nature and culture are separate categories is a Western invention. It is an artefact: indigenous peoples are cultural but often have no concept of 'nature' (Dwyer 1996: 157, 181). However, there are cultures which share a different belief:

> The dominant assumption in the United States is that nature and the physical world should be controlled in the service of human beings. This has contributed to massive abuse of natural resources in many parts of the world. (...) The American's formidable and sometimes reckless drive to control the physical world (...) is best expressed by the engineer's approach to the world, which is based on technology and applied to social spheres as social engineering and human resource management. (Steward and Bennet 1991: 115)

UNESCO and UNEP hold that cultural diversity and biological diversity are central to ensuring resilience in both social and ecological systems. They argue that both types of diversity are mutually dependent (UNESCO/UNEP 2002: 9, 14), because:

- Many cultural practices depend on, and are a result of specific elements of biodiversity. This applies strongly to the 350 million indigenous people, representing 4,500 of the estimated 6,000 cultures in the world.
- Biodiversity is, in many areas, developed and managed by specific cultural groups.

Examples of the latter are cultural landscapes and tropical agro-ecosystems. In indigenous societies, cultural beliefs and traditional spiritual values help in preventing over-exploitation of resources, and sustaining the ecosystems in and from which such societies live (UNESCO/UNEP 2002: 14).

The failure of many development projects can be connected to not having linked the tangible (health, economic capabilities, security, productivity) and intangible (participation, empowerment, recognition, aspiration) dimensions of development (Appadurai 2002: 17). This mutual dependency of biodiversity and cultural diversity has been framed as bio-cultural diversity (Posey 1999), a concept which originally focused on cultures of indigenous people, but meanwhile also applies to other (e.g. Westernised) cultures like in the South-African suburbs, where for example certain wild plants still have an important meaning and use, as:

> (...) even people who have migrated to urban or peri-urban areas and become involved in modern economic sectors still to varying degrees maintain certain cultural practices, including the use of wild resources for maintaining a sense of well-being and identity. (Cocks 2006)

The connection between both concepts also implies that the study of biodiversity can be helpful for the understanding of human cultures. The following example is illustrative. Without understanding the physical environment, anthropology would probably have never understood why leadership among South American *Yaruro* people is primarily ceremonial (Leeds 1969: 378–83):

> Briefly, Yaruro chieftainship is characterised by its ineffectiveness. The *capitan*, as he is called in Spanish, or *o'te-ta'ra* (elder head) in Yaruro, rarely commands. (...) The *capitans* (...) complained that no one paid enough attention to them, and the Palmarito chief complained that he was never kept sufficiently informed of events by his people.

It turns out that the logic behind this is that the ecological conditions of *Yaruro* life do not require cooperation, coordination or management. All tools and techniques demand utilisation by single persons, and hunting is a solo activity. Therefore, a stronger institutionalisation of the chieftainship is not possible.

The analogy between biodiversity and cultural diversity is, however, not unproblematic. Van Londen and De Ruijter (2011: 4–5) raise three questions. The first is that while biodiversity provides resilience in ecosystems, it is unclear if cultural diversity always produces resilience in societies. Cultural diversity can, under certain political and social conditions, also result in loss of social cohesion, and in tensions and conflicts over access to scarce resources. A counter-argument is that cultural diversity implies having multiple models with regards how to approach life. This provides resilience in the sense that alternative options are available when one approach no longer works.

Secondly, high levels of biodiversity are considered as insurance to future use of the biological gene pool, but it can be questioned if this also applies to cultural diversity. As a response, it could be argued that cultural diversity can also contribute to the 'gene pool' of ideas in a society, leading to innovation and creating new combinations of cultural approaches, which can produce resilience. The third point of critique is that culture has, other than biodiversity, an intentional dimension; it is not an expression of a group but an invitation to become a group (Barth 1969, 1994).

Van Londen and De Ruijter finally question whether or not there are cultural equivalents for concepts like ecosystem services and keystone species. Apart from the point that analogies do not need to extend to all aspects of the items being compared, it could be mentioned here that cultural services can provide mechanisms for meeting human needs, just like ecosystem services. In addition, key cultural elements that, if removed, lead to the collapse or complete change of the rest of the culture, do exist, as with, for example, egalitarian ethics.

The strong analogy and the linkages between the concepts of bio- and cultural diversity are an argument for incorporating cultural diversity in the objectives of sustainable development. Understanding sustainability processes requires an understanding of cultures, and this again necessitates understanding biodiversity. Another argument for giving cultural diversity a comparable place as biodiversity is that natural sciences have shown beyond doubt that people are part of nature, in the sense that they are taxonomically mammals and are biologically and biochemically closely related to certain other mammals.

2.3   Cultural Diversity and Governance for Sustainable Development

If the arguments that cultural diversity has to be included in the governance of sustainable development are convincing, why does it then not happen? In order to understand this, we must know more about how governance definitions deal with values and norms. In this section it is argued that the mere fact that governance is a normative concept implies that cultural diversity should be addressed in governance discussions.

2.3.1   Governance of and for Sustainability

When we discuss governance and sustainability, it can be useful to distinguish between governance *of* and *for* sustainable development. Governance *of* sustainable development refers to the governance dimension as such: which issues are relevant when we discuss the governance dimension of sustainability? This is a heuristic issue. Governance *for* sustainability is normative, prescriptive: it concerns the methods, tools and instruments specifically considered to be useful for sustainable development (e.g. Baker 2009).

    This chapter tackles both: what are the specific characteristics of sustainability governance, and could conditions be created for governance approaches which deal effectively with sustainable development challenges?

2.3.2   The Cultural Dimension of Governance

Some widely adopted definitions of governance assume a general applicability. An example is the 'good governance' agenda, used by World Bank the IMF and by development organisations. This concept has become increasingly refined. The 1997 World Development Report contains 45 criteria for good governance, whilst the 2002 Report lists 116 items (Grindle 2004).

    The problem is that there are many competing – but all hegemonic – definitions of what 'good' governance implies. They vary, for example, between European and American scholars (Robichau 2011: 116). In Anglo-Saxon countries, many political scientists prefer to present governance as the combination of small government and market-type instruments. In Scandinavian countries and the Netherlands, the 'governance is network' narrative has become popular among scholars. These different schools are based on different values and traditions. The market-oriented New Public Management movement was born in Anglo-Saxon countries which have a strong individualist and free-market culture. The network-orientation of the Netherlands can be traced back to centuries ago, when people had to work together to fight the water (Kickert 2003). Such different world views lead to different problem definitions and to different interests of actors (Jachtenfuchs 1994). Dixon and Dogan (2002: 191) emphasise the incompatibility of these views: Hierarchical, network and market governance:

> derive their governance certainties from propositions drawn from specific methodological families, which reflect particular configurations of epistemological and ontological perspectives. They have incompatible contentions about what is knowable in the social world and what does or can exist – the nature of being – in the social world. Thus, they have incompatible contentions about the forms of reasoning that should be the basis for thought and action.

It is difficult to grasp why the cultural dimension of governance has been neglected for so long. One possible reason is that it does not fit into the dominant paradigm of the post-war Western world, rational choice theory (Geva-May 2002: 388). In the USA, cultural diversity has been a societal (and scholarly) taboo. This absence of a cultural dimension in governance is not problematic per se when the objects of governance are clear-cut and predictable. If it works well in France, there is no reason why French railways should not be decided upon centrally in Paris. In addition, the extensive stakeholder dialogues around decisions about fast railway tracks in The Netherlands are not necessarily an example of 'best practice' which works everywhere. However, when we discuss governance for an inherently value-laden object like sustainable development, the use of standardised governance approaches may provoke unnecessary implementation problems. Sustainability governance therefore requires a definition which is based on the idea that governance is inherently normative, and that values and traditions are part of effective governance. The following definition means that different cultural versions are a possibility, and that the definition itself is only normative in one respect: it is a definition in which relations are fundamental. Governance is:

> ... the totality of interactions of governments, other public bodies, the private sector and civil society, aiming at solving societal problems or creating societal opportunities. (Meuleman 2008)

Another, similarly neutral, definition of governance is:

> a collection of normative insights on the organisation of influence, steering, power, checks and balances in human societies. (in 't Veld 2011)

Three styles of governance are usually distinguished: hierarchical, network and market governance (Thorelli 1986, Thompson et al. 1991, Kickert 2003). These styles tend to appear in combinations which can vary over time. Each style or ideal type is consistent with specific sets of values. For example, a hierarchical governance style appreciates authority, justice and accountability, network governance links with empathy, trust and equality, and market governance prefers autonomy, competitiveness and economic value (price).

There are three problems with the application of these governance styles. Firstly, they can undermine each other. Secondly, each of them has typical failures or even perversities. Thirdly, they all have an attractive, even 'addictive' logic. The latter relates to the cultural dimension.

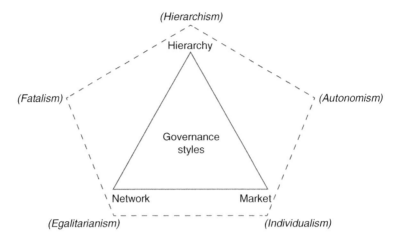

Fig. 2.2 Three governance styles and five ways of life (After Meuleman 2008)

Thompson et al. (1990: 1, 5) distinguish five human cultures or 'ways of life': Hierarchism, egalitarianism, individualism, fatalism and autonomism. In Fig. 2.2, the triangle represents the three governance styles, while the dash-lined pentagon depicts the five ways of life. Hierarchism, egalitarianism and individualism are, in terms of value orientation, congruent with hierarchical, network and market governance. The two other ways of life – fatalism and autonomism – are mixed forms. Fatalism, an attitude in the shadow of strong hierarchism, denies the possibility of coordination, which is an important function of governance, and autonomism is individualism (market governance) in its extreme form, which does not accept social responsibility.

2.3.3 Governance and National Cultures

If cultures and governance styles are so intensely intertwined, to what extent do national cultures then predict how sustainability policies are designed and implemented? Does this mean that global sustainability is a *fata morgana* – because we do not have a global culture? Although we may speculate whether nations have a dominant characteristic type of governance, many public administration studies have shown that at least national administrations can be categorised along cultural lines. For example, people in Germany and France may prefer hierarchy, while in Scandinavian countries and the Netherlands network governance may be the preferred 'default' style. Regarding the value preferences of (to a large extent business) professionals in different nations, much empirical work has been published. The sociologist Hofstede constructs and tests five indexes which may be useful for understanding different cultural approaches to sustainability: power distance, the degree of individualism, gender roles, uncertainty avoidance, and long-term orientation (Hofstede 1980, 2001). Nations with a relatively low power distance, low uncertainty avoidance and a 'feminine'[3] culture are the Netherlands and the Scandinavian countries, which also have a tradition of network governance. People living in nations like Germany and France show, compared to citizens of Scandinavian countries, a higher acceptance of power distance and a higher level of uncertainty avoidance. This may be an indicator of a more hierarchical tradition. The US, Australia and Great Britain rank 1–3 on Hofstede's list of most individualistic countries and they also score highly on 'masculinity'.[4] This correlates with the historical fact that market governance has originated in Anglo-Saxon nations.

Even when their geographical distances are small, there can be clear differences in value orientations between nations. One study, for example, shows such differences between a Western European country (the Netherlands) and four Central European countries, as well as among these four countries (Kolman et al. 2002: 87).

In the context of this chapter it is noteworthy that most environmental policies are formulated from an individualist point of view, and based on implementation through market-based mechanisms, but also include a legalist (rights-based, hierarchical) approach. A comparative study on the relations between national culture and three basic norms of governance (rule of law, corruption, and democratic accountability) in some 50 nations reveals that 'good governance' is more compatible with cultural profiles in Western European and English-speaking countries than in many other nations. The authors of this study conclude that, for example, the value of individual freedom 'runs counter to the societal emphasis on embeddedness that is common in many Asian, African and other countries' (Licht et al. 2007). In addition, Cornell and Kalt (e.g. 2005) present an extensive body of analysis with regards to Native American tribal culture and governing styles. This may well offer an interesting contrast, and may boost efforts to increase cultural diversity.

It seems therefore that it is important to recognise the cultural dimension of governance if we wish to understand why transferring a successful governance approach from one country to another in a dogmatic way, without adaptation to the national socio-politico-administrative culture and other situational factors, can result in failure (Meuleman 2010a).

The idea of 'national cultures' ought to be nuanced, however, because many countries house different groups of people with their own cultural settings. On the other hand, there are also communities who share the same values but are not geographically linked. Good examples are 'elite cultures'. There may be considerable differences between the culture of the governing elite, and the culture of the general public. For example, international climate negotiations are carried out by highly educated officers, who share a common working culture with colleagues from abroad.

---

[3] A 'feminine' culture is defined as a culture 'in which emotional gender roles overlap: both men and women are supposed to be modest, tender, and concerend with the quality of life' (Hofstede and Hofstede 2005: 401).

[4] A masculine culture is a culture 'in which emotional gender roles are clearly distinct: men are supposed to be assertive, tough, and focused on material success; women are supposed to be more modest, tender, and concerned with the quality of life' (Hofstede and Hofstede 2005: 402).

In addition, Hofstede and Hofstede (2005: 11) argue that the national level of 'mental programming of ourselves' is only one of the layers of culture, besides regional, ethnic, religious and/or linguistic affiliation level, a gender level, a generation level, a social class level, and an organisational level (socialisation by a work organisation). Finally, there are combinations of cultural values underlying governance style combinations which hardly fit with any of the ways of life which are depicted in Fig. 2.2. For example, bazaar governance as coined for the governance of Internet communities, is characterised by low levels of control (hierarchy), weak incentives intensity (market) and a kind of network governance which does not build on trust – community members seldom know each other personally and may enter or leave the network unnoticed (Demil and Lecocq 2006).

To summarise, national or regional, and even local governments may be inclined – for reasons of *opportunita*, fashion, or compliance with (e.g. World Bank) rules set by financial sponsors – to use a specific governance style mixture, regardless of the character of the societal problems to be addressed. However, we know that local, regional or national values and traditions co-determine which governance style combinations may work. Sometimes there are clear underlying national preferences for certain styles, such as a consensus style in the Netherlands and hierarchy in Germany and France, as well as market styles in Anglo-Saxon countries. In many other cases such linkages are much less clear.

What can be concluded is that without considering adapting sustainability governance to what makes sense for, and is acceptable to people, all governance attempts risk failure. The popular term 'best practice' suggests universal applicability and should therefore be exchanged for 'good practice'. When considering borrowing a successful practice, it is essential to reflect on the question of whether such an approach would work in the specific setting of values and traditions.

2.3.4   Governance and the Relations Between Values

The concept of governance is intentional, and we have defined it as a relational concept. Hierarchy needs dependent subjects, network governance requires interdependency between partners, and market governance necessitates independent relationships (Kickert 2003: 127). Hence, it is fair to assume that different governance styles also express how people consider other people's values. Five relational values which express different relation types, are (in 't Veld 2010a):

- Hegemony: "My values are superior to those of other people".
- Separatism: "I don't want to be confronted with the implications of other people's values".
- Pluralism: "Other people's values may be valuable, and I am co-responsible for protecting them".
- Tolerance: "I find my values superior to other people's values, but I abstain from interventions because of sympathy".
- Indifference: "I find my values superior to other people's values, but I abstain from interventions because I am not interested".

Table 2.1 Governance styles and relational values

Governance style	Relation to other people's values
Hierarchical governance (dependency, authority)	Hegemony or separatism
Network governance (interdependency, empathy)	Pluralism or tolerance
Market governance (independency, autonomy)	Indifference

To draw a broad typology, hegemony and separatism are related to the top-down and authoritarian thinking of hierarchical governance, pluralism and tolerance to the empathy, trust and respect of network governance, and indifference to the individualism and autonomy of market governance (Table 2.1). Hegemonic thinking is congruent with top-down governance, and cultural pluralism seems to fit better to the character of many sustainability challenges. If the complexity of a sustainability challenge leads to choosing network governance, pluralism or at least tolerance are values to be expected. However if, for a specific problem hierarchical governance is chosen as the main style, its congruency with hegemony and separatism should be taken into account: it can destroy trust and innovation power. If a market-based approach is chosen, the indifference towards values and traditions related to market governance can become a bottleneck for implementation. We have already seen that the 'wickedness' of many sustainability problems necessitates a strong network governance touch in the sustainability governance mixture. This suggests that for sustainable development at least tolerance, but even more so pluralism are probable and should be expected to be productive approaches. Earlier I have argued that governance is a normative concept, and that different governance styles can be linked to different value systems and traditions. This implies that governance can be both a hindrance and an opportunity to increase the role of cultural diversity in sustainability. How can the latter be promoted?

2.4   Cultural Diversity as an Opportunity?

The following paragraphs discuss the arguments in favour of linking cultural diversity to sustainability governance. In addition, the tension between diversity and unifying concepts will be addressed.

2.4.1   Cultural Diversity as an Opportunity for Sustainability Governance

Cultural pluralism is often seen as threatening sustainable development, especially social sustainability. The dominant attitude therefore is, and has been, assimilation of cultural and ethnical minorities (often euphemised as 'integration') (Verweel and

De Ruiter 2003). This policy of assimilation has created social tensions between
different cultural groups in many European countries. Verweel and De Ruijter
(2003: 217–219) presents three arguments against the integration principle:

- An appeal for fundamental, shared values and standards cannot work in our
  contemporary plural societies.
- Our cultural diverse societies have become less recognisable and predictable;
  institutions turn out to be unreliable.
- Having the same cognitions, standards and values are not functional conditions
  for society and communication.

If it is considered important that sustainability governance is grounded in
cultural values as drivers for social transformation, an alternative approach could
be to not focus on *communality* – commonly shared values – but on *compatibility*
(De Ruijter 1995). Communality is often politically framed as 'integration', for
example in Western immigrant policies. In reality, this is a rather hegemonic
assimilation. The compatibility principle recognises that there are (in principle
valuable) differences, which may cause tensions and incompatibilities:

> We should not remove differences, which is both impossible and unnecessary, but regulate
> and hence both recognise and appreciate these differences. Since power also implies
> inequality, it should also include organising power effects. (De Ruijter 1995: 222)

De Ruijter argues that this means, for example:

> stimulating contacts between groups with different identities, without asking these people
> to develop a common system of basic conditions.

He adds that this requires participation, which in turn, requires – at least in
Western nations – entrance to the job market, which again requires education,
including learning the social competence to deal with diversity. The latter, De
Ruijter concludes, requires the capability to deal with uncertainty, unknown
situations, limited means and one's own shortcomings.

A simple example, which many will have experienced, is the fact that different
cultures have a different notion of time (or concrete: being in time for an agreed
meeting). Some people consider it crucial to be exactly in time for a meeting, and
become impatient when others are delayed. The latter, for example, is considered to
breach the value of politeness. Others, who do not put such an emphasis on being
punctual, could be offended if people do not wait until they arrive. Their solution in
such cases is multi-tasking: if a meeting does not take place at all, or not at the
agreed moment, they switch to other useful activities without a problem. The
background of these differences could be found in profoundly different time
conceptions: linear time (e.g. Western) versus circular time (e.g. African) (Du
Pisani, 2006). Striving for compatibility could, in this example, mean creating a
compromise that includes a time window of flexibility (to which the 'multi-taskers'
should comply), and that the 'impatients' move a little towards multi-tasking.

We could ask if striving for compatibility is a one-solution-fits-all approach.
Earlier I have concluded that such panaceas are extremely rare. In nations in which
(cultural) diversity is suppressed and central solutions to sustainability and

environmental challenges are imposed, this can be successful if people are willing to accept this hierarchical and authoritarian approach, be it because of a strong underlying hierarchical value system, or because of fatalism. Although some in the environmental community dream of the effectiveness of such 'Chinese governance', these dreams contradict my argument that copy-pasting governance from one culture to another can run the risk of failure.

We have seen that framing cultural diversity as an obstacle to sustainability has been a welcome political excuse to do nothing, and continues to be the dominant view, partly because of the lack of inter- and transdisciplinary research. However, if sustainability can be linked to existing values and traditions, chances are higher that ownership develops. Ownership is a crucial condition with regards to, for example, the introduction of 'forest diplomacy' which is more based on partnerships (Hoogeveen and Verkooijen 2010), and is an essential value for multiple inclusion of actors in problem-solving (as described in configuration theory; Termeer 1993). Moreover, considering cultural diversity as an opportunity for sustainability governance may prevent the destruction of already existing sustainable practices.

The argument that social ownership is crucial, and implies co-production of situationally appropriate solutions rather than ready-made packages (GIZ 2010), has gained more influence in the field of development cooperation than in sustainability governance. In order to create a better balance, a different view of cultures in sustainability governance is required. Cultural change is a requisite of tackling the great sustainability challenges: we cannot rely on engineering, entrepreneurship and professional politics alone (Leggewie and Welzer 2009).

To conclude: Building sustainability on cultural diversity and investing in compatibility of values and practices rather than on assimilation, can lead to a rich variation of solutions to similar problems, instead of current governance practice in which centrally proposed solutions are often accepted in some cultures and rejected in others.

2.4.2 Unity and Diversity?

Besides the challenge of optimising the opportunities and minimising the hindrance of cultural diversity to sustainability governance, the tensions between *diversity* as expressions of pluralism and its 'enemy' *universalism* should be addressed in sustainability governance. This problem is not only typical for sustainable development. Three contrasting approaches are relevant:

- Universalism departs from the idea that cultures are not equivalent. According to this conviction, some cultures are superior to others, and therefore economic and cultural imperialism are legitimate (Procee 1991).
- Cultural relativism makes diversity central and chooses tolerance as the main relational value to cope with power differences between cultures, for example. Its advocates consider it taking 'a neutral vantage point' which 'calls for

suspending judgement when dealing with groups or societies different from one's own' (Hofstede and Hofstede 2005: 6).

- Pluralism also approaches cultures as equivalent and gives diversity a central place, but instead of tolerance, it chooses an active exchange between diverse cultures. This 'interactive diversity' (Procee 1991) considers cultural uniqueness as a possibility to learn.

Private companies, more than (national) governments, have also discovered the innovation power of cultural diversity. Referring to De Bono's advocacy of lateral thinking and other non-linear approaches to innovation, the Canadian CEO Singer (2008) holds that culturally diverse organisations can be more innovative when their 'cultural intelligence' is used well. She argues that multinational companies like Proctor and Gamble have come to understand this, and have developed the philosophy that diversity outperforms homogeneity. Companies who fail to understand that relationships are much more important in Asian cultures than in the USA where the focus is on the contract, on money and individual recognition, face problems. Many northern European companies have failed in India because they did not adapt their strategies to Indian norms and values (Majlergaard 2006). The challenge to turn cultural diversity from a hindrance into an asset is the reason why IBM more than 30 years ago asked the sociologist Hofstede to investigate national cultural differences by interviewing IBM employees. In addition, corporate governance theorists have stressed the usefulness of national (or other) cultures. Licht (2001) even concludes that national cultures can be seen as 'the mother of path dependencies' in corporate governance systems (Licht 2001).

If we conclude that a pluralist approach is most in line with the cultural dimension of sustainable development, how can we reconcile unity and diversity or plurality? A first observation is that, although it seems difficult to combine universalist common values with a pro-active attitude towards cultural diversity, this has been an attractive paradox for Western politicians. Since 2000, the motto of the European Union is 'United in diversity', reflecting the idea that the EU is an ambitious common project of people who recognise the richness of their continent's different cultures and traditions. More than two centuries before, in 1781, the American Congress adopted the motto *E pluribus unum* (One from many) as the motto of the USA. However, it is not only the West that feels attracted to this paradox. The principle of 'unity in diversity' is also the foundation of Hinduism, and is considered to bind India and its 1652 languages and dialects together (Satheye 2001).

Mottos like 'unity in diversity' and 'E pluribus unum' may have an important symbolic meaning, but have at the same time little practical use. There are quite different views on the unity-diversity divide. Pro-diversity advocates argue that:

- Sustainability problems differ so much (geographically and culturally), that we need different strategies in North and South, between men and women, and poor and rich (Oswald Spring 2010).
- Acceptance of (sustainability) governance depends on the match with (local/ regional/national et cetera) cultures (Meuleman 2010a).
- The current (global), and not very successful sustainability governance system has a uniformist bias.

Pro-unity advocates argue that some environmental problems (e.g. climate change), economic and social problems cross borders and require universally binding policy principles. Sustainable development requires that the human rights are extended to, for example, the right to food (De Schutter 2010).

The most successful example, which is relevant to social sustainability, of a universalist approach to cultures is the 1948 Universal Declaration of Human Rights. Indeed this declaration has (since 1993) been supported by 171 nations. In only 30 articles this Declaration formulates more than 60 universal rights. The Declaration itself is not legally binding, but has been the inspiration for other declarations which are legally binding, and for national constitutions. The seemingly valid option of an impact-rich universal set of values could be interpreted as supporting a continuation of the current centralist and universalist focus of sustainability governance.

However, the Universal Declaration has been, and still is disputed, because of its presumed incompatibility with certain cultures. This is not the place to elaborate on it further, but some of the arguments contesting the viability of universal human rights should be mentioned:

- Many sustainability challenges are of the 'wicked' type, which makes them extremely difficult to tackle with centralist (in terms of governance styles: hierarchical) processes and instruments, except maybe in nations with an accepted authoritarian regime.
- Preserving a broad perspective of diversity (including biophysical, cultural, economic, technological, social, institutional and cognitive issues) can be a suitable way to reduce socio-ecological vulnerability (Cazorla-Clariso et al. 2008) and create more social resilience.
- It can be questioned whether or not contemporary human rights do reflect a global consensus, or if the current set of human rights is congruent with all cultures. For example, they are 'overtly egalitarian in their aim to secure equal rights for everyone, regardless of social station or level of achievement' (Kao 2011: 172).

One area to draw examples from could be the impacts of the rights of indigenous peoples on sustainable development. The rights-based approach seeks to protect indigenous peoples' self-determination and preserve their traditional life styles and culture, which are often assumed to be more sustainable than industrial lifestyles. However in practice one can often observe that indigenous peoples' life styles are neither sustainable per se nor does the rights-based approach seem to help in preserving their traditions. Instead, the combination of traditional rights to lands and resources and a broad interpretation of the right to self-determination sometimes leads to highly disruptive lifestyles (e.g. hunting with skidoos and AK47 rifles). These lifestyles are of course also expressions of culture and a possible consequence of diversity. Another negative outcome could be that rights-based diversity leads communities to resist adapting to external influences, such as climate change. Most indigenous communities in Canada's north, for example, will not be able to survive in a warmer climate.

It seems that each discussion on cultural diversity leads to the question of whether there are also universal values, and if yes, how do they relate to the premise of diversity? The question is whether this is a problem, because such a never-ending discussion can also be beneficial. However, the issue of how to make trade-offs between unity and diversity in sustainability governance should be addressed, as there is a dominant coalition in scholarly and political environmental and sustainability communities pushing the 'unity' side of the equation.

An observation based on a comparative investigation of European nations dealing with these kinds of dilemmas in sustainable development governance is, that if there is a lack of success, then there is a tendency to move towards the other end of the pole (Niestroy 2005: 13). Taking this line of thought, it might be possible to formulate this as a heuristic governance rule: The dominance of centralism in sustainability governance should lead to the assumption that moving towards more diversity would lead to better results. It does not however have to imply a breakdown of the universal 'acquis' of, for example, human rights.

In addition, a contribution to dealing with the dilemma between univeralist and pluralist approaches could arise from adopting a formula like Kant's categorical imperative (already mentioned in Sect. 2.1.2) for universal values: 'everybody should act in a way that the maxim of this behaviour could become a maxim applicable to all'. Such a governance formula for sustainability governance could be framed as in the guidelines on cultural impact assessments under the Convention on Biological Diversity (CBD 2004; the *Akwé: Kon* Guidelines), which seek to bridge the environment/human rights divide in a very concrete way:

> Governments should encourage and support indigenous and local communities, where they have not already done so, to formulate their own community-development plans that will enable such communities to adopt a more culturally appropriate strategic, integrated and phased approach to their development needs in line with community goals and objectives.

2.5 Towards a Culturally Sensitive Metagovernance for Sustainable Development

How can cultural diversity, while respecting the need to have some common values, be reconciled with different approaches to sustainability governance? This section seeks an answer using the emerging concept of metagovernance: combining and managing governance style combinations which take into account the differences between value systems and traditions in different regions and for different communities.

2.5.1 Sustainability Metagovernance

I have so far argued that sustainable development cannot be promoted with a one-style-fits-all governance approach. The consequence of the failure of standard recipes (be it of the hierarchical, network or market type or a specific combination like the World Bank's 'good governance' or the 'governance-as-network-management' paradigm) should be to investigate whether it is possible to design and apply governance approaches which allow variation in time and place. For this purpose Jessop (1997) coins the concept *metagovernance*, or governance 'beyond' governance.

Metagovernance can be defined as 'producing some degree of coordinated governance, by designing and managing sound combinations of hierarchical, market and network governance, in order to achieve the best possible outcomes' (Meuleman 2008). A 'metagovernor' aims to prevent or mitigate governance style conflicts, and understands how to combine governance style elements into a productive approach. He or she also knows when and how to switch from one style to another. This seems very ambitious, but there are experienced public managers who do this by intuition and find it nothing special. Two quotes, the first from a leading manager in the Dutch Environment Ministry, and the second from a police manager of one of the largest cities in the Netherlands, can illustrate this (Meuleman 2008: 146, 214):

> In a complex and constantly changing environment a Ministry has to be flexible, always problem-oriented and impact-sensitive, and ask itself: does our governance approach deliver the expected results?
>
> We are chameleons: We switch between styles depending on the situation at hand. People in our organisation have a sense for this. When an incident occurs, they know that there is no time for discussion. Nobody asks "Why?", "Shouldn't we involve other parties?", "Isn't this too expensive?". After the incident, network and market governance elements reappear.

Biermann et al. (2010) define a similar overarching concept, namely a 'global governance architecture', which concerns the meta-level of governance for global climate change. However, this approach still has a hierarchical bias. A more participatory approach is proposed by Hoogeveen and Verkooijen (2010), who develop a portfolio approach for forest governance, which could be useful for broader sustainability issues. They suggest moving from negotiating grand agreements towards negotiating and then managing 'portfolios of instruments and the provision of the convening space in which they can operate and be nurtured coherently' (ibid. 2010: 154).

If we accept that it is impossible to determine which governance approach is *in general* the most successful, it makes no sense to design standardised approaches. What can be standardised, however, are mechanisms which increase the chance that successful governance emerges in a certain situation. Such mechanisms should take into account the existing preferences of powerful actors (governmental or non-governmental), as well as the cultural and administrative history of a location.

The concept of metagovernance provides such a mechanism. In order to make sustainability governance culturally sensitive, permanent and systematic attention is required to translate or adapt possible solutions into ones which work well in a given cultural setting. This I would call culturally sensitive sustainability metagovernance.

Several questions arise with regards to the concept of metagovernance. The first is which dimensions of governance should be involved. Should the focus be on institutions and transformation processes, or also on leadership styles, core values, preferred instruments, and so on? The answer lies in the situational core of metagovernance: metagovernance implies taking a bird's eye perspective, and takes into account all dimensions which are relevant in a given – or framed – case, and all three prototypical forms of these dimensions. Take the example of the dimension of leadership: Grint (2010) presents the example that each classical governance style can lead to an addiction: advocates of hierarchy can become addicted to command and neglect all 'wicked' problems; egalitarians who prefer consensus tend to turn any problem into a 'wicked' problem, and individualists (often advocates of market governance) tend to turn every problem into a standard problem which is subject to 'the correct understanding of cause and effect'.

Another question is who determines the relevant governance dimensions. Some argue that metagovernance is a return to the central state: a new and hidden type of hierarchical steering; metagovernance is steering at a distance, but still steering, and will have some centralising effects (Peters 2011: 9). Moreover, metagovernance studies are said to rely on a central role of states (Glasbergen 2011: 194). A solution to this problem has been advocated by Sørensen (2006: 100) and Aagaard (2009). They position the metagovernor as a 'hands-on', neutral actor who takes direct part in the policy process, but has no formal authority on behalf of the other actors. This requires that such a metagoverning process manager is trusted by all relevant parties. It can be questioned whether this is feasible for global governance issues, because in the global arena trust is a scarce value, as could for instance be observed during the first two preparatory sessions of the UN Rio 2012 conference in 2010 and 2011. Indeed discussions at this conference between the developing nations (united as the 'G77') and the group of richer nations were tense.

I would propose a different approach, based on two considerations. Firstly, in line with the situationality of metagovernance, a metagovernor may prefer to force other actors to comply (hierarchical governance, using coercion, a 'stick'; or convince actors of self-regulation and competition mechanisms, which are principles of market governance, using an incentive, a 'carrot'). He or she can also start a process with other actors on the basis of mutual dependency and voluntary cooperation (network governance; aiming at mutual gains). Mixed forms have also been described, such as co-opetition (e.g. Teisman 2001), a neologism originating from game theory, which describes cooperative competition. The term co-opetition expresses a combination of network and market governance. A state actor therefore may not have to choose a 'steering' approach.

Secondly, metagovernance may be considered as a process approach which can be used by *all* involved actors, in their own way. Certainly, a governmental actor who looks beyond old or new orthodoxies such as Weberian 'steering', New Public Management inspired 'rowing' or a consensus-searching network governance model, acts as a metagovernor. Having said this, there is no reason why a business actor or a civil society organisation should not also think beyond orthodoxies (Glasbergen 2011). If such actors embrace the metagovernance philosophy, a competition might emerge between different actors with regards to who takes the lead, but it is at least an informed governance competition which may increase the number of policy options in a given policy theme. Such a multi-actor metagovernance means, however, that building governance capacity includes training in metagovernance. Moreover, such meta-capacity building must be culturally informed. For example, in consensus-oriented societies such as the Netherlands and the Scandinavian countries, training would have to emphasise the typical failures of network governance (never-ending talks, risk of manipulation) and present insights into how to overcome these failures through introducing structures and rules. These structures and rules could pertain to issues such as determining the width of a network, the patterns of relationships of the members of a network, and determining the main issues on the network's agenda (Schvartzman 2009). Other compensation measures are ensuring transparency and introducing values from hierarchical governance like legitimacy and reliability. An interesting example of such a 'bottom-up metagovernance' is the successful 2008 action to clean the forests of Estonia from garbage in 1 day. The organisers, a group of citizens, used network governance as the key mechanism to motivate 50,000 citizens, but added efficiency (market governance) and an almost military (hierarchical) operation mechanism.[5]

Metagovernance therefore does not have to be a new shape of central steering. Sustainability metagovernance includes the necessity of metagoverning roles of more than only governmental actors in the sustainability debate. This poses an important design question for conferences such as UNCSD (Rio) 2012.

In the next section, we will briefly discuss institutions, as a crucial governance dimension, and question the common assumption that they always epress dominant, hegemonic approaches; we will also suggest how the mechanisms of culturally informed sustainability metagovernance could look.

2.5.2   Cultural Diversity and Institutions

The meaning of institutions has been phrased as 'using rules and tools to cope with the commons' (Ostrom 1990: 219). This section discusses the cultural dimension of institutions relevant for sustainable development. Contrary to the common practice of considering institutions and organisations as synonyms, the term institutions should here be broadly defined as sense-making arrangements, which are the rules of the game. These rules realise values in society and produce meaning. Such a broad definition includes interpersonal societal structures, organisations, mechanisms and orientations. Some consider the institutional dimension so important for sustainability governance that they propose it as the fourth dimension of sustainable development (Spangenberg 2005: 28–29), after the environmental, the social and the economic dimension. However, we have seen that culture can also be seen as the fourth dimension of sustainability (Nurse 2006:45), and there may be more candidates queuing. Therefore, here the institutional dimension of sustainable

---

[5] A short video summarising this endeavour can be seen on http://bit.ly/DZmMg.

development will be considered as one of the essential governance dimensions of sustainability: the rules dimension of decision-making and implementation.

The 'institutional framework for sustainable development' was one of the two *foci* of the UNCSD 2012 Rio conference. The reason why institutions were chosen as the focus lies in the weak implementation record of sustainability policies. An analysis by the Mapping Global Environmental Governance Reform project shows that:

> an impressive institutional machinery has actually been built, but also that the overall state of the global environment seems not to have improved as a consequence of this. (Staur 2006)

Which governance institutions and mechanisms could generate change? Here we must heed Machiavelli's warning to avoid wishful thinking and start with the world as it is. It is pointless to preach to consumers about abandoning their cars and plane travel, or to admonish companies to give priority to sustainability. Economic activity is deeply embedded in economic and social institutions, and companies are constrained by corporate governance, capital markets, competition, and the wider consumer culture.

The current institutional framework for sustainability governance is a patchwork or mosaic which is often labelled as 'fragmented'. This is not a neutral label: behind the term 'fragmented' lies the assumption that integration and centralisation are the most appropriate principles for institutional design. However, in the diverse, globalised world of today, other design principles might be necessary. While recognising that existing institutions will be defended by governmental and other stakeholders who have been part of their establishment, a first step can be:

> ... the willingness to discuss diverse world views, and to recognise that the situation in twenty-first-century society can no longer be adequately represented by institutions and values from times gone by. (Verweel and De Ruijter 2003: 15)

Besides being 'fragmented', the framework of (inter)governmental institutions for sustainable development and climate change contains many political players who are relatively weak, such as in environmental and sustainability policies like the UNEP and UNCSD.[6] The gap between the knowledge of threats and the adequacy of institutional response seems at its largest in environmental policy (Weiss and Thakur 2010: 215). However, decision-making and implementation concerning environmental and climate change problems is, at least to some extent, based upon well-defined legally binding commitments. In sustainable development it is mostly declaratory, defined by corner stones like the 1992 Rio declaration, or the Millennium declaration.

The institutions we focus on in this section are those which shape or obstruct a successful governance of sustainability in diverse cultural settings. Like other institutions, they follow different logics, according to the logic of the governance

---

[6] UNEP: United National Environmental Programme; UNCSD: United Nations Commission for Sustainable Development.

style (or – combination) in or for which they are designed. If we define governance as the relationship between government, civil society and business when solving societal problems and creating societal opportunities, three different institutional logics should be distinguished through the logic of hierarchies, the logic of networks and the logic of markets (Meuleman 2008):

- The logic of hierarchies: Hierarchical governance produces centralised institutions, which work on the basis of authority, with rules and regulations and imperatives. Institutions are, for example, legally based agreements. This aligns with classical representative forms of democracy, but also with authoritarian types of ruling. Decisions are made top-down. Government is central. Blueprinting is an engineering term which aligns well with the hierarchical logic. Hierarchical institutions are often best suited for dealing with emergencies and disasters, and for control tasks.
- The logic of networks: Network governance tends to produce more informal institutions in which trust and empathy are key values. Examples are covenants and Internet communities. This shares a logic with deliberative forms of democracy. Decisions are made together. Government is a partner in society. Network institutions have proven to be able to lead to ways out of 'wicked', complex and disputed problems.
- The logic of markets: Market governance aims at small, decentralised government, and at using market types of institutions such as contracts, incentives and public-private partnerships as well as other hybrid organisations. Decisions are made bottom-up, through mechanisms like the invisible hand of the market. Government is a societal service-provider. Market institutions with their focus on autonomy and efficiency are best for routine problems.

Key characteristics of the three prototypical institutional logics are compiled in Table 2.2. As argued in Sect. 2.5.1, the logic of metagovernance implies that situationally effective combinations of the three prototypical logics should be made. It is therefore impossible to describe an optional institutional framework for sustainable development, even when such a framework would include different institutions at different levels of government.

Most of the current institutions for climate change mitigation policy – a key sustainability challenge – are based on the hierarchical logic, and part of the intergovernmental discussions on renewal of sustainability governance show similar premises (Meuleman 2010b). Six implicit hierarchical premises can be observed. The first is a preference for centralised negotiations and institutions, such as the UNCFFF climate conventions, the Convention for Biological Diversity and the UNCSD sustainability summits. The second is the conviction that in the end, governments should be the only decision makers, whereas other actors are also able to make relevant decisions and take responsibility. Thirdly, there is a broadly shared belief that only legally binding decisions are effective. This can be illustrated by the first question on the cover of a recent edited book, written by a team of 30 leading experts from the European Union and developing countries, which is: 'What is the most effective overall legal and institutional architecture for successful and equitable climate politics?' (Biermann et al. 2010). The framing of

Table 2.2 Key characteristics of institutional logics (After Meuleman 2008)

	Hierarchical logic	Network logic	Market logic
Culture/Way of life	Hierarchism	Egalitarianism	Individualism
Theoretical background	Rational, positivism	Social constructivism	Rational choice theory
Primary virtues	Reliable	Flexible, discretion	Cost-driven, maximising advantage
Motives	Minimising risk	Satisfying identity	
Roles of government	Ruler of society	Partner in a network society	Delivery of societal services
Actor perception	Subjects	Partners	Customers, clients
Organisational orientation	Top-down, formal, internal	Informal, reciprocity, open-minded	Bottom-up, suspicious, external
Aim of stocktaking actors	Anticipation of protest/ obstruction	Better results and acceptance	Finding profitable contract partners
Organisation structure	Line organisation, centralised	Soft structure, few rules/regulations	Decentralised, autonomy
Flexibility	Low	Medium	High
Roles of knowledge	Supporting authority	Shared good	Competitive advantage
Type of knowledge	Authoritative	Agreed knowledge	Cost-effective
Coordination via	Imperatives	Diplomacy	Competition
Control through	Authority	Trust	Price
Communication style	Giving information	Organising dialogue	Influencing, PR
Leadership style	Top-down	Coaching, support	Empowering
Relation type	Dependent	Mutual dependent	Independent
Typical values	Legitimacy, accountability, justice	Community, empathy, harmony	Self-determination and -realisation
Affinity with problem types	Crises, disasters, legal issues	Complex, wicked problems	Routine, non-sensitive issues
Typical failures	Ineffective, red tape	Never-ending talks	Market failures
Typical perversions	Abuse of power (e.g. clientelism)	Abuse of trust (manipulation)	Abuse of money (corruption)
Preferred type of instruments	Laws, regulations, compliance	Consensus, agreements, covenants	Service, contract, product

their question reveals an assumption that there is one solution to the problem of climate change, and it is a legal one.

The fourth premise is the preference for (mono)disciplinary, 'authoritative' science, which suggests that practical knowledge is not relevant, and that scientific authority is a given thing. Both climate mitigation and adaptation policy processes are technology based, large-scale and top-down (Ayers et al. 2010: 271). This presumes a stable, clear and predictable world. However, the design and implementation of climate change policy takes place in a changing world, in which we not only bear witness to a range of interrelated global environmental problems, but also to turbulent economical and geopolitical changes. Although the IPCC process seems modern because of its network-type consensual process within the science community, the results – quite paradoxically – seem to be weighed by politicians and the media on the parameter of classical scientific authority: science should produce the truth and nothing but the (one) truth. At this point it is worth referring to the statement of a Dutch Environment minister on 3rd February 2010, on Dutch television: 'I will not accept any more mistakes from the IPCC. As a politician I must be able to have blind trust in what science says'.

The fifth premise which relies on the assumption of predictability underlying hierarchical governance is the promotion of 'best practices'. Notwithstanding the current implementation deficit of sustainability governance, many seem to believe that simply copying/pasting successful approaches to sustainability from one situation to the other, is a guarantee for success.

Finally, the language of sustainability governance is often centralist. There seems to be a preference for monolithical concepts which make the world less complex: *The* climate, *the* economy, *the* democracy, *the* culture – as if there are not many different forms which should be taken into account. The concept of 'planetary boundaries' is a similar monolithical term (see also Schmidt 2012, in this volume). Some even go one step further and use capitals for their monolith: *the* 'Earth System'. The current institutional framework is often labelled as 'fragmented', which implicitly means that integration is always better. Indeed, Biermann et al. (2010: 309) conclude after having framed climate governance as heavily fragmented, that 'a strongly integrated climate architecture appears to be the most effective solution'. Finally, in (global) governance jargon there seems to be a preference for centralised 'coordination' instead of decentralised 'cooperation' or 'collaboration'.

It is difficult to see how such an approach can provide successful answers to the complexity of climate change and related problems like hunger, water crises and migration. As long as climate change policy is considered as a top-down, state-run operation, it is bound to fail (Leggewie and Welzer 2010). Even with the inclusion of non-state actors, as in the model proposed by the German WBGU (2011), it is questionable whether central steering models are implementable, as they deny the complexity and plurality of our times, and of Beck's 'second modernity'.

However, all existing governance institutions are embedded in the current system, and thus it is naïve to simply specify 'ideal' governance institutions that would, for example, create a high global price for carbon, mandate clean production systems, and empower non-financial stakeholders (Levy 2011: 84).

We have seen that the different assumption values behind hierarchical, network and market governance are reflected in different institutional logics. When – as this chapter argues – there is a move away from the unsuccessful dominance of hierarchical governance, towards more network governance and some market governance, this should be reflected in the logic of new institutions: the institutional framework for sustainable development should, on all levels and scales, recognise that a shift is necessary from a primarily hierarchical approach towards a more horizontal logic: more partnerships, new alliances, voluntary agreements, exchange of practices, capacity building, and so on. There are already numerous examples.

Firstly, the idea that we live in a networked world, in which the measure of power is connectedness (Slaughter 2009: 94), has highlighted the enormous possibilities of large networks which the ICT revolution has made possible. Multinational companies have already shown the way. Companies like Procter and Gamble, Boeing and IBM have switched from hierarchical strategy formation to network forms, with a system of peer production, suppliers becoming partners, and by use of social network options of the Internet (Slaughter 2009: 97).

Small networks could also be used more often. New informal approaches have begun to develop. The government of the Netherlands, for example, has decided after the disappointment of the 2009 Copenhagen climate conference, to apply a more varied approach to climate governance, which is not directed against the traditional central (UNFCCC) approach, but complements this. One of the innovations which have emerged in the aftermath of the Copenhagen conference is the establishment of the so-called Cartagena Dialogue. This is a parallel process to the formal UNFCCC negotiations in which approximately 20 nations participate, looking for new ways forward and concrete action, in an informal way. It is a network form of governance, based on mutual trust and partnership.

The next example is bilateral cooperation between nations, with the additional involvement of non-governmental partners. The Netherlands, for example, supports Columbia with knowledge of the TNO applied research institute, to develop an emission registration system. This is an investment in good relations, with very concrete results, and departing not from the premise of what should be done, but what can be done.

Another innovative approach is the new direction in behavioural economics (see for example Thaler and Sunstein (2008) which introduces voluntary institutions. The architecture of choice is then designed in such a manner that people behave voluntarily in the way the architects feel is desirable. It is about creating new behavioural defaults. This type of institutional setup is of course subject to the reproach of manipulation. However, if these architects' activity is accepted in a democratic process then maybe these institutions would be sustainable themselves and at the very least, would be relatively cost-effective. At this point it is worth thinking about the development of benign markets. This opens up another set of fascinating ideas about institutional rearrangements which might be utilised for sustainable development in different cultural contexts. A good example is the proposal to replace the pollution model of minimising greenhouse gas emissions with a mutual gains approach based on the right of universal access to clean, low-carbon energy services (Moonaw and Papa 2012). This is analogous to the ecosystem services approach in environmental policy.

A last example which reflects a more horizontal orientation of institutional design involves covenants between governments or other public authorities and private companies, in which common targets are set, and are in use in the UK and the Netherlands. Such covenants combine the voluntary attitude of market governance, the network governance principles of mutual dependency and trust, and are supported by a formal, legal framework (hierarchical view). Such forms of co-regulation could also be useful in a multi- or international setting, for example for climate governance (Telesetsky 2010).

To conclude, it seems important to investigate connectivity strategies between institutions, and their implications for institutional change. We should also direct more analytical power to the legal system, and learn from 'jailed institutions'. Finally, in line with the 'and', not 'or' argument of second modernity mentioned earlier, it might be wise to consider institutional redundancy as useful rather than inefficient.

2.5.3 Principles for Culturally Sensitive Sustainability Metagovernance

If there is a strong focus on centralised governance, how can sustainability governance then become more varied? Ostrom (1990: 14) points out that 'getting the institutions right' is a difficult, time-consuming, and conflict-evoking process. It is a process 'that requires reliable information about time and place variables as well as a broad repertoire of culturally acceptable rules'.

As there is not one singular approach, we should concentrate on which governance principles could be useful for the design of effective institutions and transformation processes for sustainability. Besides the often-mentioned principles of *reflexivity, resilience, transparency* and *inclusiveness*, others might have additional value. The first is that *problem-orientedness* should be a point of departure: any governance design should start with a transdisciplinary analysis of the problem in its context. The well-structured internal Impact Assessment procedure of the European Commission applies for all its proposals and could be a good example. In addition, *temporality* is important: The terms 'time' and 'place' in Ostrom's quote above refer to what I would call, respectively, the temporality and the locality principle. This specifies that governance has the potential to be diverse, and that governance for sustainability is multilevel, -scalar and -actor. Another principle is therefore *locality*: the focus on hierarchical governance leads to a concentration on administrative areas and scales, and thus to neglecting the need of exceeding such barriers when dealing with certain societal problems. The feasibility of governance also depends on what works best given the physical borders of a certain problem; such borders may be very different from, for example, national borders. Water systems are a good example: for historical reasons national and regional borders often follow the course of rivers. Water management should take into account the whole catchment area of a river. To overcome this problem, by 1950 on the scale of the Rhine basin, European countries and regions had already established the International Commission for the Protection of the Rhine (ICPR), an example of a governance arrangement which follows the geography of a problem.

Ostrom's references to conflicts and cultural acceptance point at the cultural dimension. This could lead to a principle which has been framed as *culturality*

(Abdallah-Pretceille 2006: 497); culturality refers to the understanding of cultural phenomena 'based on dynamics, transformations, fusion and manipulations'. Abdallah-Pretceille argues that the variety of cultural fragments and cross-cultural exchange has become more important than cultures in their entirety: they help to make sense of what happens in our contemporary globalised world. Therefore, an institutional framework for sustainability governance could profit from a cultural assessment: what are key values linked to both the objectives of sustainability and of the *problematique*, and how can they be reconciled? Simply put, how can they be made compatible? This could result in different approaches in countries with an individualist value pattern and in nations with a collectivist culture. In this context, it is important that we understand that some of the well-known models of value patterns, like Maslow's hierarchy of human needs, have a strong Western bias. In Maslow's model, individualist self-actualisation represents the top of the pyramid, whereas in collectivist countries like China, the basic need is 'belonging', and self-actualisation concerns societal needs (Gambrel and Cianci 2003).

Another principle could be *polycentricity*, a concept introduced by Ostrom et al. in 1961. The basic idea is that 'any group of individuals facing collective problems should be able to address that problem in whatever way they best see fit' (McGinnis and Walker 2010: 294). Polycentricity shares decentralism and self-regulation with market governance. It is not anti-state, but 'building a polycentric system (...) acts as a spur to national and international regimes to get their act together!' (Ostrom 2010, about climate change governance).

The last principle I propose is *historicity*. Institutions are mortal, but they can survive a long time, much longer than the objective or the policy theory from which an institution originates. There is a gradual dialectic dynamic of funding values, in which these values are destroyed by non-intentional effects of formalisation. This is the inherent curse of formalisation. Taking into account the historical experience, or historicity, is therefore an important principle. Changing the underlying mental model of current institutions requires an understanding of the mental model on which they are based (Stahl-Role 2000: 28).

Such principles might help to decide what should be done in a specific situation with regards to top-down, bottom-up, or a combination of the two.

2.5.4 Actor Perspectives

> Princes and governments are far more dangerous than other elements within society.
> (Niccolo Machiavelli)

When Machiavelli wrote about state power, he did not think that non-state actors were very influential in determining the course of society. Of course, this has changed considerably. Non-state actors have both informal and formal influence and cannot be neglected when governance challenges are discussed. Public access to information and to decision-making processes has become a right in many countries, and at the UN level, nine 'major groups' have been distinguished who are invited to be involved in the debate on sustainable development.

Therefore, the question which should be answered is that of which role cultural diversity plays in different actor arenas that are relevant for sustainable development. In political science, four types of actors are often distinguished: political and administrative decision-makers, business actors, civil society organisations, and science representatives. We could distinguish a fifth type: boundary workers, whose task it is to link and translate between the other types.

Beauty is in the eye of the beholder, as are other frames of reality. Different societal actors therefore have different opinions on how to design sustainability governance. Concerning governmental actors, we should realise that classical bureaucracies modelled on the Weberian ideals are usually fuelled by hierarchical values. Hofstede (1980) argues that the cultures of political and administrative cultures are influenced by national cultures. For example, low-trust societies with a high preference for uncertainty reduction usually have developed large bureaucracies. The Weberian model has dominated the political style for decades. Its hegemony was only tested during the 1980s, when in Western countries, but was also disseminated to developing countries by the World Bank and IMF. With this, the New Public Management movement became an extremely successful campaign to undermine public support for government (Lipsky 2008). The result has been a heightened public distrust in government; governments 'were denigrated as a set of failed institutions inherently incapable of responding to critical social needs' (Lipsky 2008: 143). This is one of the reasons why political and administrative actors have become dependent on other actors. If the weakness of the state is not compensated by a broad 'social pact' between state and non-state actors, emergencies such as the Katrina hurricane disaster in the USA and other environmental crises, cannot be dealt with optimally. Governmental actors must therefore find productive relationships with other actors. This is not only the case for disaster prevention and management, but maybe even more so for 'wicked' problems such as the vast number of sustainability challenges. The value-laden character of such problems requires an understanding and preparedness in order to deal with cultural diversity in a constructive way.

Although environmental concerns are the cornerstone of sustainable development, the economic dimension is still dominant. It has been argued that mainstream notions of sustainable development do not challenge neo-liberal economic hegemony because they share hegemonic premises and growth as well as efficiency, which are part of the sustainable development discourse (Nurse 2006: 35). From a business perspective, cultural diversity is relevant in corporate strategies, and with regard to dealing with globalisation and the role of the private sector in economic development as part of the sustainability challenge. The private sector is increasingly engaged in sustainable development. Frontrunner companies have united in, for example, the World Business Council for Sustainable Development, and call for national governments and international organisations to stimulate sustainable innovation, create level playing fields and punish free riders and other laggards.

Civil society organisations are, more than governmental actors, broadening their institutional perspective with soft, informal structures, often using social media as their communication platform. They profit from the increasing self-organisation and capacity of social media through the Internet. At the same time, the existence of established civil society organisations can also be threatened by the rise and fall of instable, one-issue, Internet-based pressure groups. A metagovernance perspective on sustainability governance might offer established civil society organisations competitive advantage. However, they could choose to not only lobby and influence governmental and business actors, but to also take more responsibility for action themselves, by, for example, establishing alliances with private companies and/or public-sector organisations. The future roles of science in the context of sustainability governance have already been adequately addressed in this volume by Jungcurt (2012).

Finally, if the effect of sustainability governance depends on its contextuality, this implies that there is a need for institutions whose remit is to translate and transform different visions, knowledge and problem perceptions into situationally working variations. Such boundary work organisations, such as national sustainable development councils, have no power themselves, but can be hugely influential with regards to the institutional framework. Their main functions are giving policy advice, acting as agent and facilitator, as well as communication and stimulating involvement (Niestroy 2007).

2.5.5 Sustainability and the Unpredictability of Crowds

A roar of grief and rage rose over the city and boomed, relentless, obsessive, sweeping away any other sound, beating out the great lie. *Zi, zi, zi,* He lives, he lives, he lives! A roar that had nothing human about it. In fact, it did not rise from human beings, creatures with two arms and two legs and minds of their own; it rose from a monstrous, mindless beast, the crowd, the octopus that at noon, barnacled, with clenched fist, distorted faces, contracted mouths, had invaded the square of the orthodox cathedral, then stretched its tentacles into the nearby trees, jamming them, submerging them, implacable as the larva that overwhelms and devours every obstacle, deafening them with its *zi, zi, zi*. (Fallaci 1981)

It is useful for sustainability governance to categorise societal actors into classical clusters as demonstrated above and because this is the way in which they are usually organised. However, social reality can also be strongly influenced by ad hoc social groups with their own, particular behaviour. Basten (2010) introduces the term 'public' for a temporary community which only exists around a special event or emergency, and has an action perspective. This can be the burial of a famous and loved person like the British Princess Diana, or that of the Greek activist Aleksis Panagoulis, who is the hero in the above citation.

Publics can have two faces. One is a wise one: according to Surowiecki (2004), there is a lot of wisdom in crowds. If the success of certain collective actions is extrapolated, it may be that unpredictable 'publics' will play an important role in the future of sustainable development.

The other face of such 'publics' is a darker one. The description of the 'octopus' by Oriana Fallaci is an example of Wallace's (1970: 234) madness of crowds. Examples are 'mass panics, group delusions and illusions, mass hysterias, and mob violence'. He argues that such events depend either on a specific situational context (a threat, combined with a limited escape route feeds panic), or 'the dissociation effect on the individual of repetitive mass suggestion in a crowd', which can lead to, for example, a lynch mob. Interestingly, Wallace observes that such phenomena are in a way a-cultural: a 'mad crowd consistently violates culturally accepted norms'. Social media are twenty-first century creatures, which probably also have this dual character: wisdom and madness.

2.6   Transgovernance: One Step Beyond...

In this chapter I conclude that sustainability governance should be culturally informed, and that this requires the situationality and multi-perspectivity of a metagovernance attitude: combining governance style elements together could function well in a specific situation. I would argue, however, that one step further is recommendable. Due to the challenges and constraints of an emerging knowledge democracy, the second modernity concept (not 'or' but 'and': plurality), and the wickedness of sustainability problems, an awareness is needed which goes beyond the metagovernance method. The bird's eye view, which is typical for metagovernance, could be useful for sustainability governance in many ways.

We do not need a new paradigm or a new orthodoxy, but should develop a sensitivity and capability which we have framed in this volume as transgovernance. The well-known quote attributed to Einstein 'We can't solve problems by using the same kind of thinking we used when we created them' seems appropriate here. The challenge is therefore to get politicians and scientists out of their 'comfort zone' into trying new approaches. Transgovernance implies looking beyond classical governance styles and towards a culturally sensitive metagovernance for sustainable development. It is an approach beyond disciplinary scientific research, towards more transdisciplinarity, and beyond borders formed by states and other institutions, towards trans-border approaches. Transgovernance is beyond conventional means for measuring progress, towards new and more interactive measuring methods, and beyond linear forms of innovation, towards open innovation. Last but not least, it is an approach to governance which is beyond cultural integration or assimilation, and towards searching for compatibility.

2.7   Summary and Conclusions

Against the background of the multiple tensions between old and new forms of politics, science and media – the emerging knowledge democracy – we have seen that sustainability governance is a double normative construct (both terms are normative). It is currently characterised by a dominance of centralism as design strategy, a neglect of the complexity and 'wickedness' of the challenges, and ignorance of the cultural dimension. The first two characteristics belong together: centralist thinking about solutions to societal problems tends to lead to the political construction of central, simplified problems to which classical hierarchical governance approaches can be applied. The centralist focus is visible in the belief that solutions to climate change or biodiversity should in the first place be legally binding. Such a belief is underpinned with globalised research and knowledge, based on reductionist, monolithical frames (the climate, the biodiversity, the economy, the media), and on challenges which, to a large extent, possess the characteristics of 'wicked problems'. Each of these problems is unique, value-laden and reflexive.

Initial conclusions are that the wickedness of many sustainability challenges implies that sustainability governance will depend largely on the success of non-hierarchical governance approaches; the usefulness of additional legal constructions depends on the context. Wicked problems are value-laden, as are the terms 'governance' and 'sustainable development'. Therefore, values and traditions, and hence the cultural dimension, must be included in sustainability governance. The cultural dimension is, for example, embedded in the notion of 'glocalisation', which points at a twinning of globalisation and localisation. Globalisation may have made cultures increasingly ambivalent, ambiguous and paradoxical, but the counter reaction, localisation, is equally important for effective governance: identity provides security, which is a condition for relating to other cultures with an open mind.

In this chapter we have concentrated on the cultural dimension of governance for sustainable development. The key question is formulated as: How can cultural diversity contribute to sustainable development (meta)governance, and what can be done to prevent it from becoming a hindrance? We start with the latter. The centralist bias in sustainability governance is congruent with a widely shared conviction among decision makers that cultural diversity is an obstacle. Different strategies are applied to mitigate the perceived problem, such as considering cultural diversity as a taboo, and promoting cultural assimilation (often euphemised as 'integration'). In social and natural sciences, inter- and transdisciplinary research on the cultural dimension of sustainable development is lagging behind. Political science, anthropology, sociology and ecology have continued to study only parts of the puzzle. This has hindered insight into the broader picture, the consequences being that political decisions regarding sustainability on all governmental levels may be ill-informed with knowledge about existing values, traditions and practices, and therefore also ill-informed about the possibility of implementation.

This lack of knowledge does not prevent national governments who lag behind with regards to the implementation of sustainable development objectives, from playing the cultural card as an argument for not being overly ambitious. Besides lack of knowledge, another important reason for considering cultural diversity as a hindrance to sustainability is a hegemonic attitude towards other cultures. Because values are what you believe in, it is only logical that people consider their own culture 'better' than those of other people. However, this is not the point. Our conclusion is rather that cultural hegemonism is almost a guarantee for further stagnation of the implementation of sustainable development strategies; what is needed is investment in cultural pluralism.

Policy makers and scholars alike have found the analogy between cultural diversity and biodiversity attractive. It is, however, a problematic analogy: concepts like risks, resilience, (eco)system services and keystone species may have very different meanings in biological and social systems. Nevertheless, sustainability could profit from the introduction of cultural diversity besides biodiversity as an objective of sustainability. However, as this is based on the conviction that humankind is part of nature, this is a normative, not scientific consideration.

In addition, we can conclude that, as biodiversity (or broader, nature) and cultures are interlinked in many ways, understanding sustainability processes requires an understanding of cultures, which in turn necessitates understanding biodiversity.

After having discussed the relationship between cultural diversity and sustainable development, a third normative dimension is introduced, governance. Governance is defined as a relational concept. In order to link cultures and sustainability, it is crucial to note that three – usually combined – governance styles can be distinguished: hierarchical, network and market governance. Each of these prototypical styles is consistent with specific cultural values and has its own institutional logic.

National or regional, and even local governments may be inclined to use a specific style mixture regardless of the character of the societal problems to be addressed, as well as of local, regional or national values and traditions that co-determine which governance style combinations may work well. Sometimes there are clear underlying national preferences for certain styles, such as a consensus style in the Netherlands, hierarchy in Germany and France, and market mechanisms in Anglo-Saxon countries. In many other cases this is much less clear. However, without adapting sustainability governance to what makes sense for, and is acceptable to people, all governance attempts risk failure. The term 'best practice' suggests universal applicability and should be replaced by 'good practice'.

We have seen that framing cultural diversity as an obstacle to sustainability has been a welcome political excuse to do nothing, and continues to be the dominant view, partly because of the lack of inter- and transdisciplinary research. In order to turn this around, a different view of cultures in the context of sustainability governance is required. Building sustainability governance on cultural diversity and investing in the compatibility of values and practices rather than in assimilation, will lead to an increased variety of solutions to similar problems. This is superior to the current practice in which centrally proposed solutions are accepted in some cultures and rejected in others.

Each discussion on cultural diversity leads to the question of whether there are also universal values, and if yes, how are they related to the premise of diversity? The paradoxical situation which we apparently want, is expressed both in the European Union's and Hindu motto 'Unity in diversity' and in the 'E pluribus unum' of the USA. The message may be that there are merits in this being a never-ending discussion. The question of how to make the trade-off between unity and diversity in sustainability governance is relevant. This is because, as we have seen, there is a dominant coalition pushing the 'unity' side of the equation. It can be concluded that the dominance of centralism in sustainability governance leads to the assumption that moving towards more diversity does not have to imply a breakdown of the universal 'acquis' of, for example, human rights.

The next question is how to design and manage sustainability governance approaches which are situational. The first conclusion here is that if we accept that it is impossible to determine which governance approach is generally the most successful, it makes no sense to design standardised approaches. What can be standardised, however, are mechanisms which increase the chance that successful governance emerges in a certain situation. Such a mechanism is 'governance beyond governance', or metagovernance. In order to make sustainability governance culturally sensitive, permanent and systematic attention is required to translate or adapt possible solutions into ones which work well in a given cultural setting. This can be called culturally sensitive sustainability metagovernance.

Some have argued that metagovernance is a new form of hierarchy, because it implies someone (from the government) who coordinates the governance process. However, this does not have to be the case. There is no reason why business actors or civil society organisations should not also think beyond orthodoxies and act as metagovernors when they are involved in sustainability governance.

The different assumptions and values behind hierarchical, network and market governance are reflected in different institutional logics. This chapter argues that when there is a move away from the unsuccessful dominance of hierarchical governance, towards more network governance and some market governance, this should be reflected in the logic of new institutions. The institutional framework for sustainable development should, on all levels and scales, recognise that a shift is necessary from a primarily hierarchical towards a more horizontal logic: more partnerships, new alliances, voluntary agreements, exchange of practices, capacity building, and so on.

The same applies to the organisation of transitions or the 'management' of societal transformation. Sustainable transitions/transformations require dynamic mixtures of different governance logics, adapted to place-based values and traditions. In addition, leadership should be situational: sometimes steering, sometimes rowing, and sometimes surfing the waves.

I do not claim to have found a general recipe or a panacea for sustainability governance in a cultural context, but it seems that metagovernance as a mechanism, and a tool beyond standardised governance, can be useful. The bird's eye view, which is typical for metagovernance, could be useful for sustainability governance in many ways. We do not need a new paradigm, or a new orthodoxy, but should develop the sensitivity and capability to apply *transgovernance*. This is, as with metagovernance, a method rather than a prescription, and implies looking beyond classical governance style and towards a culturally sensitive metagovernance for sustainable development; beyond disciplinary scientific research, towards more transdisciplinarity; beyond borders formed by states and other institutions, towards trans-border approaches; beyond conventional means to measuring progress, towards new and more interactive measuring methods; beyond linear forms of innovation, towards open innovation; and beyond cultural integration or assimilation, towards looking for compatibility.

Acknowledgements   I would like to thank Maria Ivanova, Deborah Rogers, Harald Welzer and the TransGov project team members for their invaluable critical comments on earlier versions of this chapter, and Madelon Eelderink for her critical editorial comments.

References

Aagaard P (2009) Ideas and institutions in global governance – the case of microcredit. Paper presented a governance networks: democracy, policy innovation and global regulation, Ringsted, Denmark

Abdallah-Pretceille M (2006) Interculturalism as a paradigm for thinking about diversity. Intercult Educ 17(5):475–483

Appadurai A (2002) Diversity and sustainable development. In UNESCO/UNEP (ed) Cultural diversity and biodiversity for sustainable development. UNESCO/UNEP, Nairobi pp 16–20

Ayers J, Alam M, Huq S (2010) Global adaptation governance beyond 2012: developing-country perspectives. In: Biermann F, Pattberg P, Zelli F (eds) Global climate governance beyond 2012: architecture, agency and adaptation. Cambridge University Press, Cambridge, pp 270–285

Baker S (2009) In pursuit of sustainable development: a governance perspective. Paper presented at the 8th international conference of the european society for ecological economics (ESEE), Ljubljana, 29 June–2 July 2009 http://bit.ly/ulfG6s. Accessed 1 Dec 2011

Barnett J, Campbell JL (2010) Climate change and small island states: power, knowledge and the south pacific. Earthscan, London

Barth F (1969) Introduction. In: Barth F (ed) Ethnic groups and boundaries. Universitetsforlaget, Oslo

Barth F (1994) Enduring and emerging issues in the analysis of ethnicity. In: Vermeulen H, Govers C (eds) The anthropology of ethnicity. Beyond 'ethnic groups and boundaries'. Het Spinhuis, Amsterdam

Basten F (2010) Researching publics. In: In 't Veld RJ (ed) Knowledge democracy: consequences for science, politics and media. Springer, Heidelberg, pp 73–85

Beck U, Bonss W, Lau C (2003) The theory of reflexive modernization: problematic, hypotheses and research programme. Theory Cult Soc 20(2):1–33

Bennet JW (1990) Ecosystems, environmentalism, resource conservation, and anthropological research. In: Moran EF (ed) The ecosystem approach in anthropology: from concept to practice. University of Michigan Press, Ann Arbor, pp 435–458

Biermann F, Pattberg P, Zelli F (2010) Global climate governance beyond 2012: architecture, agency and adaptation. Cambridge University Press, Cambridge

Bronfenbrenner M (1973) Equality and equity. Ann Am Acad Pol Soc Sci 409(9):9–23

Cazorla-Clarisó X, Cañellas-Boltà S, Domingos-Abreu A (2008) Unity in diversity: perspective for long-term sustainability in Europe. In: EEAC Working Group SD (2008) sustaining Europe for a long way ahead: making long-term sustainable development policies work. Theme papers. EEAC, Brussels

CBD (2004) The Akwé: Kon guidelines. Secretariat of the Convention on Biological Diversity, Montreal

Cocks M (2006) Biocultural diversity: moving beyond the realm of 'indigenous' and 'local' people. Human ecology (C_2006) doi:10.1007/s10745-006-9013-5

Demil B, Lecocq X (2006) Neither market nor hierarchy nor network: the emergence of bazaar governance. Org Studies 27(10):1447–1466

De Schutter O (2010) The right to food. Report of the Special Rapporteur on the right to food, New York, 11 Aug 2010

De Ruijter A (1995) Cultural pluralism and citizenship. Cult Dynam 7(2):215–231, Sage: London

Dixon J, Dogan R (2002) Hierarchies, networks and markets: responses to societal governance failure. Admin Theory Praxis 24(1):175–196

Du Pisani J (2006) Sustainable development – historical roots of the concept. J Integr Environ Sci 3(2):83–96

Dwyer PD (1996) The invention of nature. In: Ellen R, Fukui K (eds) Redefining nature: ecology, culture and domestication. Berg, Oxford, pp 157–186

Fallaci O (1981) A man. Random House, New York

Ferro N (2009) The Chinese path to sustainable development, a critical analysis of reality and propaganda. SusDiv working paper 13.2009. Fondazione Eni Enrico Mattei, Milano

Florida R (2005) The world is spiky. The Atlantic monthly. http://bit.ly/mCuLvp Accessed 27 Nov 2011

Friedman Th (2006) The world is flat: a brief history of the twenty-first century. First updated and expanded edn. Farrar, Straus and Giroux, New York

Gambrel PA, Cianci R (2003) Maslow's hierarchy of needs: does it apply in a collectivist culture. J Appl Manag Entrepreneur 8(2):143–161

Geva-May I (2002) From theory to practice: policy analysis, cultural bias and organizational arrangements. Public Manag Rev 4(4):581–591

GIZ (2010) Diversity – driver for development. Deutsche Gesellschaft für Technische Zusammenarbeit GmbH, Eschborn

Glasbergen P (2011) Mechanisms of private meta-governance: an analysis of global private governance for sustainable development. Int J Strateg Bus Allian 2(3):189–206

Grindle M (2004) Good enough governance: poverty reduction and reform in developing countries. Govern Int J Pol Adm Inst 17(4):525–548

Grint K (2010) The cuckoo clock syndrome: addicted to command, allergic to leadership. Eur Manag J 28(4):306–313

Gould PR (1969) Man against his environment: a game theoretic framework. In: Vayda AP (ed) Environment and cultural behaviour: ecological studies in cultural anthropology. The Natural History Press, New York, pp 234–251

Hall S (1991) The local and the global: globalization and ethnicity. In: King AD (ed) Culture, globalization and the world-system. Macmillan, London, pp 19–39

Haque MS (1999) Sustainable development under neo-liberal regimes. Int Pol Sci Rev 20(2): 197–218

Hofstede G (1980) Culture's consequences: international differences in work-related values. Sage, Beverly Hills

Hofstede G (2001) Culture's consequences: comparing values, behaviours, institutions and organisations across nations. Sage, Newbury Park

Hofstede G, Hofstede GJ (2005) Cultures and organizations: software of the mind, Revised and expanded 2nd edn. McGraw-Hill, New York

Hoogeveen H, Verkooijen P (2010) Transforming sustainable development diplomacy: lessons learned from global forest governance. Wageningen University, Wageningen

Hulme M (2010) Problems with making and governing global kinds of knowledge. Glob Environ Chang 20:558–564

In 't Veld RJ (2010a) Kennisdemocratie: opkomend stormtij. SDU, The Hague

In 't Veld RJ (ed) (2010b) Knowledge democracy: consequences for science, politics and media. Springer, Heidelberg

In 't Veld RJ (2011) Transgovernance: the quest for governance of sustainable development. IASS, Potsdam

Jachtenfuchs M (1994) Theoretical reflections on the efficiency and democracy of European governance structures. In: Conference paper, European community studies association, 2nd World conference, Brussels, 5–6 May 1994

Jessop B (1997) Capitalism and its future: remarks on regulation, government and governance. Rev Int Pol Econ 4(3):561–581

Jungcurt S (2012) Knowledge management for international sustainable development governance in the context of the knowledge democracy. In: Meuleman L (ed) Transgovernance: advancing sustainability governance. Springer, Heidelberg, pp 332–354

Kao G (2011) Grounding human rights in a pluralist world. Georgetown University Press, Washington, DC

Kickert WJM (2003) Beneath consensual corporatism: traditions of governance in The Netherlands. Public Adm 81(1):119–140

Kolman L, Noorderhaven NG, Hofstede G, Dienes E (2002) Cross-cultural differences in Central Europe. J Manag Psychol 18(1):76–88

Cornell S, Kalt JP (2005) Two approaches to economic development on American Indian reservations: one works, the other doesn't. Harvard project on American Indian economic development and the native nations institute for leadership, management, and policy on behalf of the Arizona Board of Regents

Leeds A (1969) 1969 ethics report criticized. Newsletter of the American Anthropological Association 10(6). Reprinted in: Weaver T (ed) (1973) To see ourselves: anthropology and modern social issues. Scott, Foresman, Glenview, pp 49–50

Leggewie C, Welzer H (2010) Another "great transformation"? Social and cultural consequences of climate change, J Ren Sust Energy 2(3): Online: http://dx.doi.org/10.1063/1.3384314

Leggewie C, Welzer H (2009) Das Ende der Welt, wie wir sie kannten. Klima, Zukunft und die Chancen der Demokratie. Fischer Verlag, Frankfurt am Main

Lévi-Strauss C (1958) Structural anthropology. The Penguin Press, London

Levy D (2011) Private sector governance for a sustainable economy: a strategic approach. In: Najam A (ed) Beyond rio+20: governance for a green economy. Frederick S. Pardee Center, Boston University, Boston, pp 83–90

Licht AN (2001) The mother of all path dependencies: toward a cross-cultural theory of corporate governance systems. Del J Corp Law 26(1):147–205

Licht AN, Goldschmidt C, Schwartz SH (2007) Culture rules: the foundations of the rule of law and other norms of governance. J Comp Econ 35(4):659–688

Lipsky M (2008) Revenues and access to public benefits. In: De Jong J, Rizvi G (eds) The state of access: success and failure of democracies to create equal opportunities. Brookings Institution Press, Washington, pp 137–147

Majlergaard FD (2006) Release the power of cultural diversity in international business. Paper presented at the summit for the future on risk, Amsterdam, 3–5 May 2006

McGinnis MD, Walker JM (2010) Foundations of the Ostrom workshop: institutional analysis, polycentricity, and self-governance of the commons. Public Choice 143(3, 4):293–301

Meuleman L (2008) Public management and the metagovernance of hierarchies, networks and markets. The feasibility of designing and managing governance style combinations. Dissertation. Springer/Physica-Verlag, Heidelberg

Meuleman L (2010a) The cultural dimension of metagovernance: why governance doctrines may fail. Public Organ Rev 10(1):49–70

Meuleman L (2010b) Metagovernance of climate policies: moving towards more variation. Paper presented at the Unitar/Yale conference. Strengthening institutions to address climate change and advance a green economy, Yale University, New Haven, 17–19 Sept 2010

Moonaw W, Papa M (2012) Creating a mutual gains climate regime through universal clean energy services. Medford, Fletcher School, Tuft University

Moran EF (1990) Ecosystem ecology in biology and anthropology: a critical assessment. In: Moran EF (ed) The ecosystem approach in anthropology: from concept to practice. University of Michigan Press, Ann Arbor, pp 3–40

Niestroy I (2005) Sustaining sustainability. RMNO/EEAC/Lemma, Utrecht

Niestroy I (2007) Stimulating informed debate – sustainable development councils in eu member states. a compilation of tasks, capacities, and best practice. EEAC, Brussels

Nurse K (2006) Culture as the fourth pillar of sustainable development. Commonwealth Secretariat Malborough House, Pall Mall, London, pp 32–48

Ostrom E (1990) Governing the commons: the evolution of institutions for collective action. Cambridge University Press, Cambridge

Ostrom E (2010) Nested externalities and polycentric institutions: must we wait for global solutions to climate change before taking actions at other scales? Econ Theory published online 6 August 2010

Ostrom V et al (1961) The organization of government in metropolitan areas: a theoretical inquiry. Amer Pol Sci Review 55:831–842

Oswald Spring U (2010) Towards a sustainable peace in the Anthropocene. Paper presented at the annual meeting of the Theory vs. Policy? Connecting Scholars and Practitioners New Orleans, Feb 17, 2010. Online http://bit.ly/vZhfkY

Parillo VN (1994) Diversity in America: a sociohistorical analysis. Sociol Forum 9(4):523–545

Peters BG (2011) Steering, rowing, drifting, or sinking? Changing patterns of governance. Urban Res Pract 4(1):5–12

Posey DA (1999) Cultural and spiritual values of biodiversity: a complementary contribution to the global biodiversity assessment. In: Posey DA (ed) Cultural and spiritual values of biodiversity. UNEP and Intermediate Technology Publications, London, U.K., pp 1–19

Procee H (1991) Over de grenzen van culturen :voorbijuniversalisme en relativisme. Boom, Amsterdam

Rittel HW, Webber MM (1973) Dilemmas in a general theory of planning. Pol Sci 4:155–169

Robertson R (1995) Glocalisation: time-space and homogeneity-heterogenity. In: Featherstone M, Lash S, Robertson R (eds) Global modernities. Sage, London, pp 25–44

Robichau RW (2011) The mosaic of governance: creating a picture with definitions, theories, and debates. Pol Stud J 39(1):113–131

Schmidt F (2012) Governing planetary boundaries – limiting or enabling conditions for transitions toward sustainability? In: Meuleman L (ed) Transgovernance: advancing sustainability governance. Springer, Heidelberg, pp 211–230

Schvartzman Y (2009) Revising meta–governance: an analytical model for comparing the impact of state intervention on network structures in the Swedish and the Danish organic food industry. Paper presented at the conference. Governance networks: democracy, policy innovation and global regulation, Roskilde, 2–4 Dec 2009

Secretariat of the Convention on Biological Diversity (2004) Akwé: Kon voluntary guidelines for the conduct of cultural, environmental and social impact assessment regarding developments proposed to take place on, or which are likely to impact on, sacred sites and on lands and waters traditionally occupied or used by indigenous and local communities. CBD Guidelines Series, Montreal

Singer R (2008) Cultural diversity as a source of economic growth. Paper presented at Metropolitan Review of Toronto, OECD Mission, 30 Apr 2008

Slaughter A (2009) America'sedge. Power in the networked century. Foreign Affairs 88(1):94–113

Sørensen E (2006) Metagovernance: the changing role of politicians in processes of democratic governance. Am Re Public Adm 36(1):98–114

Sovacool BK (2011) Tailor renewable energies to local culture. Published 23 Feb 2011 on http://bit.ly/e6gXKw (Science and Development Network)

Spangenberg JH (2005) Die Ökonomische Nachhaltigkeit der Wirtschaft: Theorien, Kriterien und Indikatoren. Edition Sigma, Berlin

Stahl-Role S (2000) Transition on the spot: historicity, social structure, and institutional change. Atl Econ J 28(1):25–36

Staur C (2006) Preface. In: Najam A, Papa M, Taiyab N (eds) Global environmental governance: a reform agenda. IISD, Winnipeg, pp iii–iv

Stewart and Bennet (1991) American cultural patterns: a cross-cultural perspective. Intercultural Press, Yarmouth, Maine

Surowiecki J (2004) The wisdom of crowds: why the many are smarter than the few and how collective wisdom shapes business, economics, society and nations. Anchor Books, London, New York

Teisman GR (2001) Ruimte mobiliseren voor coöperatief besturen, over management in netwerksamenlevingen. Inaugural speech, Erasmus Universiteit Rotterdam, Rotterdam

Telesetsky A (2010) Climate co-regulation: creating new international governance. Paper presented at the Unitar/Yale conference. Strengthening institutions to address climate change and advance a green economy, Yale University, New Haven, 17–19 Sept 2010

Termeer C (1993) Dynamiek en inertie rondom mestbeleid: een studie naarveranderingsprocessen in het varkenshouderij-netwerk. Gravenhage, Vuga

Thaler R, Sunstein C (2008) Nudge. Yale University Press, New Haven

Thompson M, Wildavsky RE, Wildavsky A (1990) Cultural theory. Westview Press, Boulder

Thompson G, Frances J, Levačić R, Mitchell J (eds) (1991) Markets, hierarchies and networks: the co-ordination of social life. Sage, London

Thorelli HB (1986) Networks: between markets and hierarchies. Strateg Manage J 7(1):37–51

UNESCO/UNEP (2002) Cultural diversity and biodiversity for sustainable development. Background documents to the UNESCO/UNEP, Johannesburg, Roundtable held on 3 Sept 2002

VanDeveer SD (2011) Consuming environments: options and choices for 21st century citizens. In: Najam A (ed) Beyond Rio + 20: governance for a green economy. Frederick S. Pardee Center, Boston University, Boston, pp 43–52

Van Huijstee MM, Francken M, Leroy P (2007) Partnerships for sustainable development: a review of current literature. J Integr Environ Sci 4(2):75–89

Van Londen S, De Ruijter A (2011) Sustainable Diversity. In: Janssens M et al (eds) The sustainability of cultural diversity. Edward Elgar, Cheltenham, pp 3–31

Verweel P, De Ruijter A (2003) Managing cultural diversity. J Today 2:1–20

Von Barloewen C, Zouari S (2010) Die Kultur-und Religionsgeschichtlichen Voraussetzungen: Zur Anthropologie der Globalisierung und der Weltzivilisation

Wallace AFC (1970) Culture and personality, 2nd edn. Random House, New York

WBGU (2011) World in transition – a social contract for sustainability. WBGU, Berlin

Weiss T, Takur R (2010) Global governance and the UN: an unfinished journey. Indiana University Press, Bloomington

World Commission on Environment and Development (1987) Our common future. United Nations, New York

Chapter 3
Growth: A Discussion of the Margins of Economic and Ecological Thought

Alexander Perez-Carmona

Abstract In the late 1960s a debate about the long-term feasibility and desirability of economic growth as a one-size-fits-all economic policy emerged. It was argued that economic growth was one of the underlying causes of ecological and social problems faced by humanity. The issue remained strongly disputed until the inception of the Sustainable Development discourse by which the debate was politically settled. Nevertheless, given that many ecological and social problems remain unsolved and some have become even more severe, there are renewed calls for the abandoning of the economic growth commitment, particularly in already affluent countries. This chapter summarises the growth debate hitherto and examines two alternatives, the steady-state economy proposed by Herman Daly and economic de-growth proposed by Serge Latouche. In spite of recent disputes between the Anglo-Saxon steady-state school and the emerging continental de-growth school, it is argued, consistent with recent contributions on the issue, that steady-state and de-growth are not mutually exclusive but *inevitably* complements. The steady-state has the advantage of comprehensive theoretical elaboration, while de-growth has the advantage of an attractive political slogan which has re-opened the debate on the issue. Latouche is also a social thinker who gives a voice to the critiques of economic growth contained in the notion of development from outside Europe and the United States. The steady-state economy, and de-growth are held by some analysts to be beyond what is politically feasible. Although this argument is valid, it fails to recognise that past desirable societal changes were made possible through reflexive societal processes conducive to collective action and institutional change. It is concluded that the debate must ultimately rest in the physical quantities that a given economy needs for the 'good life' in the long run, how to decide on these quantities, how to achieve them, and how to maintain an approximate global steady-state. Finally, some recommendations for further research along with some reflections on the potential role of scholars are provided.

A. Perez-Carmona (✉)
Research fellow at the Institute for Advanced Sustainability Studies, Potsdam, Germany,
e-mail: alexandrop@gmx.net

L. Meuleman (ed.), *Transgovernance*,
DOI 10.1007/978-3-642-28009-2_3, © The Author(s) 2013

3.1 Introduction

Is there something new under the sun? This is the question asked by Historian John Robert McNeill (2000) drawing on verse 1: 9–11 from the Ecclesiastes. The verse, probably written in the third or the fourth century B.C., gave a negative answer to this question. Although there is much to learn from the verses of the Ecclesiastes, McNeill claims that there is indeed something new under the sun. True, the ubiquity of wickedness and vanity may have remained as much as part of human life today as it was when the Ecclesiastes was written, yet the place of humankind within the natural world is not what it once was. The global magnitude and devastating impact of the human scale on the rest of the biosphere is something truly new under sun. In contrast, while the magnitude and impact of the human scale on the biosphere is new, the intellectual debate on it is not. It has been taking place for the last 40 years, albeit with dissimilar intensity. What is meant by the impact of the human scale on the biosphere? There are two physical interrelated magnitudes: the size of the human population and the size of man-made capital 'population'. These populations live off the biosphere; broadly speaking they take its resources, transform them, and return them back to it in the form of waste and pollution. These physical magnitudes are commanded by non-physical magnitudes such as preferences, knowledge (e.g. non-embodied technology) and the social institutions that govern production and distribution (e.g. markets, the state, et cetera). Both physical and non-physical magnitudes are parts of what is called the economic system.

On the scale of the population there exists a wide consensus that it cannot perpetually grow, for the planet does not physically grow and there are limitations to its ecological functions to support not only but particularly human life. Today, this debate seems beside the point because some of the most populous countries, after years of population control policies, already have fertility rates even below replacement level. Globally, fertility rates are slowly tending towards stabilisation. In contrast, on the general scale of a given economy the consensus that it cannot perpetually grow for the same reasons that population cannot do it is still absent. Economic growth – measured as Gross Domestic Product (GDP) at constant prices – as an all-encompassing economic policy remains firmly established. Challenging economic growth started in the late 1960s, when some economists and natural scientists began to understand that the pursuit of perpetual economic growth was physically impossible. It will eventually end. Ignoring this physical impossibility, they argued, would bring a wide array of evils, that is, it would make ecological problems more intractable, it would make the abusing of other sentient beings unavoidable, and it would further exacerbate all kind of social conflicts at different levels. What is new under the sun would intensify what was not new. At the international level, these arguments remained however, largely without political implications through the inauguration of the Sustainable Development discourse. Currently, and in view of the fact that our growing global economy has already overshot the carrying capacity of the planet, there are renewed calls articulated mainly from social thinkers in Western Europe for 'de-growth'.

The aims of this chapter are (1) to provide a summary and analysis of the growth debate hitherto and (2) to scrutinise and compare alternative policy proposals. The structure of the chapter is the following: summary and analysis of the growth debate from the late 1960s until present are dealt with in the next section. In Sect. 3.3, I describe the theoretical underpinnings, the basic model and some policy recommendations for institutional change in order to achieve and eventually to manage a steady-state economy. The steady-state economy was conceived by one of the founding fathers of Ecological Economics, Herman Daly. In Daly's conception, the optimal scale of the economy replaces economic growth as the overall goal of macro-economic policy. In Sect. 3.4, I explore the ideas of the principal intellectual figure behind the emerging de-growth movement, Serge Latouche. He argues for a cultural change that would, physically speaking, de-grow the economies of rich countries in order to 'make room' for development in poor countries, while at the same time severely criticising the very notion of 'development'. In Sect. 3.5 a comparative analysis of Daly's and Latouche's ideas are provided. Conclusions and prospects for the social sciences are dealt with in Sect. 3.6.

3.2   The Growth Debate: Its Sources and Contours

The discussions in this section will be set against two backgrounds: (1) the prevailing economic doctrine alone with some relevant events, and (2) the global ecological footprint metric (see Fig. 3.1). Two prevailing economic doctrines can be distinguished in this period. First, Keynesianism which was adopted and largely implemented after the great depression of 1930 as well as during the post-war period in the West. It lasted until the early 1970s. The application of the ideas of J. M. Keynes constituted incidentally the beginning of an active pro-growth policy after the great depression and the split of economics between macro- and micro-economics.[1] The 1970s saw the end of the convertibility of the dollar to gold (1971), high oil prices (1973–1986), a stock market crash, and an economic crisis (1973–1975) in two core countries, the United States (US) and Britain. Following this, a political window of opportunity was seized by a revitalised laissez-faire or neo-liberal intellectual movement prominently represented by F.A. von Hayek and M. Friedman. Neo-liberal doctrines were partially implemented in the West, but even more in its zones of influence and later worldwide after the collapse of the Soviet Union (1991). This phenomenon was later labelled as 'globalisation'.[2] After the preceding economic crisis (2008–2009), there was a temporary renaissance of

---

[1] The terms economic growth and growth will be interchangeably used.

[2] Neo-liberal proponents cling to their classical intellectual precursors. Policy implementation attempts took place between 1870 until 1930. The central tenets were unregulated markets (including labour markets), unregulated international trade, stable currency and capital mobility.

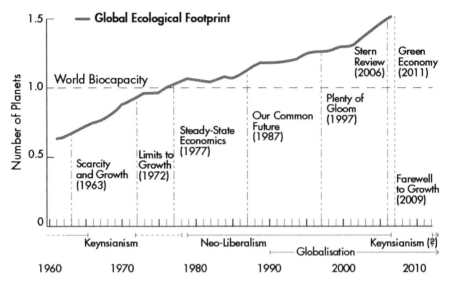

Fig. 3.1 Global ecological footprint (1960–2007), eight relevant publications and two macro-economic doctrines (Ewing et al. 2010. Modified by the author)

Keynesianism including its 'greened' version that had been proposed to come to grips with the ecological predicament. In recent times, however, Keynesianism seems to have been reduced to a minor option given the current multi-crisis of high oil and commodity prices, US fiscal problems, and the Eurozone debt crisis, in which austerity forces seem to have won the overhand. This information is placed on the x-axis of Fig. 3.1.

The global ecological footprint is an aggregate index which measures the ability of the biosphere to produce crops, livestock (pasture), timber products (forest), fish, to host built-up land, as well as to uptake carbon dioxide in forests.[3] Carbon dioxide emissions are the largest portion of humanity's current footprint. The ecological footprint is less controversial than other ecological metrics.[4] Figure 3.1 depicts the ever rising global ecological footprint in the period 1961–2007. Humanity started to overshoot the world carrying capacity, or 'biocapacity', roughly in 1975. By 2007 it was 'using' 1,5 planets.[5] In that year, the last one in which the metric was estimated, half of the ecological footprint was attributable to just 10 countries, whereby the US and China alone were using almost half of the earths' biocapacity with 21% and

---

[3] The ecological footprint is a metric developed in the early 1990s by William Rees and Mathis Wackernagel. For an extensive explanation of the metric see Wackernagel and Rees (1996).

[4] It is less controversial in the official sense since it has been endorsed by the United Nations Development Programme (UNDP 2010) and the United Nations Environment Programme (UNEP 2011).

[5] It is of course impossible to use planets that do not exist. Excessive carbon dioxide emissions are in reality accumulating in the atmosphere.

24% ecological footprint respectively (Ewing et al. 2010: 18).[6] Figure 3.1 also shows eight publications which are milestones in the economic growth debate. It is around the message of these publications that the discussion will be centred, whereby *Steady-State Economics* (1977) and *Farewell to Growth* (2009) will be dealt with in greater detail in two separate sections.

3.2.1   Scarcity, Pollution and Overpopulation

The origins of the economic growth debate lie in the late 1960s and the early 1970s, when a bundle of ecological concerns articulated primarily by natural scientists converged in rich countries.[7] A general public preoccupation with pollution and the political backdrop of the 'environmental revolution' was the book *Silent Spring* (1962) authored by Rachel Carson. Concerns about scarcity emerged with the dramatic increase in world population. This concern was epitomised by Paul Ehrlich's book *The Population Bomb* (1968). While the environmental discussion was primarily framed by natural scientists and the emerging political activism of the late 1960s, the most important social discipline was also taking position in that debate: economics.[8] In the US the think-tank Resources for the Future was established in 1952, which, in line with governmental concerns on potential shortages of raw materials published *Scarcity and Growth* authored by Barnett and Morse (1963). This study turned into economic orthodoxy (Daly 1991: 40, Dryzek 1997: 46). The emphasis of the study was to show that resource scarcities do not impair economic growth. The authors revised the classical economic doctrines of resource scarcity and compared them with what they called the contemporary 'progressive world' (Barnett and Morse 1963: 234). They concluded that techno-logical innovation, resource substitution, recovery and discovery of new resources

---

[6] It must be also mentioned that these countries have within their borders a great portion of the global biocapacity, namely 10% in the US and 11% in China.

[7] For reasons of convenience, I will split the world into 'rich' and 'poor' countries in the conventional terms of per-capita income. In other cases, I will use also the notions 'core' and 'peripheral' in the sense of material and discursive bargaining power of the latter with respect to the former. Rich or core countries are those located in North America, Western Europe, and the countries Australia and Japan. Poor or peripheral countries are the rest. When necessary, I will mention countries, regions or more recent categories grouping countries such as 'emerging countries'.

[8] An economic system means, stated in its simplest terms, how a given human group attempts to stay alive, that is, how it acquires food (energy), build housing, organise labour, and how is what is produced distributed among the members of the group. Given the overwhelming importance for human affairs of economic systems, it follows that the social discipline that studies it, must be equally of overwhelming importance, in this case economics. A step further is the distinction between the dominant school in economics, that is, neo-classic economics, which is also some-times labeled as 'economic orthodoxy', and the less influential schools, such as Ecological Economics, Economic Anthropology, Old-Institutional Economics, and so on.

made Malthus' and Ricardo's doctrines basically obsolete. These mechanisms would function not only better within the free-market system, but also more rapidly as to broaden the availability of resources, even making the definition of 'resource' uncertain over time. Therefore: 'A limit may exist, but it can be neither defined nor specified in economic terms [...]. Nature imposes particular scarcities, not an inescapable general scarcity' (Barnett and Morse 1963: 11). With respect to pollution, economists were borrowing from the thought of its welfare economists' precursors. The concept of externality, already familiar from the writings of Cecil Pigou in the 1920s and Ronald Coase in the 1960s fitted nicely into pollution issues (Pearce 2002). The economist's mission became the design of allocation mechanisms capable of realising foregone costs and benefits. As leading environmental economists Baumol and Oates in their textbook observed: 'When the 'environmental revolution' arrived in the 1960s, economists were ready and waiting' (1988: 1). For economists, doubts about the feasibility or desirability of economic growth were not raised. Beyond economic orthodoxy, human ecologists were further drawing attention to the world's population increase, mainly territorially restricted to poor countries,[9] while the expanding environmental movement was concluding that the mounting ecological problems were rather caused by 'consumerism', and more broadly by wasteful lifestyles. As wasteful lifestyles became synonymous with the pursuit of economic growth, the 'antigrowth' movement was born (Pearce 2002: 60).

However, it was not only the emergent environmental movement which perceived economic growth as the problem. The position of economists concerning the link between economic growth and the natural environment also began to show fissures. The discussion did not focus only on the concepts and relationships between a given set of assumptions, but also on the assumptions which themselves sustain the superstructure of macro-economic theories which made possible the belief in perpetual economic growth. In the list of economic assumptions nature was missing. 'Land' had been long since reduced to merely an input factor, deprived of all environmental functions and any traditional social meaning; and the newly re-emphasised 'externality' was seen rather as an exceptional case, therefore constituting a half-hearted ad hoc recognition of the sink function of nature in the economic process. As historian McNeill (2000: 335) put it: 'if Judeo-Christian monotheism took nature out of religion, Anglo-American economists (after about 1880) took nature out of economics'. The expansion of ecological problems was caused by the fact that economists were living in the 'cowboy economy' of the 'illimitable plains and also associated with reckless, exploitative, romantic, and violent behavior', while humanity were rather approaching the 'spaceman economy' in which the earth was a 'single spaceship, without unlimited

---

[9] A trend that continued with Hardin's controversial piece *Lifeboat Ethics* (1974) and Catton's insightful *Overshoot* (1980).

reservoirs of anything, either for extraction or pollution' (Boulding 1966: 6).[10] In the spaceship economy, perpetual economic growth was physically unfeasible and given the ensuing social and ecological costs of post-war growth not even desirable, as British economist Mishan (1967) reasoned. Mishan condemned what he called the 'growthmania' suffered by his fellow economists and professional politicians. However, comprehensive explanation and modelling of growth-related problems would only be offered in the following years.

Ayres and Kneese (1969) published their *Production, Consumption and Externalities* in which they showed, partially consonant with the arguments already made before by Kapp (1950) that externalities were not exceptional cases but rather an inherent part of the economic process. This seminal article would in due course give birth to the discipline Industrial Ecology. Similarly, *Environment, Power and Society* (1971) authored by pioneer ecologist Howard T. Odum attempted to frame the relationship between human and natural systems in terms of matter and energy analysis, equally showing the inherent production of waste/pollution which necessarily returns to the natural environment. This work would bring a number of young ecologists into what later would be called Ecological Economics. Nicholas Georgescu-Roegen a former student of Schumpeter published *The Entropy Law and the Economic Process* (1971), a book in which he explained from a historical perspective, the weak spot of economic orthodoxy in handling the issues of depletion and waste/pollution. In the formative years of economics as a science, it borrowed the mechanistic/circular outlook from Newtonian physics; hence the economist was failing to account for irrevocable linear processes occurring to energy/matter in the process of economic transformation. From this perspective, Herman Daly, a student of Georgescu-Roegen, proposed the *stationary-state economy* (1971) which he felt should replace the growth-policy as an overall societal objective. Georgescu-Roegen on the other hand would later insist on a de-growth policy.

This fertile intellectual activity and debate between 1966 and 1971 took place mainly in the limited arena of academia. Projecting the discussion beyond this was the achievement of a team of natural scientists at the Massachusetts Institute of Technology who published in March 1972 a small report entitled *The Limits to Growth*.

3.2.2   Understanding the Whole

According to the scientist team, the failure of adequate political responses to tackle environmental and resource problems were due to a lack of understanding of the

---

[10] It is widely recognised that the picture of the earth taken by Apollo 8 in 1968, the 'earthrise', gave a massive boost to the environmental movement. The 'earthrise' made it possible to conceptualise the earth as a beautiful, fragile, floating in the middle of nowhere, and especially, finite planet. It is remarkable that Kenneth Boulding introduced the spaceship analogy 2 years before the earthrise picture shaped public imaginary.

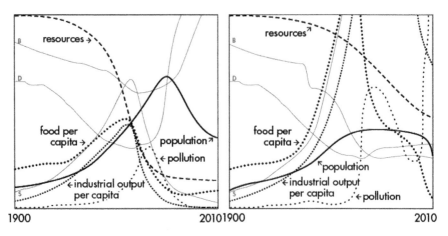

Fig. 3.2 Two scenarios of the world model: (**a**, *left*) standard run, and (**b**, *right*) comprehensive use of technology (Meadows et al. 1972)

human system as a whole: 'we continue to examine single items in the *problematique* without understanding that the whole is more than the sum of its parts' (Meadows et al. 1972: 11, italics in original text). Using the new system dynamics methodology developed by Jay Wright Forrester and the computer model World3, the authors of *Limits to Growth* (LtG) examined the interaction of five key subsystems of the global system: population, industrial production, food production, pollution and natural resources. They assumed that population and industrial production were growing exponentially, in a world with absolute fixed available resources. The time scale of the modelling ranged from 1900 until 2100. As the team abundantly emphasised, the world model was not intended to make exact predictions (Meadows et al. 1972: 93, 94, 122) given the extreme complexity and uncertainties involved in the real world. Their aim was rather to understand the global system's behavioural tendencies and to offer a plausible answer to the question: are our current growth policies leading to a sustainable future or to a collapse?

Figure 3.2a, b show two scenarios of the world model from a total of seven. Figure 3.2a plots historical values from 1900 to 1970 until 2100. It assumes no major changes in historical socio-economic relationships. It is the 'standard run' which illustrates that the world is 'running out of resources' in the first decade of the twenty-first century, while population collapse occurs in the middle of it. As industrial output increases exponentially, it requires an enormous input of resources. Resources becoming scarce led to a rise in prices which conversely left less financial capital to be re-invested for future growth. Ultimately, investment did not keep up with depreciation and the industrial base fell along with agricultural systems which became dependent on industrial outputs such as fertilisers, pesticides, and especially, energy sources for mechanised agriculture. Population continued to increase for approximately two decades and finally started to decline when the death rates were driven upward by a lack of food and health services. The team ran five more scenarios in which the initial assumptions made in the standard run were additively relaxed. Nonetheless, in each case the population inevitably collapsed during the twenty-first century due to an ever rising pollution, food shortages, and so on. Figure 3.2b plots an aggregate scenario of several technological and political responses to shortages. Technology is being implemented in every sector: nuclear power, recycling, mining the most remote reserves, withholding as many pollutants as possible, pushing further yields from the land and having 'perfect birth control'.[11] Population collapse has simply been delayed by several decades. In this scenario, three crises hit simultaneously, food production drops because of land erosion, resources are depleted by a prosperous population holding an average income per capita of close to the US level, pollution rises, drops and then rises again dramatically causing a further decrease in food production.

The study was presented at a perfect time, as the first United Nation Conference on the Human Environment was held in June of that year 3 months after the study was released. Nevertheless, the policy goal of stabilisation which the team proposed and which happened to resemble Daly's idea of the *stationary-economy* (zero-growth) advanced 1 year before was largely dismissed. According to Beckerman (1972), delegates of poor countries made it profusely clear that they were not going to accept any policy arising from the study of some uncertain planetary limits that would hamper their future development. Henceforth, international relations could continue to operate under the frame of development set out by the US president Truman in his inaugural address of 1949, that is, actively reducing trade barriers and making the benefits of industrial progress available 'for the improvement and growth of underdeveloped areas' (Truman 1949). Additionally, LtG was unanimously rejected by leading economists (Beckerman 1972, Kaysen 1972; Solow 1973, 1974; Beckerman 1974). The common argumentative line was that technological progress and the market mechanism could prevent scarcity and pollution from constituting a substantial limitation on long-term economic growth. In essence, their way of looking at the problem was identical to that established by Barnett and Morse a decade before. Cole et al. (1973) re-ran the world model, yet they eliminated absolute limits of resources and let them increase *pari passu* with population and consumption, assuming additionally total control of pollution. They claimed if 'the rates of (technological) progress are increased to 2% per annum collapse is postponed indefinitely' (Cole et al. 1973: 118).

The emerging economic heresy also contributed to the LtG debate. They were particularly emphatic about the incongruences and fallacies committed by their orthodox colleagues (Daly 1972: 949–950, Georgescu-Roegen 1975: 363–366, Mishan 1977). Georgescu-Roegen, for example, was impressed by the fact that many of the critiques made by economists on the methodology employed in LtG,

---

[11] Perfect birth control meant parents wishing and voluntarily having just two children from 1975 onwards.

was the very same which they themselves routinely used.[12] They condemned LtG for having used the assumption of exponential growth; nonetheless, economists themselves have always suffered from 'growthmania'. Economic plans have been designed with the explicit aim of obtaining the highest rate of growth possible and the very theory of economic development is firmly anchored in exponential growth models. Furthermore, some of them used the very same argument of exponential growth – but applied it to the 'increase' in technological progress in order to criticise LtG. This argument besides being circular, is fallacious on other grounds. Technology is a non-physical entity – unless it is embodied in capital – that as such cannot (exponentially) grow as a population does. Georgescu-Roegen concluded that economists proceeded according to the Latin adage: *quod licet Iovi non licet bovi* – what is permitted to Jupiter is not permitted to an ox (1975: 365).

Six years later, after LtG's release, Daly published his *Steady-State Economics* (1977). The book was a collection of essays which dealt with logical inconsistencies made by pro-growth proponents, and expanded on physical and economic motives for a stationary but developing economy. Chapter 4: 'A Catechism of Growth Fallacies' dealt with 16 fallacious arguments. Four of them were of particular significance to reproduce here given their endurance: (1) becoming rich through economic growth is the only way to afford the costs of cleaning up pollution: as Daly noted, this statement skips the relevant question of *when* economic growth will start to make a nation poorer and not richer. The problem is that economists do not attempt to compare costs and benefits of growth, apparently because it is tactility implied that growth is always 'economic'. (2) Growth is necessary to combat poverty: Daly argued that in spite of the growth of the preceding years in the US, there was still poverty. The benefits of the reinvested surplus which generates growth go preponderantly to the owners of the surplus, who are not poor and only some of the growth dividends 'trickle down'. For growth-economists, Daly further reasoned, growth has become a substitute for inequality concerns. Yet, with less inequality, less growth and consequently less ecological pressure would be required. (3) Growth can be maintained by further shifting the economy to the service sector: Daly argued that after adding the indirect aspects of services activities (inputs to inputs to inputs, that is, Leontief's input–output-analysis), we will likely find out that they do not pollute or deplete less significantly than industrial activities. Casual observation shows that universities, hospitals, insurance companies, and so on, require a substantial physical base. The reason why employment in the service sectors has grown relative to total unemployment is because of the vast increase of productivity and total output of industry and agriculture which conversely has required more throughput given

---

[12] According to Levallois (2010) the comprehensive review of LtG's critique made by Georgescu-Roegen, which I am only partially reproducing here, was an outcome of the contacts that Georgescu-Roegen entertained with the group. He eventually entered the Club of Rome but abandoned it later, apparently disillusioned with the club's fascination for computer-based models and appetite for public relations. He did not fail to mention the latter in his review.

the increased scale. (4) Oil is not recycled because it is still uneconomic to do so; humankind is less worried about the environment because it is currently not totally dependent on it, and nature imposes no inescapable scarcities: According to Daly these arguments can only be made given economist's illiteracy in basic natural sciences.

Notwithstanding these arguments – which were largely ignored – orthodox economists contributed to producing the general impression that LtG was simply *pessimistic*, and *predicting* something alone with the reaffirmation that technological progress would cope with all sorts of ecological problems. In contrast, LtG did contribute to popularising the sustainability debate which was emerging at that time by selling millions of copies and being translated into 30 languages (Meadows et al. 2004: x), even influencing the opinion of leading politicians in Europe. Sicco Mansholt, the president the European Commission (1972–1973) read LtG and concluded that growth in Europe should not only be stopped but even reduced, and replaced with another 'growth', that is, the growth of culture, happiness and well-being (Mansholt 1972).

In the late 1970s, the US was re-entering another economic crisis and successive efforts were focused on monetary policy in order to fight inflation at the cost of employment creation, thus risking a deeper recession. Almost simultaneously humankind was entering a global era of planetary overshoot (Fig. 3.1). The oil embargo imposed by the Organisation of Petroleum Exporting Countries (OPEC) upon rich countries in 1973 helped to trigger not only an economic stagflation but also a debate on energy dependency. Subsequently, an energy policy embracing (1) nuclear power and (2) energy efficiency measures was discussed and partially implemented. As industrial growing economies need correspondingly increasing amounts of energy, and a part of the energy must be produced at home instead of being imported from countries located thousands of kilometres away, the vital but visible nuclear reactors rapidly produced a social response which had been in gestation years before: rejection. In 1969 physicist Starr had already proposed a risk-benefit analysis by means of 'historically revealed social preferences' (Starr 1969: 1232) with favourable results for nuclear power and speculated on the causes of the irrational risk *perception* by the lay public which was generating the opposition.[13] Later on, the social conflict was renamed the not-in-my-backyard syndrome (NIMBY), elevated into an analytical concept, and extended to all kinds of facility siting conflicts. Nonetheless, after the Three Mile Island incident of 1979, it was evident that the risk aversion and the *nimbysm* of the lay public could not be

---

[13] As Otway (1987) explained, risk perception studies appeared as the public entered decision-making over technological risks, therefore turning upside down the fiduciary trust in public servants issuing the licenses and even more, antagonising the deep-grained notion of technological progress. As risk perception studies did not bring the expected results, communicative risk studies emerged in an attempt to bring public opinion in line with experts' assessments. It must be mentioned, however, that communicative risk studies turned out to be useful in dealing with, for example, occupational and natural risks.

entirely dismissed as irrational. On the issue of energy efficiency and conservation policies, two energy economists were raising doubts about the effectiveness of such policies. They were resuscitating Jevons' conclusion made more than a 100 years ago that, contrary to common expectations, energy efficiency improvements would lead to more energy consumption, that is, such policies would 'backfire' (Brookes 1979, Khazzoom 1980). Hence, alone with the revival of the pessimism of the so-called Neo-Malthusians, the pessimism of Neo-Jevonians also came about. By 1980, another pessimist report was released in the US, the *Global 2000 Report for the President* that, as the title implies, did not look as far ahead as LtG. The major finding was that:

> If present trends continue, the world in 2000 will be more crowded, more polluted, less stable ecologically, and more vulnerable to disruption [...]. Despite greater material output, the world's people will be poorer in many ways than they are today. (Quoted by Dryzek 1997: 28)

Georgescu-Roegen would have certainly said *because* 'of greater material output'. Nevertheless, the timing for pessimistic antigrowth positions could not be worse, for an era of exuberance would begin which could not handle the pessimism of the preceding years. In the core countries of the West, the US and Britain, a new formula for economic growth was proposed, (allegedly) away from state interventionism, and thus strong labour unions would be put in place: neo-liberalism. The optimism of the new era found its place in the ecological debate concerning economic growth through what would be later called 'cornucopianism'.

3.2.3   The Sustainable Development Discourse

In congruence with the rising optimistic era of neo-liberalism but acknowledging that there were real ecological issues at stake, the United Nations (UN) created the World Commission on Environment and Development. The commissions was established in order to investigate the links between the deterioration of ecological systems and economic growth in 1983, the same year in which the newly formed Green Party in West Germany managed to win enough votes to trespass the election threshold for federal parliament. The world commission was the follow up of the conference held in 1972, and it is better known by the name of its chairwoman, Mrs. Brundtland. The commission delivered the report *Our Common Future* in 1987, roughly a year after the optimism of infinite energy supply was shattered anew by the disaster of Chernobyl.

On the political consequences of conceptual ambiguities and the strong anthropocentrism of the report enough attention has been drawn.[14] For the aims of this

---

[14] For a good overview on the diverse interpretations from different theoretical perspectives and policy implementation see Sneddon (2000) and Hopwood et al. (2005).

chapter it is useful to highlight the origins of these ambiguities and the ambiguities specifically in relation to growth. If Sustainable Development (SD) was to have a chance of future implementation, it had to have an appeal of political acceptability in order to initially bring different interests to the table of negotiation. Nevertheless, and according to political scientist Dryzek (1997: 124), as it was recognised that sustainable development would become the global dominant discourse, powerful actors, mainly big businesses, made sure to cast it in terms which were favourable to them. Ultimately, sustainable development was politically successful, but it achieved this by sacrificing substance: 'lots of lobbyists coming together, lots of blurring going on – inevitably, lots of shallow thinking resulting' was the judgment of historian Donald Worster (1993: 143). To be sure, the difficulties lay in putting together the relatively well-framed 'sustainability' and 'development'. Sustainability was at the bottom an ecological concept traceable to the German enlightenment. What is to be *sustained* is the environment, although mainly for human purposes.[15] On the other hand the notion 'development', as previously noted, was established by the emerging leader of the West in 1949.[16] Given the ecological debate of the preceding years in the US, and the increasing appeal of the notion 'qualitative growth', that is, more leisure for family and hobbies during the 1970s and 1980s in Germany and France among others,[17] it was evident that the general economic policy goal of growth was at stake. The question to be solved was then: how to maintain the perpetual economic growth policy if the planet has ecological limits?

Although, as noted before, the report was (inevitably) a product of political bargain, it is necessary to understand how the report coped with the dilemma,

---

[15] The concept appeared in Germany in the late eighteenth and early nineteenth century. As Germany's economy depended in essential ways on its forests that were rapidly declining, scientists were consulted to give advice. They started to talk about *managing* forests as to attain a *sustained-yield* so that periodic harvests would match the rate of biological growth (Worster 1993: 144). Southern notions of sustainability, however, had given forests a less anthropocentric meaning.

[16] However, at this time the official meaning of development had undergone several changes. Development meant practically projecting the US model of society onto the rest of the world, but in the late 1960s and at the beginning of the 1970s too little advancements in this direction could be attested. As Sachs (1999: 6) explained: 'Poverty increased precisely in the shadow of wealth, unemployment proved resistant to growth, and the food situation could not be helped through building steel works'. Hence, in the 1970s and 1980s the meaning of development was broadened as to include justice, poverty eradication, basic needs, woman issues, and of course, ecological problems.

[17] During the 1980s, the Green Party and the Social-Democratic Party of Germany had been advancing a change in the stability-act enacted in 1967 that basically reflected Keynesian doctrines of high employment through steady-growth and balanced terms of trade. The reform of the stability-act should aim rather at 'qualitative growth' in the sense explained above and ecological balance. In France, during the 1970s, the demand for more leisure was famously made by philosopher André Gorz (For the former insight I thank Dr. Angelika Zahrnt, and for the latter one, Dr. Giorgos Kallis).

especially the arguments pertaining to needs and ecological limits so central to the growth debate. The emphasis was first placed on poor countries, who were after all the ones to be aided with their development. Here, essential needs were defined in conventional terms: food, clothing, shelter, and jobs. It was also accepted that beyond them, the poor have the legitimate aspirations for an improved quality of life (WCED 1987: 43). When the report switches into the realm of the rich, needs become perceived, socially and culturally determined what possibly drives up levels of consumption. Therefore, it is reminded that in the context of sustainability, values encouraging 'consumption standards within the bounds of the ecological possible and to which all can reasonably aspire' (ibid.: 44) are required. Although reaching ecological limits can be slowed through technological progress 'ultimate limits there are' (ibid.: 45). Since sustainable development also involves equity, equitable access to the constrained resources ought to be granted before the 'ultimate limits' are reached. From these premises relating to frugality, equity and time-bounded growth because of ultimate limits the conclusion was however:

> The Commission's overall assessment is that the international economy must speed up world growth while respecting the environmental constraints. (ibid.: 89)

How to speed up world growth, that is, economic growth for both rich and poor countries, *while* respecting ecological limits? The solution advanced was a change in the *quality* of economic growth, but not in the sense advanced in Europe years before. Qualitative growth meant rather that growth must become less energy/ matter intensive and more equitable in its impact (ibid.: 52). On this general recommendation some comments are needed, for the official environmental discourse became locked in sustainable development until the present.[18]

First, the report was advising something that one of the main drivers of global economic growth, the manufacturing sector, had been doing since the industrial revolution, namely becoming less energy/matter intensive. In Canada, the US and Germany, energy intensity (ratio of energy use to GDP) declined after about 1918, in Japan after 1970, in China around 1980 and Brazil in 1985. The US used half as much energy and emitted less than half as much carbon per constant dollar of industrial output in 1988 as in 1958. For the world as a whole, energy intensity peaked around 1925 and by 1990 had fallen by nearly half (McNeill 2000: 316). However, these global happy trends of 'dematerialisation' and 'decabornisation' obscured the trends in industrial expansion. In fact, industry had been *too* successful in this domain, inasmuch as when consumers were not able to cope with what manufacturing industries were putting on the market, it started to *produce* consumers at home and to lobby for free trade abroad – a foreign policy already practiced by the first industrial nation Britain. What was happening entered the intellectual radar of economist John K. Galbraith (1958), who resuscitated the forgotten Say's law: a growing supply creates its own growing demand. Yet his

---

[18] The following analysis is not intended to diminish the advances made in other realms that have been guided by the SD discourse.

arguments found little response from his colleagues, who two decades before restricted the boundaries of the study of economics as being unresponsive about the inquiry on the origins preferences.[19] The social-engineered cultural change partially accomplished by advertising techniques was investigated in the US by Vance Packard (1960). He described the birth of easy-credit and the general inculcation of self-indulgence in the management of money, as well as the commercialisation of virtually every aspect of life, and the technique of built-in 'progressive obsolescence'. Progressive obsolesce was introduced by both lowering standards of quality by design and psychologically outmoding products after a given time.[20] Growth became de facto a self-contained policy rather than a mean to achieve a societal goal, since the 'private economy is faced with the tough problem of selling what it can produce' (Packard 1960: 17). What is important to highlight from this process is what Packard and Galbraith troubled at that time, namely that the consequences for social welfare were neglected, let alone the political and ecological consequences of which Packard was not unmindful. The topic would be discussed years later by Erich Fromm (2007) and Fred Hirsch (1977), yet all of these growth caveats had little incidence in the Brundtland report.

Second, the fact that becoming even more efficient leads to an increase of throughput (input + output) went rather unnoticed. This was presumably because the revival of the Jevons' paradox was accomplished a couple of years before it became irrelevant at the political level as oil prices returned by the mid-1980s to their customary level. Third, the rationale that already rich countries must further pursue economic growth by consuming even more was that of helping poor countries with their economic growth as they are 'a part of an interdependent world economy' (WCED 1987: 51). The alternative that poor countries could create their own markets by selling necessities to each other instead of selling 'even more extravagant luxuries to the jaded and harried rich' (Daly 1991: 151) or allowing for import-substitution as had been put forward by Latin American economists in the 1970s and practiced with some success in the region, was entirely neglected. The mainstream doctrines of economic development that prevailed at the time in which the Brundtland report was embedded did not permit this. The policy of perpetual

---

[19] Given the insurmountable problems of direct measurement of utility (happiness), definitional confusions between utility and *usefulness*; and the embracing of the logical-positivism of the Vienna circle in the 1920s, economists decided to focus upon market revealed preferences. The formation of preferences and their ends were declared beyond the scope of economics (see Cooter and Rappoport 1983 and Bromley 1990). An interesting account of the debate on commensurability and comparison of values between von Mises, von Hayek and Otto Neurath, a member of the Vienna Circle, can be found in Martínez-Alier (1987: 211–218).

[20] According to Strasser (1999) the term was coined in the 1920s by Christine Frederick, a US household economist. Progressive obsolescence was an attempt to introduce what was already common in the upper-classes regarding clothing. Strasser also explained that the idea had a great appeal for businesses, and transferring progressive obsolescence from clothing to other commodities was pioneered by the automobile industry, which at that time was worried with a saturated middle-class. With progressive obsolescence the 'throwaway society' was born.

economic growth for the entire planet remained virtually intact in spite of discussions regarding the issue in the preceding years. Indeed with SD, the intellectual debate was politically settled (Du Pisani 2006: 93) – with one single exception: population growth. The report mentioned as a 'strategic imperative' the realisation of a 'sustainable level of population' (WCED 1987: 49). The *combination* of free trade and population control policies in poor countries were indeed, mildly put, suspicious.

After the UN Conference on Environment and Development (UNCED) held in Rio de Janeiro in 1992, sustainable development became gradually operationalised. The firmly established 'qualitative growth' has made it possible to talk ever since about 'patterns' of consumption and production, and to carefully avoid *less* consumption and production. This is despite the fact that during the earth summit which endorsed the Agenda 21, it was argued that global ecological problems arose as a result of profligate consumption and production in rich countries.[21] When the report was launched, the global economy required roughly 1,1 planets, hence, humanity had started to live from the natural capital, and not from its income. By the publication date of the report there was of course no ecological footprint metric, but LtG had been around for 15 years. Additionally, just 1 year before the report's publication, a group of natural scientists had published another study showing that humans were already appropriating 25% of the global potential product of photosynthesis (terrestrial and aquatic), and that when only terrestrial photosynthesis was considered, the fraction increased to 40% (Vitousek et al. 1986).

By 1989 the Washington consensus was formulated and the receipt was applied to poor countries which had previously become over-indebted; partially as a result of the pressures to reinvest the so-called 'petro-dollars' gained from the OPEC embargo in the 1970s which flooded development banks. The Washington consensus contained items such as the redirection of public spending from subsidies into pro-growth services, namely primary education, health care and infrastructure; trade liberalisation and privatisation of state enterprises; in short, the well-known Structural Adjustment Programs of the International Monetary Fund (IMF). In the same year, the Berlin wall fell and the process of German reunification began, thus shifting attention away from the previous discussions of reforming the Keynesian stability-act (1967) for the purposes of 'qualitative growth' – in the West German sense. After the Soviet Union collapsed in 1991 and the 'end of history' was proclaimed, neo-liberal doctrines conquered not only the Soviet Union but also

---

[21] Recently, the nineteenth session of the UN commission on sustainable development concluded in disappointment as governments were unable to establish a consensus to produce a final outcome text. Apparently, one of the main reasons for the lack of consensus was the failure to agree over the 10-Year Framework Programme on Sustainable Consumption and Production. To this shortcoming, the UN secretary general Ban Ki-Moon stated: 'Without changing consumption production *patterns* – from squandering natural resources to the excessive life-style of the rich – there can be no meaningful realization of the 'green economy' concept'. (Anon 2011). *Rio + 20 Expectations Unclear as CSD 19 Ends on Sour Note* [online]. International Centre for Trade and Sustainable Development. Available from: http://bit.ly/pClJCd. Accessed 24 May 2011. Emphasis supplied).

its former influence's zones as to transform them into a more efficient growth machines than they had been previously (McNeill 2000: 334). The world entered the era of *globalisation* institutionally rounded up in 1995 when the World Trade Organization (WTO) emerged out of the culmination of the Uruguay Round of negotiation of the General Agreement on Tariffs and Trade.

By 1992 the World Bank (WB) published its World Development Report entitled *Development and the Environment* embracing without conceptual difficulties as the following anecdote shows: during a session in which the schematic representation of the economy was being discussed, the WB's chief economist Lawrence H. Summers refused to draw a larger box around the smaller box representing the economy.[22] The larger box would represent the natural environment as suggested by Herman Daly, who was serving as senior economist at the WB's environment department. Why refuse something so simple and evidently true? As Daly explains, it was because of the subversive iconographic suggestion that the economy could not grow in perpetuity given the limits that the environment imposes. Moreover, 'a preanalytic vision of the economy as a box floating in infinite space allows people to speak of 'sustainable *growth*' – a clear oxymoron to those who see the economy as a subsystem' (Daly 1996: 7. Italics in original text).

3.2.4   Between 'Cornucopians' and Cautious Optimists

According to Dryzek (1997: 30–31), the fact that an economist of Kenneth Arrow's intellectual calibre and reputation co-authored a paper stating that the resource base is finite and that there are 'limits to the carrying capacity' (Arrow et al. 1995: 108) is an effect of the field of Ecological Economics pioneered by Kenneth Boulding, Nicholas Georgescu-Roegen and Herman Daly. The authors focused on unravelling the fashionable claim that economic growth and free trade (export-led growth) in poor countries (development[23]) are in the long run beneficial for the environment, a claim that, as noted before, had already been made in the 1970s. During the 1990s it came to be known as the Kuznets' curve hypothesis. It postulated an inverted U-shaped curve which described the relationship between per-capita income and indicators of natural and resource quality, that is, when a poor country becomes rich through export-led growth, only then will its population start to become

---

[22] In the same year of the WB's publication Summers attracted international attention through an internal memo that was leaked to the public. Using impeccably the doctrine of comparative advantage, he suggested that many poor countries were 'underpolluted' and that dirty industries should be encouraged to move to them (for a retrospective analysis see Johnson et al. 2007).

[23] The differentiation of economic growth and development gained support in some sectors of the development community during the 1990s (see for example Sen 1999), while other sectors where rejecting the notion outright (see for example Escobar 1992, and Sachs 1992).

preoccupied with environmental quality. As Arrow et al. explained, the Kuznets' curve hypothesis had been shown just for a selected set of pollutants, yet orthodox economists have conjectured that the curve applies to environmental quality in general. Moreover, they were neglecting the export of pollutants from rich to poor countries effectively done by offshoring highly polluting industries, the purposeful policy implementation to reduce environmental impacts in rich countries and finally, that sometimes environmental concerns are not only about increased demands for environmental 'quality', as the resilience of ecosystems upon which communities depend can be irreversibly damaged.[24]

Two years after the article appeared, and 10 years after the launching of the Brundtland report, the influential British magazine The Economist published in its Christmas special edition an article with the title *Plenty of Gloom* (Anon 1997). The article attempted to show their readers by means of time-series graphs the predictive errors made in the past by Malthus, concluding that there was no reason to believe in their modern proponents. The article was important as it epitomised reasonably well another persuasive position going beyond the trend set by Barnett and Morse in 1963.[25] The so-called 'cornucopians', famously represented by economist Julian Simon. The cornucopian rationale is the following: minerals, food production have been made plentiful in the past, standards of living and life expectancy have been risen, and technological substitution has taken place many times. By extrapolating these past trends into the future, in which the basic metric of scarcity are market prices, it is concluded that the reason for growth pessimism is without substance. For example, on the issue of oil which is the 'master resource', Simon stated that we will never run out of it (Simon 1996: 179). His argument was however, subtler and the phrase misleading. In his view, it is not the oil that is important, but its service: energy. Indeed, the service of energy can be delivered by other sources rather than oil (substitution). As we will never run out of oil (energy), and energy will become increasingly cheap as in the past, it

> ... would enable people to create enormous quantities of useful land. The cost of energy is the prime reason that water desalination now is too expensive for general use [...]. If energy costs were low enough, all kinds of raw materials could be mined from the sea. (Simon 1996: 162)

All of this is possible because the 'ultimate resource' is after all human inventiveness (technology), which is 'unlimited'. Prominent orthodox economists such as Beckerman never went so far as Simon, but Beckerman had also been using time-series in order to show that there is little reason to attend the warnings of natural scientists and derailed economists – the former ones have been wrong too many times (Beckerman 1974, 1995). Beckerman additionally disdained the sustainability

---

[24] The argument that only rich countries are preoccupied with the environment was also refuted by Martínez-Alier. He coined the term 'environmentalism of the poor' (Martínez-Alier 1995).

[25] In a subsequent study called *Scarcity and Growth Reconsidered* (1979), Barnett reaffirmed his position. Nevertheless, many others authors including Georgescu-Roegen and Daly commented on the issue.

discourse for being 'morally repugnant' (Beckerman 1995: 125). He argued that needs are subjective, and poverty is the contemporary world malady to be tackled – certainly through economic growth, for the entire world, and using the standard instruments of neo-classic economics to tackle scarcity and ecological problems.

The Economists' article presented a set of figures taken from the Food and Agriculture Organisation and the WB, showing declining price of metals and food. It was argued that despite the fact that the world population almost doubled from 1961 to1995, food production had more than doubled, even resulting in falling food prices. Other tragedies predicted but which turned out to be wrong, according to the magazine were rising cancer rates because of pollution, forest decline in Germany in the 1980s caused by acid rain and famines due to population increase. Later, the journal of Environment and Development Economics called for a response to the Economist's article. It was attended by 12 scientists: 9 environmental economists, 2 ecologists and a climate scientist. They responded in the Policy Forum section of the journal and argued about the absence of markets and property rights on environmental services but also about the complexity and uncertainty in socio-ecological systems, and the non-linearity of numerous ecological processes. I will go into some detail regarding two arguments which reflect, in my view, the gained influence of Ecological Economics and Industrial Ecology upon Environmental Economics. The arguments are: (1) the problem with time series statistics versus processes and (2) the 'Heisenberg Principle' (Portney and Oates 1998: 531) which is at work when a prediction is made.

1. Time series statistics versus dynamic processes

Using time series to show that natural scientists were wrong is a weak argument because it does not take into account the natural resource-base upon which production depends (Dasgupta and Mäler 1998). In agriculture, for example, increased food production (green revolution) had been achieved by monocultures, pesticides, fungicides, soil depletion, and so on (Krebs 1998). Hence, the question to be asked is not only if we can produce more food, but what are the long term ecological/ social consequences of doing so in the way it is done. On scarcity, Dasgupta and Mäler (1998) pointed out that price can be a very bad indicator. In fact, prices can decrease while the resource in question also becomes scarcer.[26] Krebs (1998) argued that for predictive purposes, the understanding and modelling of underlying dynamic processes are more promissory than simple time series statistics.

2. Heisenberg principle

Portney and Oates (1998) and Polin (1998) stated that the act of observing and forecasting social events is likely to affect the outcome. Hence, the previous predictions made by natural scientists raised awareness of looming problems,

---

[26] For a discussion of this paradox see Daly and Cobb (1994: 450) and for more general discussion on scarcity and prices see Norgaard (1990), Daly (1991: 265–256), and Wackernagel and Rees (1997: 13–14).

namely exponential population growth, ozone layer depletion, the effects of acid rain on German forests, and so on. The raised awareness was conducive to political action which prevented the prediction from coming true and which stopped damaging activities. Levin (1998: 527) affirmed that 'the greatest reward for one predicting catastrophe is to stimulate the implementation of measures that invalidate the predictions'.

These answers were very significant, and as far as I know, The Economist did not refute them – although it might have shaped opinion more effectively than the responses of a scientific journal with a specific and limited audience. As one of the main targets of ridicule was LtG, several scientists' responses sadly repeated the distortions made years before, for example, on the alleged predictions that LtG made (Hammitt 1998: 511, Perrings 1998: 491), and the supposed failure of taking into account technical change (Portney and Oates 1998: 530). On predictions, the following is one of the many phrases written by the LtG's authors:

> This process of determining behaviour modes is 'prediction' only in the most limited sense of the word ... these graphs are *not* exact prediction [...] They are only indications of the system's behavioral tendencies. (Meadows et al. 1972: 92–93. Italics in original text)

With regards to the fact that technical change was not taken into account (Fig. 3.2b). Finally and as previously mentioned, Krebs (1998) maintained that the understanding and modelling of underlying dynamic processes is superior to simple time series statistics. Nevertheless, Krebs failed to give proper recognition or to defend the LtG team who inaugurated these types of studies.[27]

The attention on LtG also raised the central question concerning economic growth, since after all, LtG's central tenet is that economic growth (and population growth) is in the long run simply impossible and a failure to recognise that would be calamitous. The only comment in this direction was made by environmental economists Dasgupta and Mäler (1998: 505) who expressed that:

> By concentrating on welfare measures, such as GNP and life expectancy at birth, journalists, political leaders and, frequently, even economists, bypass the links that exist between population growth, increased material output, and the state of natural-resource base.

They argued later that environmental problems are sometimes correlated by 'some people' with wrong sorts of economic growth. On the other hand, Kneese (1998) expressed gratitude to the magazine for reminding the readers that the impacts of economic growth on natural resources can and have been cancelled by technological progress. He explained that with endogenous growth theory, national

---

[27] I bring LtG to the end-1990s again because the widespread idea that LtG was 'refuted' contributed to several issues being left unattended for many years. Presumably, this widespread perception also meant that the two last updates published in 1992 and 2004 correspondingly were largely ignored. More recently, Turner (2008) published an analysis of 30 years of historical data (1970–2000) and concluded that they compared favourably with the key features of the 'standard run model' reproduced in Fig. 3.2a.

economics do not growth like balloons, for efficiency in the use of energy/matter prevents them from doing so. Similarly, Kriström and Löfgren (1998: 525) asserted that endogenous growth theory 'promises us permanent growth, due to constant returns to capital'. It may be worth reminiscing that endogenous growth theories simply attempt to account for the origins of technological progress which was previously treated as given, that is, 'exogenous' to the neo-classical growth models. However, exogenous or not, it does not handle the issue of scale or the Jevons' paradox already mentioned, resulting in an impact on the natural environment and related social conflict. When this discussion was taking place, the global economy was already necessitating 1,2 planets, from which the largest share was what The Economist's author dismissed as the 'mother of all environmental scares': global warming.

3.2.5   Climate Change

From the 1990s on, the focus of the debate on ecological problems shifted progressively from depletion to pollution, more specifically to greenhouse gas emissions (GHG) causing an increase in global average temperature.[28] Climate change was put on the international political agenda at the Earth Summit in 1992 when the United Nations Framework Convention on Climate Change (UNFCCC) was created. The ultimate objective of the convention (article 2) was the 'stabilization of greenhouse gas concentrations in the atmosphere at a level that would prevent dangerous anthropogenic interference with the climate system', and in line with sustainable development it re-affirmed the objective of 'sustainable economic growth' within the context of the 'open international economic system' (UNFCCC 1992). The convention acknowledged several principles, such as the precautionary principle, the protection of the climate system on the basis of equity, the necessity that rich countries take the lead in combating climate change, and a consideration of the circumstances of developing countries. After 5 years of negotiations, the Kyoto Protocol with legally binding commitments was agreed in 1997. Thereafter, a political process of ratification began. The protocol included three international mechanisms in order to facilitate its implementation: International Emissions Trading, Joint Implementation Mechanism and the Clean Development Mechanism. According to Munasinghe and Stewart (2005: 2) these mechanisms were

---

[28] An emphasis was also set to the state of ecosystems and development/poverty. It resulted in the release of the Millennium Ecosystem Assessment in 2005 and in the Millennium Development Goals in 2000. It is however my belief, that climate change has been more at the centre of public attention in rich countries than the bad shape of ecosystems and global poverty with reference to the disposition for real action at the international level. The reason might be that climate change is logically related to the most sensitive geostrategic concerns of rich and emerging countries: energy.

developed to specifically satisfy the conditions required by the US, yet the progress initially made suffered a reverse when the US government refused to sign the Bonn agreement – an extension of the Kyoto protocol – in July 2001. Two months later the US suffered a terrorist attack and the attention of the entire West shifted away from the climate change issue.

The visibility of the subject was again given a massive boost in 2006, when British economist Lord Nicholas Stern published his *Stern Review*. That the attention on climate change was brought back to the forefront by an economist indicated once again the extraordinary power of the profession.[29] As Jackson (2009: 11) put it: 'it's telling that it took an economist commissioned by a government treasury to alert the world to things climate scientist [...] had been saying for years', namely that humanity is at crossroads. Climate change is a global and serious threat – and there is no doubt that it is anthropogenic. Climate studies have been compiled by the Intergovernmental Panel on Climate Change (IPCC) created in 1988. It has delivered four comprehensive reports thus far: 1990, 1995, 2001 and 2007. The following information is taken from the synthesis of the last IPCC report (IPCC 2007a).

Global atmospheric concentrations of greenhouse gases (GHG) such as carbon dioxide (CO_2), methane (CH_4), nitrous dioxide (N_2O) and halocarbons have clearly increased since 1750 (pre-industrial times) as a result of an expansion in 'human activities', whereby halocarbons did not even exist in pre-industrial times. For example, the global atmospheric concentration of carbon dioxide, the most important anthropogenic GHG, increased from 280 parts per million (ppm) to 379 ppm in 2005. The major growth in GHG emissions between 1970 and 2004 has come from energy supply (fossil fuels), transport and industry. There is convincing evidence that the rising levels of GHGs emissions have a warming effect on the climate because of the increasing amount of heat energy (infrared radiations) trapped in the atmosphere: the greenhouse effect. In fact, the earth has become warmer since around 1900 by 0.7°C and it will continue to do so for the next two decades at a rate of 0.2°C for a range of emission scenarios, and 0.1°C per decade even if the concentration of GHGs is kept constant at 2,000 levels.

Increases in temperature estimates depend on specific emission trajectories for stabilisation which have been provided by the IPCC since 2001. They show, for example, that a doubling of pre-industrial level of greenhouse gases is *likely* to raise global average temperature by between 2°C and 4.5°C, with a best estimate of approximately 3°C, and that it is *very unlikely* to be less than 1.5°C.[30] Presently

---

[29] Climate change was arguably not the only factor for a revived preoccupation with the topic. Since 2003 oil prices had been on the rise.

[30] The IPCC used three different approaches to deal with uncertainty which depended on the availability of data and experts' judgment. Uncertainties' estimations concerning the causal link between increased concentration in the atmosphere of GHGs and the rising of temperature consisted of expert judgments and statistical analysis. Likelihood ranges were then constructed to express assessed probability of occurrence from *exceptionally unlikely* <1% to *virtually certain* >99%. In this paragraph: *likely* >66%, and *very unlikely* <10%.

neither adaptation nor mitigation can avoid climate change and expected impacts at all. Adaptation is necessary in both the short and the long term to the warming which will occur even for the lowest estimated stabilisation scenario: 445–490 ppm CO_2e.[31] Indeed, this will increase global average temperature by 2.0°C and 2.4°C. The stabilisation of GHGs' concentrations in the atmosphere would need to peak and decline thereafter, and the lower the stabilisation level chosen, the faster the peak and the decline will occur. By now, humanity has years rather than decades to stabilise emissions of GHGs.[32] The expected impacts of global warming are unevenly distributed according to sectors and regions. In the following paragraphs a summary of expected effects taken from the working group II (IPCC 2007b) is provided.

Ecosystems: The resilience of many ecosystems is likely to be exceeded this century. Climate change will lead to increased flooding, drought, wildfire, pest outbreaks, ocean acidification, land use change, pollution, and overexploitation of natural resources. With an increase in global temperature which exceeds 1.5–2.5°C, 20–30% of plant and animal species assessed thus far are likely to be at risk of extinction. In Latin America increases in temperature and associated decreases in soil quality and water availability are projected to lead to gradual replacement of tropical forest by savannah in Eastern Amazonia. In Asia, climate change will compound the pressures on natural resources associated with rapid urbanisation and industrialisation. In both Polar Regions, specific ecosystems and habitats are projected to be vulnerable as climatic barriers to species invasions are lowered.

Food: Globally, the potential for food production is projected to increase in some regions by an increased local average temperature in the 1–3°C range. Above this range food production will decrease. In seasonally dry and tropical regions, crop productivity will decrease for even small local temperature increases (1–2°C). It will augment the risk of malnutrition and weaken political efforts to attain food security, whereby Africa will be especially affected. By 2030, production from agriculture and forestry is projected to decline in Southern and Eastern Australia, and over parts of eastern New Zealand because of increased drought and fire. Similar projections are made for Southern Europe.

---

[31] The totality of GHGs is usually converted into CO_2 equivalents (CO_2e).

[32] A '2-degree goal' was agreed by G8 leaders in Italy in July 2009. They committed to cutting their GHGs emissions by 80% by 2050. Nevertheless, they left the baseline year vague. On December 2009, the fifteenth conference of the parties (COP15) took place in Copenhagen resulting in a non-binding agreement (Copenhagen Accord). Later, Annex-I-countries, roughly speaking rich countries, submitted their quantified emission targets for 2020 with baselines which ranged from 1990 (EU) to 2000 (Australia) and 2005 (US and Canada). One year later, at the COP16 in Cancun rich countries agreed on a *Green Climate Fund* worth USD 100 billion a year by 2020. The declared purpose of the fund was that rich countries assist poor countries in financing GHGs emissions' mitigation and adaptation. How the Green Climate Fund will be raised is still an open question. The overall assessment of the COP16's achievements depends of course on whether the analysts use political criteria or rather criteria oriented to the mitigation and solution of the climate problem.

Coasts: Settlements located in coastal and river flood plains will be severely affected as sea level is expected to rise due mainly to the thawing of the Greenland ice sheet. In the meantime, gradual sea level rise is expected to exacerbate inundations, storm surge, and erosion, therefore threatening vital infrastructure, and facilities which support the livelihood of island communities. Coastal areas, especially the heavily populated regions in the South, East and South-East Asia, will be at the greatest risk due to increased flooding from the sea and rivers.

Health: The health of millions of people is projected to be affected because of increased malnutrition, deaths, diseases and injury driven by extreme weather events such as floods and higher concentrations of ground-level ozone in urban areas. Some health benefits from climate change are projected in temperate areas, such as fewer deaths from cold exposure. However, it is anticipated that these benefits will be outweighed by the negative health effects of rising temperatures. In Europe and North America climate change is also projected to increase health risks due to heat waves and the frequency of wildfires.

Water: Climate change will exacerbate current pressures on water resources from population growth and land use change such as urbanisation. Many semi-arid areas such as the Mediterranean Basin, Western US, Southern African and North-eastern Brazil will suffer a decrease in water resources. Runoff from changes in precipitation and temperature will increase by 10–40% by the mid-century at higher latitudes. Drought-affected areas are projected to increase in extent, with the potential for adverse impacts on multiple sectors such as agriculture, water supply, energy production and health. In Southern Europe, climate change is projected to worsen conditions due to high temperatures and drought in a region already vulnerable to climate variability.

It is worthwhile to mention that many causal chains are not completely understood by climate scientists. For example, the understanding of important factors driving sea level rise is limited, hence, the IPCC does not provide a best estimate for sea level rise, in part because sea level projections do not include uncertainties arising from carbon cycle feedbacks which can amplify the warming effect. Warming amplifying effects are, for example, that natural carbon absorption will be further weakened as severe increases in global temperature could be caused by the liberation of methane from peat deposits, wetlands and thawing permafrost. It means that some effects in their likelihood and magnitude can be underestimated. An increase in the global average temperature of more than 5°C would lead to major disruption and large-scale movement of population. Catastrophic events of this magnitude are difficult to capture with current models as temperatures would be so far outside human experience. What is already well understood is that past and future anthropogenic GHG emissions will continue to contribute to the warming and sea level rise for more than a millennium because of the time scales required for the natural removal of the gases from the atmosphere. Although the prospects of climate change are appalling, let alone the limited capacity of the relevant political actors at the international arena to deal with it, even more appalling is that the warming of the atmosphere is not the only sharpened ecological problem which humanity is facing. Indeed, other problems are plentiful and include ecosystem liquidation, unprecedented biodiversity loss, the collapse of fish stocks, water scarcity, loss of productive soil and impoverished communities. These ecological and social problems will simply become more acute through climate change.

The magnitude and urgency of the problem is evident; a notion which was conveyed by Stern. Nonetheless, his message was one of hope. Taking as the target the stabilisation of carbon emissions in the atmosphere at 550 ppm CO_2e, it would cost approximately 1–2.5% of annual GDP (Stern 2007: 227). The cost is modest ($1 trillion by 2050) with respect to the level and expansion of economic output expected over the next 50 years which is likely to be over 100 times this amount (Stern 2007: 265). He argued that in order achieve that target, strong policy would be required as to redirect research and investments in green technologies away from carbon intensive technologies, especially in the area of energy provision. Unfortunately, Stern took as the target the stabilising of carbon emissions in the atmosphere at 550 ppm CO_2e, yet the IPCC's Fourth Assessment Report showed 1 year later that a 450 ppm CO_2e will be needed if climate change is to be restricted to an average global temperature increase of 2°C. In fact, the target may be even more punishing. Jackson (2009: 83–84) explained, drawing on two articles published in the journal Nature, that 350 ppm target offers the best hope of preventing dangerous climate change. Stern could not have known this writing 3 years before and using largely IPCC's information published in 2001 – even though there was already an international 350 ppm movement and the European Union (EU) had already proposed the 450 ppm goal.

When Stern published his review in 2006, the global economy already required almost 1.5 planets, yet a discussion on the causality's direction between economic growth and ecological obliteration so fervently debated prior to the Brundtland report was completely absent in Stern's work. Economic growth was Stern's default assumption for the entire globe. Finally, some of Stern's ideas would be eventually brought to the international political arena after a global shock, which instead of slowly worsening environmental conditions, expeditiously and decisively set political forces in motion.

3.2.6   Greening the Economy

The financial turmoil caused by the housing bubble burst in the US which almost resulted in a fully-fledged global economic recession between 2008 and 2009 and which greatly shattered the food crisis of the preceding months, opened a political window of opportunity for a greened version of neo-Keynesianism worldwide. In September 2008, the Political Economy Research Institute (PERI) at the University of Massachusetts proposed a fiscal expansion of USD 100 billion (bn) which would create two million green jobs in key areas such as building retrofitting to improve energy efficiency, expansion of mass transit/freight rail, the building of a 'smart' electrical grid, wind power, solar power and biofuels (Pollin et al. 2008). A month later, the executive director of the UNEP, Achim Steiner argued for a 'Global Green

New Deal' as to redirect a substantive portion of the stimulus packages and bank bailouts prepared at the time to the green sector. The green sectors were the same areas already proposed by the PERI but adding ecosystem 'infrastructure' and sustainable agriculture (Nuttall 2008). A month later, a group of investment advisors of the Deutsche Bank revealed the 'green sweet spot' for green investment formed from the junction of three factors: climate change, energy security and the financial crisis (DB 2008). Finally, in January 2009, the US president raised the development of a 'green economy' to the top of the US political Agenda (Goldenberg 2009). Since the EU had for a long time been making active use of fiscal policy to 'decarbonise' their economies so as to meet their emission targets,[33] a green consensus among rich countries was achieved. From the global stimulus plans worth nearly USD three trillion, over USD 430 billion went to the green sector (almost 16% of the total), primarily for energy efficiency (buildings, rail, and so on), water infrastructure and renewables (Robins 2009). In absolute terms, the green stimuli in China and the US took the lead, with USD 221 billion and USD 112 billion respectively. Yet, the real green new deal took place in South Korea, with more than 80% of the total stimulus package (USD 38 billion) allocated for the green fund (ibid.).

In the following years, as the dust of the economic crisis temporarily settled, the idea of the green economy turned into a firmly established notion in the official environmental discourse through the *Green Economy Report: Towards a Green Economy* (UNEP 2011). In this report, the UNEP broadened the focus on green investments in energy efficiency as to include the main *raison d'etre* of SD: development and poverty. It also added many important elements of Ecological Economics in all the green-investment scenarios such as investment in natural capital, eco-taxation, shifting away subsidies from harmful industry, and so on. The topic played a central role during the United Nations Conference on Sustainable Development (Rio+20) in June 2012. Despite the fact that the definition of the green economy is as broad as the definition of SD,[34] the authors of the report made a concise statement about why so little has been achieved in the years since the inception of the sustainability discourse. Their answer was: 'there is a growing recognition that achieving sustainability rests almost entirely on getting the economy right' (UNEP 2011: 16), and getting the economy right means in this new context of Keynsianism active state intervention in order to achieve *sustainable* or, by now, *green* growth.

Although laissez-faire proponents condemn this shift to green neo-Keynesianism, the authors of the report explain that markets' instruments alone cannot deal with pervasive externalities such as climate change in order to globally achieve an economy less dependent on fossil fuels. On the other hand, green technologies also need public

---

[33] Germany has been the forerunner with the enactment of the Renewable Act from the year 2000. The government introduced feed-in tariffs encouraging the deployment of onshore and offshore wind, biomass, hydropower, geothermal and solar facilities.

[34] The 'green economy [is] one that results in improved human well-being and social equity, while significantly reducing environmental risks and ecological scarcities' (UNEP 2011: 16).

procurement so as to protect them against the brutal competition of the market. Many technologies and public facilities which are taken for granted today, contrary to neo-liberal beliefs, have been created and built under the tutelage of state such as aviation, internet, roads and schools. It also seems clear that poor countries, especially the largest and rapidly growing ones such as China and India must be locked into an energetic path different from fossil fuels so as to meet their energetic requirements. Indeed, this is vital if humanity is to have a chance to tackle at least global climate change – whether this is doable given the gigantic and increasing energetic requirements, price uncertainties and the changing geo-strategic game remains an open question.

By and large the report has historical relevance. It captures the changes in the direction of environmental policy which had been taking place within the borders of global players such as members of the EU and China, later joined by the US out of a financial crisis and with a president less hostile to spend taxpayer money for green investment. These factors might explain the swiftness with which the green economy became environmental mainstream discourse. To climb up to this status sustainable development has taken almost 20 years, while the green economy made it in just 3 years.[35] The question that arises and which will be shortly examined is whether this response is adequate in view of the truly civilisational shift needed to cope with a worsened ecological and social crisis.

First at all, the report maintains the growth commitment for the globe, after all growth is also the goal of Keynesianism.[36] Keynes made stimulated public or private demand-driven growth a policy objective in the past century after 1945 (or before, in Roosevelt's New Deal) as a mean to overcome the vicissitudes of the Great Depression. However, Keynes himself saw it as a time-limited policy and not intended to be a perpetual endeavour as implied since the Harrod-Domar growth models of the 1950s.[37] Second, the authors of the report maintain that the 'funda-mental' reason for the social and ecological crisis is 'the gross misallocation of capital' in the last two decades (UNEP, 2011: 14). Certainly, subsidising heavily polluting industries or failing to respect the regenerative capacity of ecosystems has been a grave mistake. However, it hardly follows that the fundamental reason for the ecological and social crisis is because of the misallocation of capital in the recent past. The general preoccupation with both ecological problems and even less with poverty did not start with the inauguration of sustainable development, for this

---

[35] The authors of the report assert that the green economy is not meant to replace SD (UNEP 2011: 2).

[36] The rationale of Keynsianism is that fiscal stimulus funded by deficit spending will create employment, employment will generate income, income will generate private spending and savings, income will spur consumption and savings investment, and consequently employment. With the revenues raised from a reinvigorate economy the government will pay off the debt. The whole purpose of the mechanism is economic growth.

[37] See in particular his essay *Economic Possibilities for our Grandchildren* written in 1930 (Keynes 2009).

was a response to the joint-effects of these problems within the constraints of the political possible. An alternative fundamental reason would be that ecological and related social problems exist because of the metabolism of the industrial economy, and the economic policy of perpetual economic growth largely driven by the search of profits and rents in a non-growing planet. Third, the projections of the report reach as far as 2050. Assuming that through green investments – which are absolutely necessary – and further improvements in energy/matter efficiency we maintained global growth until 2050 what will happen thereafter? It is highly probable that humanity will end up simply doing the same or even more of the things which became cheaper because of the very same improvements in energy/matter efficiency. This is the Jevons' paradox which has been mentioned several times in the last sections and which now requires more elaboration.

William S. Jevons in his *The Coal Question* (1865) was concerned about Britain losing her economic dynamism and worldwide position because of a foreseeable depletion of coal reserves. On the one hand, while other countries were living on the annual regular income from harvest, Britain was living on capital which would not yield interest as it was being turned into heat, light and power, that is, that capital was disappearing forever (Martínez-Alier, 1987: 161). On the other hand, he doubted that gains in technical efficiency with regards to the use of coal would lead in the future to less coal consumption as was argued at that time:

> It is wholly a confusion of ideas to suppose that the economical use of fuel is equivalent to a diminished consumption. The very contrary is the true [...] new modes of economy will lead to an increase in consumption. (Jevons 1865. Quoted by Polimeni et al. 2008)

The topic remerged almost 100 years later after industrial economies had largely switched from coal to oil and later on to nuclear power for electricity as a result of partial oil-demand destruction caused by the OPEC's embargo during the 1970s and early 1980s. The article of Khazzoom (1980) elicited a renewed interest on the issue as he explained that some mandated standards for energy saving would even 'backfire' (Khazzoom 1980: 35). From then on an enlargement of the Jevons' paradox, which has been renamed as the rebound effect has been taking place. Theoretical and empirical studies have attempted to trace, for example, micro- to macro-economic effects. Nonetheless, the results of these studies remained unconvincing. For example, increased energy/matter efficiency would make a given commodity cheaper, what conversely would free household's income which would be spent on either more consumption of the same product or on other products in case of low-demand elasticity. Eventually it will pull up economic growth, and economic growth will mean, *ceteris paribus*, more resource extraction (inputs) and waste/pollution (output). The unconvincing part of this argumentative line is related to the insurmountable empirical task of following income effects up to the macro-economy, also aggravated by the different theoretical growth-approaches and the terminology used (see the following reviews Herring (1999), Biswanger (2001), Alcott (2005) and Jenkins et al. (2011)). However, and as already shown when discussing SD, the Jevons' paradox seems not to be a paradox at all. It was after all a major component in the pattern of development of the West – at least in its own terms.

The authors of the report fully recognised the Jevons' paradox in the green investment scenarios for the manufacturing sector (UNEP, 2011: 257–258), energy efficiency in buildings (ibid.: 357–361) and green cities (ibid.: 461, 479). Nevertheless, the policy implications which have followed from its recognition are by and large inconclusive. In the context of increasing energy efficiency in buildings the report could only simulate power demand and not overall energy use due to a lack of data. Power demand accounts only, according to the report, for roughly 30% of total energy used in buildings. In spite of the partial but highly positive results of the simulation, it is stated that 'economic growth in the green investment scenarios, approximately offsets the savings in power demand' (ibid.: 357). This is the Jevons' paradox. However, policy implications are left rather inconclusive. It is simply stated that it 'highlights the importance of accompanying new technologies with appropriate behavioral and institutional change' (ibid.: 357), without specifically mentioning what kinds of behavioural and institutional changes are needed.

In the context of green cities, an example of a current green community in Britain is given in which households have achieved 84% of energy reduction and decreased 36% of their ecological footprints. Nevertheless, it is specified (in a footnote) that although the residents of the community have reduced their footprint on site:

> A lot of their ecological impact is made outside of it, in schools, at work, and on holiday . . . [they also] fly slightly more frequently than the local average, presumably due to their higher average income. (ibid.: 461)

This is the Jevons' paradox. The authors argued that these limitations do not undermine the achievements of the local development, which is utterly correct. They finally suggested the need for 'scaling up energy efficiency measures in wider urban settlement systems' (ibid.: 461). The problem is that scaling up efficiency measures will necessarily culminate in efficiency measures for the entire world, that is, from what is called relative decoupling (energy/matter efficiency gains) to absolute decoupling. That is precisely what is proposed for the manufacturing sector. In the context of manufacturing, or green investment scenarios, the report states that overall emissions, energy and material use have been growing in spite of efficiency gains. Figure 3.3 depicts a global trend in increasing resource extraction, population and GDP, while the use of materials has markedly declined (increased efficiency) in the period 1980–2007. The dilemma is settled by stating that 'what economies world-wide need is absolute decoupling of the environmental pressure with resource consumption from economic growth' (ibid.: 257). Absolute decoupling will imply that worldwide total resource extraction is held constant, while GDP still increases, as the report maintains the growth commitment. This conclusion may have the following problems. First, resource extraction as depicted in Fig. 3.3 is an aggregate of metal ores, industrial and construction minerals, fossil fuels and biomass. Resource extraction could be limited in one of these sectors because of substitution effects caused by scarcity. However, this would increase resource extraction in other sectors which conversely may still increase overall resource extraction. This is at least the pattern which the historical evidence has shown so far. Second, and provisionally setting aside increasingly political and

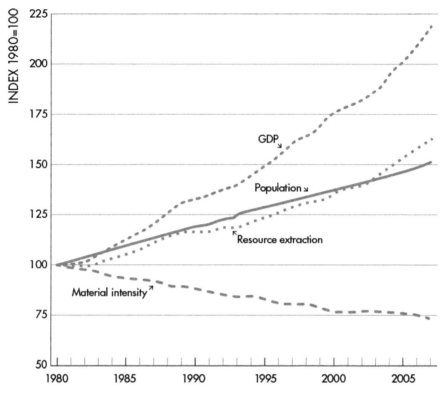

Fig. 3.3 The Jevons' paradox (UNEP 2011. Modified by the author)

ecological conflicts associated with extractive industries, the problem seems not to be that the earth's crust does not contain enough minerals to maintain customary growth levels in the long run, but in waste/pollution. In other words, *currently* and only *physically* speaking, the problem does not lie in the input-side but in the output-side of the global economy. This observation does not disclose any recondite truth. Georgescu-Roegen stated, or rather prophesised 40 years before we became so concerned with issues such as climate change, that because:

> Pollution is a surface phenomenon which also strikes the generation which produces it, we may rest assured that it will receive much more official attention that its inseparable companion, resource depletion. (1975: 377)

Thus, it can be argued that once absolute decoupling is achieved, then waste/ pollution problems will be gradually solved, but before this can be concluded, policy instruments facilitating absolute decoupling should be discussed and proposed. This is what is largely left inclusive in the report. A proposal would be to restrict the quantities of the resources according to the more stringent ecological or social necessity, and to let market prices fulfil their function. This proposal will be examined in detail in Sect. 3.3.5. However, it can be stated in advance that the chances for its implementation are rather low – as any other alternative whose implementation necessarily requires international governance structures dealing with constraints.

Foreseeable political difficulties at this level are perhaps an *approximate* explanation of why the report left largely unresolved the Jevons' paradox and even included as a major finding that the 'trade-off between economic progress and environmental sustainability is a myth' (UNEP 2011: 622). Industrial ecologist Robert Ayres, who was one of the chapter coordinators of the report, stated a couple of years ago that:

> None of the important economic actors, whether government leaders or private sector executives, has an incentive compatible with a 'no growth' policy. No economic growth is evidently not a politically viable proposition for a democracy, at least in a world with enormous gaps between poverty and wealth. But 'no growth' is an imperative as regards extractive materials, energy and pollution emissions because economic activity is based on a material function. (Ayres 2008: 290)

And yet, unviable policy proposals do not transform theory and evidence into a myth.

3.2.7 Wither Economic Growth?

Over the last 40 years economic growth has not only been assiduously cherished, but it has been elevated from time to time to a truly panacea: unemployment, development/poverty, overpopulation ('demographic transition'), and even ecological degradation ('environmental quality') have been claimed to be solved by economic growth, nay, by export led-growth.

Of course the problems of unemployment could be, at least partially tackled in rich countries by working-less/work-sharing. Poverty in rich countries could also be overcome by using other instruments such as a basic citizen income, and to effectively tackle the gap between rich and poor which is increasing even in Western Europe (Jackson 2009). The citizens of the poorest countries in the world could also be relieved from this malady by a global minimum wage, or if it is held to be an illusion opposed not only due of ideological concoctions but also due of foreseeable implementation problems, then at least by a better distribution of the gains of economic growth that hardly anyone claims they do not need. As economist Andrew Simms (2008: 49) observed:

> During the 1980s, for every $100 added to the value of the global economy, around $2.20 found its way to those living below the World Bank's absolute poverty line. During the 1990s, that share shrank to just 60 cents. This inequity in income distribution – more like a flood up than a trickle down – means that for the poor to get slightly less poor, the rich have to get very much richer. It would take around $166 worth of global growth to generate $1 extra for people living on below $1 a day.

From this perspective, claiming for more export-led growth as a mean for development and poverty alleviation is misguided and it has been long before recognised as such.

On the problem of population, China for instance, did not wait for the effects of a 'demographic transition' which should automatically happen once she becomes rich through growth, instead she preferred active top-down population policy. Contrastingly, the poor and working class in Western Europe and the US, and in some countries of Latin America, were practicing a century ago what Martínez-Alier and Masjuan (2008) called 'bottom-up neo-Malthusianism'. This was a popular movement which helped to bring down fertility rates in Western Europe against the pro-population growth policy of the state.[38] Respected demographer Carl Haub explained that 'well organized family planning campaigns are much more important than economic growth' (Hickman 2011). On the other hand, and although population growth still constitutes a problem for development in some poor countries, the truly global ecological problem is overconsumption in rich countries and its increasing emulation in emerging ones, as the very same author of the *Population Explosion* Paul Ehrlich maintains nowadays.[39] Frugality or sufficiency (less consumption) is still a necessary condition for environmental sustainability as it was acknowledged 40 years before. Indeed, it is increasingly accepted today by social scientists who in the recent past have focused primarily on technological progress (Weizäcker et al. 2009: 346). Based on the same rationale, there is a call to draft the 'Millennium Consumption Goals' (Assadourian 2011) and to implement, in line with democratic traditions and environmental justice, the 'One Man – One Vote – One Carbon Footprint' (Töpfer and Bachmann 2009).

However, since 'growthmania' is still in place and ecological problems continue to rise as expected in a world subjected to the laws of thermodynamics and ecological limits, the afore-mentioned scattered proposals are barely taken seriously by the social agents who matter: decision-makers in rich and by now emerging countries. The only way to maintain the growth commitment is to forcefully presuppose that only technological progress will drastically reduce the impact of growth on the biosphere. Technology is still 'the rock upon which the growthmen built their church' (Daly 1972: 949) in spite of recent historical evidence showing that technological progress can bring severe risks (EEA 2001), that it makes societies prone to fall into 'progress traps' (Wright 2005),[40] and that therefore, technological faith encompasses a great deal of utopianism which must be denounced as such (Jonas 1979: 9). As will be shown later, these caveats do not

---

[38] The arguments for voluntary population control were women's freedom, relieving pressure on wages ('womb strike'), anti-militarism, impeding migration overseas and the natural environment. Not surprisingly governments at that time harshly repressed the movement on grounds of religion and national interests (See Martínez-Alier and Masjuan 2008).

[39] Ibid.

[40] The notion of 'progress trap' coined by anthropologist Ronald Wright means that the problems created by technology can usually be solved only by more technology, and the new problems created by the latter must be solved by even more technology, and so *ad infinitum*. He also explained how 'too much progress' can be made. For instance since the Chinese invented gunpowder, there has been great progress in the making of bangs, but 'when the bang we can make can blow up the world, we have made rather too much progress' (2005: 5).

involve a rejection of technological progress altogether – the problem is (still) 'growthmania' and growth.

Although it is probable that the green economy will dominate the environmental official discourse for the following years, it is convenient to examine less-political realist but 'imperative' proposals which could replace economic growth.

3.3   The Steady-State Economy of Daly

3.3.1   Intellectual Foundations: Mill and Georgescu-Roegen

Classical economists were growth economists.[41] Material progress[42] was not only the source of national power – the interests of kings and merchants, but also a source of prosperity to the population at large (Arndt 1978: 7). Nonetheless, they all expected with pessimism an economic stationary-state. For Adam Smith the 'stationary [state] is dull; the declining melancholy' (Smith 1991 [1776]: 86). In the hands of Malthus the stationary-state is not only melancholic but dreadful given the propensity of humans to increase in numbers faster than the ability to produce food. Hence, population checks would inevitably arrive either by the 'vices of mankind' such as wars; and in the case it fails then by 'sickly seasons, epidemics, pestilence [...] plague [and] famine [...]' (Malthus 1998 [1798]: 139–40). The Ricardian stationary state was not attractive but at least it did not have the horror portrayed by Malthus, for it can be postponed through *laissez faire* policy, developing free trade and the exploitation of the resources in the *new* world (Hicks 1966: 260). In general, however, the normal expectation of the individual was to live on the brink of starvation, and material progress would improve the conditions of those who were already wealthy. Political economy was indeed, as Thomas Carlyle once judged it: 'the dismal science'.

It was Mill who introduced a radically different view of the stationary-state. In his view the stationary state is highly desirable and as such, it deserves to be put as an overall policy objective. His line of reasoning anticipated many of the ecological and social arguments made against the perpetual growth policy from the late 1960s up to now. He saw no reason why the natural environment should be sacrificed

---

[41] Reducing Daly's intellectual foundations to Mill and Georgescu-Roegen is an arbitrary choice for his views were also shaped by the works of John Ruskin, Frederick Soddy, Kenneth Boulding, and Irving Fisher among others. Nevertheless, as it will be shown, Mill's and Georgescu-Roegen's ideas constitute Daly's strongest foundations.

[42] 'Progress ceased to be an issue of metaphysics as understood in the middle ages, and came to be a material issue in the early eighteenth century. Material progress or 'raising standards of living' became the mean to achieve the greatest happiness for the greatest number, as the utilitarian principle proclaimed (Pollard 1968).

through the combined forces of affluence and population growth. His arguments are worth quoting at length:

> Nor there is much satisfaction in contemplating the world with nothing left to the sponta-
> neous activity of nature; with every rood of land brought into cultivation, which is capable
> of growing food for human beings; every flowery waste or natural pasture ploughed up, all
> quadrupeds or birds which are not domesticated for man's use exterminated as his rivals for
> food, every hedgerow or superfluous tree rooted out, and scarcely a place left where a wild
> shrub or flower could grow without being eradicated as a weed in the name of improved
> agriculture. If the earth must lose that great portion of its pleasantness which it owes to
> things that the unlimited increase of wealth and population would extirpate from it, for the
> mere purpose of enabling it to support a larger, but not a better or happier population,
> I sincerely hope, for the sake of posterity, that they will be content to be stationary, long
> before necessity compels them to it. (Mill 2004 [1848]: 692)

Although his advocacy for conservation was specially directed at his home country, Britain, his vision can be enlarged as to encompass today's rich countries for:

> It is only in the backward countries of the world that increased production is still an
> important object; in those most advanced, what is needed is a better distribution, of
> which one indispensable means is the stricter restrain of population. (ibid.: 691)

Mill, differing from Ricardo, viewed birth controlling measures as the most important public policy, so that population becomes the fixed factor of production, and in so doing, ensuring that a large portion of the production surplus flows to wages. With regards how to attain distribution Mills stated that:

> ... this better distribution of property [may be] attained, by the joint effect of the prudence
> and frugality of individuals, and of a system of legislation favouring equality of fortunes.
> (ibid.: 691)

Mill also addressed what Fred Hirsch 120 years later would call the *Social Limits to Growth* (1977), whose ideas Daly integrated into his model. Mill could not conceive as the most desirable state of social life the one in which the norm is: 'struggling to get on; that the trampling, crushing, elbowing and treading on each other's heels' (ibid.: 690).

The second main intellectual source of Daly's thought was the work of the mathematician and economist Nicholas Georgescu-Roegen,[43] who rigorously treated the implications of thermodynamics in the economic process. He disclosed the fallacy of misplace concreteness in which the marginalists, and later neo-classical economists have incurred by forgetting the resource base of the economy and in viewing the economic process through the lenses of Newtonian mechanics.[44]

---

[43] For a review of Georgescu-Roegen's thought see Maneschi and Zamagni (1997) and Daly (1996: 191–198).

[44] Georgescu-Roegen maintained that the fallacy of misplaced concreteness was the cardinal 'sin' of orthodox economics from which only Marx, Veblen and Schumpeter offered substantial ways to transcend it (1971: 231). The fallacy, formulated by philosopher Alfred Whitehead, consisted of 'neglecting the degree of abstraction involved when an actual entity is considered merely so far as

For the authors of the marginalist revolution,[45] the problem of land – until recently the economic term encompassing all natural resources – was abandoned, and economic growth ceased to be the central topic. They became rather concerned with the allocation of given resources (Screpanti and Zamagni 2005: 165), in spite of Jevons' energy analysis. Neglecting the role of resources in the economy was so intriguing, that, as Georgescu-Roegen observed: 'Not even wars [. . .] for the control of the world's natural resources awoke economists from their slumber' (Georgescu-Roegen 1971: 2).

On the other hand, the ambitions of the marginalists in making out of economics a scientific discipline led them to adopt the Newtonian mechanistic worldview into their modelling. Nonetheless, while the marginalist revolution was taking place in economics through the adoption of Newtonian mechanics from physics, a revolution was taking place in physics which was abandoning Newtonian mechanics. The revolutionaries were Rudolf Clausius, Robert Mayer, and Herman Helmholtz who grounded the new branch of physics thermodynamics (Georgescu-Roegen 1971: 141–195, Martínez-Alier 1987: 73–88) and from which the law of conservation of energy and the entropy law were postulated. They are correspondingly the first and the second law of thermodynamics.[46] For Georgescu-Roegen the entropy law was the most relevant physical law in economics, which leaves no room for the mechanistic view of modern neo-classical economics so clearly implied in macro-economic books' charts depicting the economic process as a circular flow of national product and income in a perfectly competitive market. Entropy means that in an isolated system, energy would move towards a thermodynamic equilibrium in which energy is equally diffused throughout the closed space.

The relation of the two thermodynamic laws and the economic process can be exemplified as follows: in the combustion chamber of the modern car engine the fuel is burnt. The resulting heat and the pressure of the gases apply force to the components of the car engine such as the pistons and the wheels. The evident result of the combustion process is locomotion: the car moves from A to B. According to the first thermodynamic law, the quantity of energy has not changed, yet a qualitative change has taken place. Before the fuel entered the combustion chamber, its chemical energy was available for producing mechanical work. After the fuel leaves the combustion chamber the chemical energy loses its quality and dissipates into the atmosphere where it becomes non-available energy, that is to say, it can no longer be used for the same purpose. This strict linearity and irrevocability from order to disorder represents the entropy law. The entropy law has enormous

---

it exemplifies certain categories of thought. There are aspects of actualities which are simply ignored so long as we restrict thought to these categories' (Whitehead 1978: 8).

[45] The figures were mainly William Stanley Jevons, León Walras and Carl Menger. For a detailed account see Screpanti and Zamagni (2005: 163–195).

[46] The third law of thermodynamics is less relevant for economics. It states that the entropy of any pure, perfect crystalline element or compound at absolute zero is equal to zero.

relevance, from the human perspective, to non-renewable resources.[47] If uranium, petroleum or coal could be re-used *ad infinitum*, scarcity would cease to be an economic problem and the resource pressures arising from a growing population and affluence could simply be solved by more frequently using the flows of the existing stocks. As much as we might believe in human inventiveness with respect to technological progress and semantics, it cannot reverse this linearity.

Georgescu-Roegen was also very clear in stating that the dictates of the entropy law happens whether or not humans are around, for the economic role of humans is simply that of 'pushing or pulling' (Georgescu-Roegen 1971: 141). In other words, the economic process consists of accelerating the transformation from low entropy energy/matter into high entropy energy/matter,[48] that is, from speeding up depletion to speeding up waste/pollution. It also follows, *ceteris paribus*, the greater the size and intensity of the economic activity the more depletion/pollution which occurs. From this perspective it is not surprising that the greatest ecological problems have been caused by industrial economies based on fossil fuels in spite of continued efforts in 'ecological modernisation'. It is worthwhile to emphasise again that Georgescu-Roegen's central point is that these physical facts are not accounted for in economics:

> Had economics recognized the entropic nature of the economic process, it might have been able to warn its co-workers for the betterment of mankind – the technological sciences- that 'bigger and better' washing machines, automobiles, and superjets must lead to 'bigger and better' pollution. (Georgescu-Roegen 1971: 19)

3.3.2 Unravelling Fallacies of Misplaced Concreteness

Drawing upon the ideas of Mill and Georgescu-Roegen, Daly further pursued the revision of economic theory disclosing and correcting further fallacies of misplaced concreteness (FMC). In the next paragraphs, I will discuss two of these fallacies which are central to understanding the theoretical tenets of steady-state economy: markets and technology.[49]

3.3.2.1 The Market

Daly fully recognised the superiority of the market-economy in allocating scarce resources among alternative uses compared to a planned economy; nonetheless

---

[47] As the earth is not an isolated system (it receives and reflects solar radiation) but a closed system (it does not exchange relevant amount of matter with the outer space), nonrenewable resources (fossil fuels and minerals) are in absolute terms finite.

[48] Georgescu-Roegen latter extended the entropy law as to include matter and proposed the fourth law of thermodynamics. It has been disputed whether a 'fourth law' can be formally enunciated. It is however, not disputed that matter inherently tends toward disorder too (see Daly and Farley 2011: 66).

[49] The following paragraphs rely heavily on Daly (1991: 281–287), Daly and Cobb (1994: 25–117) and Daly (1996: 38–44).

there are some negative features which require correction. They are (1) the tendency for competition to be self-eliminating, (2) the corrosiveness of self-interest on the moral context of the community that is presupposed by the market, (3) the existence of externalities which can be localised or pervasive, (4) an implicit amoral position on the issue distribution, and (5) the lack of defining the optimal scale of the economy relative to the natural system.

1. *The tendency for competition to be self-eliminating.*

Competition is cherished by orthodox economists on the grounds that it improves allocative efficiency, keeps profits at the normal level and avoids, at least theoretically, the emergence of monopoly which can negatively influence market prices. The slogan is 'the more buyers and sellers the better'. Nevertheless, in the middle run many firms become few firms and monopoly power increases. In addition, in the long run giant conglomerates appear with their correspondingly giant corporate bureaucracies making the market economy hardly indistinguishable from a planned economy. Within a single country this development is economically and politically damaging, and even more so within the relentless pursuit of a global integrated economy.

As explained in the last section, as the laissez-faire intellectual movement gradually gained strength, free trade and capital mobility doctrines were (selectively) re-adopted and re-implemented. In this context, the enforcement of antitrust laws of individual nations became more costly, if not impossible. One of the reasons is that the accumulation of wealth tends to increase *pari passu* with political power. Agri-business, energy provision, media-entertainment organised as transnational corporations along with financial institutions are today in a position to influence polities and politics at different levels through many direct or indirect means. It ranges from structurally having become 'too big to fail', effectively lobbying for favourable legislation, to simple unspoken and direct threats of offshoring production or capital flight. Under these circumstances, not only the credibility but even the actual functioning of representative democracy erodes.

The theoretical foundation of free trade draws from the theory of the comparative advantage as formulated by David Ricardo. However, one of the many assumptions upon which the comparative advantage was formulated was capital immobility, an assumption which was taken for granted by Adam Smith prior to David Ricardo,[50] in spite of his famous invisible hand thesis.[51] The capitalist would

---

[50] Capital immobility is certainly not the only assumption that does not hold today. Understandably Ricardo could not think of environmental costs (pollution). On the other hand, he also did not consider transport costs, the costs of specialisation, and more fundamentally, the loss of freedom of *not* to trade. For a detailed review and analysis see Daly and Farley (2011: 355–363).

[51] The often-quoted passage of the invisible hand of Adam Smith portraying the capitalist as a simple egoist who through his actions indirectly increased total wealth sometimes overlooks the very beginning of the quote: 'By *preferring the support of domestic to that of foreign industry*, he intends only his own security [...] he is in this, as in many other cases, led by an invisible hand [...]' (Smith 1991 [1776]: 351. Emphasis supplied).

not invest abroad even in view of larger profit margins, since according to Smith and Ricardo, the capitalist is primarily a member of the national community which forms his very identity. She/he would consequently avoid living under customs alien to her/him. This assumption clearly does not hold in today's globalised world of cosmopolitan money managers and global corporations. As Daly and Cobb observed: 'it is clear that Smith and Ricardo were considering a world in which capitalists were fundamentally good Englishmen [and] Frenchmen' (1994: 215).

2. *The corrosiveness of self-interest on the moral context of the community which is presupposed by the market.*

During the LtG-debate, Fred Hirsch authored *Social Limits to Growth* (1977). He believed that the growth discussion emphasising distant and uncertain physical limits was inappropriate, as it was overlooking closer and more certain limits, namely social limits. Social limits is a dual social phenomenon caused by economic growth. They are (a) the increasing importance of *positional* goods and services, and (b) the decreasing morality of individuals. As economic growth increases, affluence also increases, and with increasing affluence, individuals tend to value goods and services rather in relation to the valuations made by other individuals. In this process individuals are trapped in a spiral of social competition ('keeping-up-with-the-Jones') which conversely makes the social position attached to those goods and services 'scarce'. From this process a 'paradox of affluence' results (Hirsch 1977: 175). When the growth process is sustained and generalised the outcome is frustration instead of happiness. The other social limit is the weakening of social values. Hirsch argued that the social foundations upon which the contractual economy works such as truth, trust, acceptance, restraint and obligation are undermined by the individualistic and competitive ethos nurtured by economic growth. Both arguments are taken up by Daly and put into the box of FMC's cases. It is the fallacy of *homo economicus*. Orthodox economists abstracting from community forgot that there are also a *homo ethicus*, *homo politicus*, and more broadly the 'person-in-community' (Daly and Cobb 1994: 159).

3. *The existence of externalities that can be localised or pervasive.*

The standard market argument runs as follows: in a perfectly competitive market self-interest seeking individuals voluntarily exchanged goods and services. However, as some of the elements neglected in reality became evident to economists' experience, their existence had to be somehow acknowledged. It was noticed that many transactions between self-interest seeking individuals unintentionally affected other parties which were not involved in the exchange. This acknowledgement was integrated through the concept of externality. While Alfred Marshall was the first to draw attention to externalities, it was his pupil Arthur Cecil Pigou who developed a rigorous treatment of the issue in his *The Economics of Welfare* published in 1920. As previously mentioned, the concept gained relevance in the 1960s when concerns with environmental degradation emerged, especially those captured with the label 'pollution'. Pollution was then integrated in economic theory with the formerly introduced concept of externality. The concept externality primarily suggests that the phenomenon is external to the market, and therefore, measures to internalise them are proposed, namely Pigovian taxes/subsidies and Coasian property rights and markets. What is more important is that the phenomenon is also external to the theoretical edifice that builds on the market as an economic concept. Hence, the ad hoc introduction of the externality served to circumvent the revision of the entire theory, just as the ad hoc introduction of epicycles permitted Ptolemy to not reconsider his astronomy. However, and as Daly reasoned, when externalities are exceeding the absorption capacity of the biosphere, and threatening human life support-systems, it is time to rethink the whole theory and re-start with different abstractions.

4. *An amoral position on the issue of distribution.*

Markets criterion in the distribution of, for example income, is allocative efficiency rather than justice. People have no rights excepting the ones which they can buy according to what they can sell in the labour market. It can be seen as a sort of morality which was seen as inevitable by Malthus and Ricardo ('iron law of wages'), when they, among many other intellectuals at that time, were intellectually overwhelmed in trying to explain why Britain was becoming so wealthy while at the same time generating so many poor people. This sort of morality is however, hardly tenable within the humanistic tradition inherited to and preached by Adam Smith. For that reason, and as in the case of antitrust laws, societies have crafted institutions such as minimum wages and income tax progressivity as a societal mechanism of self-protection (Polanyi 2001). However, as in the case of antitrust laws, such social institutions have been gradually eroding in the second wave of globalisation.

5. *The lack of defining the optimal scale of the economy relative to the natural system.*

Markets do not have an 'organ' which tells us when to stop the demands made from the biosphere. This is the organ that Daly introduced. It is the notion of a macro-economic optimal scale of the economy, relative to the natural environment. The optimal scale is at the heart of the steady-state economy, and is what ultimately gives a sense to any concept of environmental and economic sustainability.

3.3.2.2 Technological Progress

Daly is not a neo-luddite, but equally not a believer in promethean gifts. He claims that the standard practice of attributing to technology all sorts of mystical faculties has its origins in 'growthmania'. The issue of technology is itself broad, so that only the relationship between scarcity, substitution and technology will be addressed.

Scarcity is the *raison d'etre* of economic thought. In production, scarcity of a given input factor is relative to the scarcity of other input factors, such as the fact that oil has largely substituted coal, aluminium has largely substituted iron and copper, and perhaps uranium will be substituted on a larger scale in the future by thorium. Nevertheless, in Daly's conceptualisation, this line of thinking is only the half-truth, and is what makes it a FMC. Resources were and are indeed substituted; however, substitution occurs within the strictly limited total of low-entropy stock. In the context of SD, orthodox economists advanced the idea of maintaining aggregate capital constant, that is, natural, man-made, human and social capital (Pearce 2002: 63–66). It implies that these forms of capital are substitutable, specifically, that natural resources can be substituted by reproducible man-made capital. The strongest position on this issue was once formulated by Nobel-prize winner growth-economist Robert Solow (1974: 11):

> If it is very easy to substitute other factors for natural resources, then there is in principle no 'problem'. The world can, in effect, get along without natural resources, so exhaustion is just an event, not a catastrophe.

In the hands of Daly, man-made and natural capital are complements and only marginal substitutes (Daly 1996: 76). The reason is plainly obvious: there are no other 'factors' apart from natural resources. Producing more of the allegedly substitute (man-made capital) requires more of what it is substituted for (natural capital). On the other hand, and as already noted, the overemphasis sometimes placed on the input-side fails to recognise that abiotic resources (fossil fuels and in general minerals) do not disappear when they are used up, they return to the biosphere as waste/pollution causing acid rain, global warming, oil spills, discarded plastics and e-waste. By now it seems that 'the sink will be full before the source is empty' (Daly and Farley 2011: 81) – as Georgescu-Roegen explained in 1971, and one of the LtG scenarios suggested in 1972.

Daly saw technological progress as necessary pertaining to what we can get out of the entropic direction of the flows arising from stocks, that is, energy/matter efficiency, but not within the paradigm of economic growth. Within the economic growth paradigm, technological progress will necessarily aggravate ecological and social vicissitudes.

3.3.3   From Social and Physical Limits to Growth Toward a Steady-State Economy

Daly departed from the pre-analytic vision that the economy is a sub-system of the larger environmental system. This pre-analytic vision implies, first, that there are physical limits to the smaller system with respect to the larger system. Since the latter does not grow, then the former cannot possibly grow beyond the physical limits imposed by the larger system. Second, since such physical limits exist, albeit not always straightforwardly knowable, it is also possible to derive a desirable (economic) limit of the smaller sub-system.[52] Therefore the question is: what is the

---

[52] It is also called sometimes the 'threshold hypothesis' enunciated by Chilean economist Manfred Max-Neef independently from Daly. The hypothesis states that 'for every society there seems to be

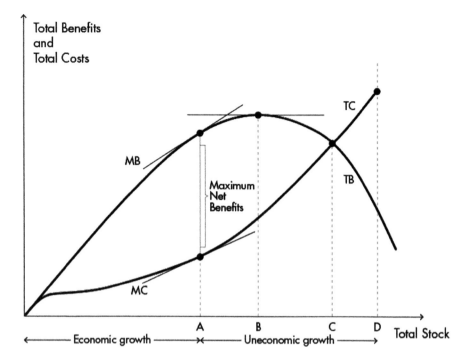

Fig. 3.4 Economic and uneconomic growth (Daly 1991. Modified by the author)

optimal scale of the economy? Concerning physical limits, and as previously mentioned in Sect. 3.1, natural scientists have been working for a long time on indexes which measure both the relative and absolute impact of the economic activities on the biosphere, such as LtG, the percentage of human appropriation of the total world products of photosynthesis, the footprint aggregate metric, IPCC estimations, and more recently, the planetary boundaries (Rockström et al. 2009). The rationale concerning the optimal scale of the economy is illustrated in Fig. 3.4. It shows the curve total benefits (TB) and total costs (TC) in relation to the total stock. Benefits decrease by each consumed unit (marginal benefits-MB), while the costs of producing a further unit of the stock increases (marginal costs-MC). Marginal benefits and marginal costs are represented by the corresponding slopes. Maximum net benefits are reached when marginal costs are equal to marginal benefits. That is at point A. At point B marginal benefits are zero, and thus there is no reason for growing beyond B even if costs are zero. C is a turning point, at which total benefits of past growth are balanced by total costs of past growth. Yet, it is economically wise to be governed by current marginal costs and benefits instead

---

a period in which economic growth (as conventionally measured) brings about an improvement in the quality of life, but only up to a point – the threshold point – beyond which, if there is more economic growth, quality of life may begin to deteriorate' (Max-Neef 1995: 117).

of past costs and benefits. At point D, the marginal costs of growth tend to be infinite, so even in the case that marginal benefits are still great, economic growth will cease. On the whole, a sensible policy recommendation would be to stop economic growth at point A. Beyond point A, economic growth ceases to be 'economic' and starts to be 'uneconomic', that is, it starts making a country poorer, not richer.

Note that this argumentative line is far from radical or even novel; the principle that economic agents should expand the scale of a given activity up to the point where marginal costs equal marginal benefits is the principle around which micro-economic theory gravitates. In macro-economics the principle of optimality is dropped, which is what Daly called the 'glittering anomaly' (1996: 60). Given the physical and economic limits to growth, Daly proposed a simple overall policy objective: the steady-state economy (SSE). The SSE is the intellectual response for a world which is no longer empty but full,[53] which strongly resembles the cowboy/spaceship analogy of Boulding.

The SSE has three important components: (1) the stock of capital composed of people and artefacts (consumer and producer goods), (2) the flow of energy/matter throughput and (3) the service. The economy, just as animals, lives from its metabolic flow, beginning with extractions from the biosphere, and ending with the return of waste/pollution back to the biosphere. Input and output are conflated into the term 'throughput' coined by Boulding, and as already explained, through-put is entropic (linear, irrevocable and irreversible). The stock of capital needs throughput because capital is also entropic. The stock of capital is composed by dissipative structures, that is, structures which decay, rot, die and fall apart. Although waste materials can be recycled by biochemical processes powered by solar energy, such recycling is external to the animal or economy whose life depends on the services provided by the natural environment. Even though the SSE is primarily a physical concept, Daly acknowledged that the purpose of the economy is the satisfaction of human needs/wants (Daly 1991: 16), or as Georgescu-Roegen called it the 'immaterial flux, the enjoyment of life' (1971: 18). This is conceptualised as the service. The SSE is defined as: 'an economy with constant stocks of people and artifacts, maintained at some desired, sufficient levels by low rates of maintenance 'throughput' (Daly 1991: 17). Hence, the service is the final benefit of the economic activity, while the entropic throughput is the final cost. The quality and quantity of services are strictly provided by the stocks and not by the flows. The relationships of the three components are depicted in the following definitional equation taken from Daly (1991: 36):

$$\frac{Service}{Throughput} \equiv \frac{Service}{Stock} \times \frac{Stock}{Throughput}$$
$$(1) \qquad\qquad (2) \qquad\qquad (3)$$

---

[53] On several occasions Daly has conceptualised Ecological Economics as economics for a 'full world'.

The ratio (3) represents the maintenance efficiency of the throughput and the ratio (2) the service efficiency of the stock. Stocks cancels out as in real life they exhaust, hence the ultimate benefit is the service efficiency of the sacrificed ecosystem caused by throughput (1). Each component requires a mode of behaviour: regarding stocks, a level must be chosen which is *sufficient* for a *good life* and is sustainable in the long run. Throughput is to be minimised, while service must be maximised. Both throughput and service are subject to the maintenance of the chosen levels of stock. If the SSE's goal is to maintain constant the stock of people and artefacts, what is the part which should not be held constant? Daly's answer was straightforward: culture, morals, knowledge (technology), distribution, mix of capital, and so on, that is, qualitative change. Here, Daly differentiated between economic *growth* and economic *development*. Economic growth is quantitative change, whereas development is qualitative change. A SSE 'develops but does not grow' (Daly 1991: 17), just as the planet does. Daly in line with Mill maintained that humankind, especially rich countries, should be more concerned with being *better* (development) than with being *bigger* (economic growth).

3.3.4   ISEW/GPI Instead of GDP

The conception of the SSE necessarily led to a proposal which would replace the most important national account used to measure economic growth: GDP. The new metric would attempt to measure human welfare, and not simply unqualified market activity. Daly and Cobb developed the *Index of Sustainable Economic Welfare* (ISEW) in 1989 which was improved 5 years later (Daly and Cobb 1994: 62–83, 443–507). It originated in an extensive range of similar studies during the 1990s up to the present. The ISEW was first tested for the US in the period 1950–1900. It was shown that from 1975 until 1985, the ISEW started to decline even when GNP[54] was rising. From 1985 until 1990 the ISEW raised slightly but much slower than GNP (Daly and Cobb 1994: 464). Instead of showing numbers and figures, I will instead discuss the conceptual differences between GDP and ISEW.

GDP is the total monetary value of the goods and services produced annually with the factors of production located in a particular region, usually the country. GDP is held to measure only market activity and not human welfare – although it is widely believed and acted upon the premise that it does.[55] This is the idea which was disputed by Daly and Cobb on the following grounds: (1) GDP considers

---

[54] At that time, Daly and Cobb (1994) were using Gross National Product (GNP). GNP measures the same as GDP, with the difference that what counts is not the location of the factors of production but their ownership (the residents of the country). GNP became outdated in the beginning of the 1990s.

[55] On the issue the United Nations System of National Accounts (UNSNA) states the following: 'GDP is often taken as a measure of welfare, but the SNA makes no claim that this is so and indeed

Table 3.1 Original items used to estimate the index of sustainable economic welfare in the US (Daly and Cobb 1994)

Items used to calculate the ISWE for the US (1950–1990)	Contribution to the ISEW
Personal consumption expenditures – A	
Distributional inequality – B	
Weighted personal consumption (A/B) – C	
Services: Household labour – D	+
Services: consumer durables – E	+
Services: highways and streets – F	+
Improvement health and education public expenditures – G	+
Expenditures on consumer durables – H	−
Defensive private expenditures/health and education – I	−
Cost of commuting – J	−
Cost of personal pollution control – K	−
Cost of auto accidents – L	−
Costs of water pollution – M	−
Costs of air pollution – O	−
Costs of noise pollution – P	−
Loss of farmland – Q	−
Depletion of non-renewable resources – R	−
Long term environmental damage – S	−
Cost of ozone depletion – T	−
Net capital growth – U	+
Change in net international position – V	+
Index of sustainable economic welfare – ISEW (Sum)	
Per capita ISEW	
Gross National Product – GNP	
Per capita GNP	

defensive expenditures and other social costs as contributions to welfare and (2) GDP is a poor measure of income and wealth. Therefore, Daly and Cobb deduct defensive expenditures and other social costs from the ISEW (Table 3.1, items I-P).

Regarding (2) the prime aim of Daly and Cobb was to produce a metric that tells us something about human welfare. Since in constructing the components of human welfare many controversial issues arise, the concept of income is preferred as it has a stronger theoretical foundation. Additionally, as it is supposed that income positively relates to human welfare, the ISWE departs from it. Two complementary conceptualisations of income are used for the ISEW, the first one is from the British economist John Hicks who explained the purposes of income and offered a workable definition. The second one is from the US economist Irving Fisher who mentioned another dimension of the income concept. For Hicks the 'purpose of income calculations in practical affairs is to give people an indication of the amount which they can consume without impoverishing themselves' and the practical

there are several conventions in the SNA that argue against the welfare interpretation of the accounts' (UNSNA 2009: 70).

purpose is 'to serve as a guide for prudent conduct'. Income is then defined as 'the maximum value which he can consume during a week, and still expect to be as well off at the end of the week as he was at the beginning' (Hicks 1948, quoted by Daly and Cobb 1994: 70). The same practical purposes of income, prudence and economic sustainability should be applied for GDP. Yet, GDP does not measure it, as it excludes capital depreciation while capital depreciation impoverishes a country. Hence, GDP does not offer a prudent guide as to avoid impoverishment. In this sense, Net Domestic Product (NDP) would be superior to GDP (NDP = GDP – capital depreciation).[56] On the other hand, NDP is also not sufficient, for it includes only man-made capital, and ignores natural capital.

The reason is that orthodox economists, as previously shown, have taught that human made capital is a near-perfect substitute for natural resources, when in fact they are complementary. Therefore, resource depletion and environmental losses are included in the ISEW (Table 3.1, items Q-T).

The notions of capital and income of Irving Fisher are of greatest importance for the SSE, and consequently for the ISEW. For Fisher, capital or wealth is the stock of physical objects owned by human beings in a period of time, and income is the flow of service in its psychic magnitudes yielded by the capital owned (Daly 1991: 32). For example, an LCD television purchased this year is not part of this year's income, but an addition to man-made capital from which psychic income flows. It implies that a proper accounting of income will only reflect the flow of services of man-made capital enjoyed in the subjective stream of people's consciousness. As previously explained, the SSE requires that man-made capital accumulation is minimised, hence expenditures on consumer durables are accounted as costs, while their services are accounted as benefits (Table 3.1, items E,F,H).

Finally, since GDP does not include the value of household labour, performed mainly by women and the welfare effects of income inequality, they are also included in the ISEW (Table 3.1, items B,C,D). The value of some public expenditures are also imputed (Table 3.1, item G). Net capital growth (increases in fixed reproducible capital minus the capital requirement, see item U) means that for economic welfare to be sustained over time, the supply of capital must grow to meet the demands of a growing population. However, it is expected, in line with the SSE, that at some point the population will stabilise.[57] Change in net international position (Table 3.1, item V) is national investment overseas minus foreign investment in the nation. If the change is positive, the nation has increased its capital assets. The final ISWE value is then divided by the population yielding ISEW per capita, and the same operation is conducted with GNP. Finally, both are compared.

---

[56] It must be mentioned the UNSNA recognises the inferiority of GDP to NDP. The problem is that not all countries make such calculations, and when they do it, it does not meet the requirements of the UNSNA. Nonetheless, it is acknowledged that NPD should be calculated (UNSNA 2009: 34). Whether the same considerations were made when Daly and Cobb were working on the issue is beyond my knowledge.

[57] The assumption of a growing population is made in the context of the US.

The ISEW, with some variations in content, and later called the Genuine Progress Indicator (GPI) has been calculated for the majority of Western European countries, Canada, Australia, Chile (for a review of the studies see Lawn, 2003) and more recently in countries of the Asia Pacific region such as New Zealand, Japan, India, China, Thailand and Vietnam (Clarke and Sardar 2005, Zongguo et al. 2007; Lawn and Clarke 2010). The frequent result of these studies has been that increasing GDP stops being economic at a certain point. The index remains either constant in spite of an increasing GDP, or begins to decline. When the index starts to decline, it simply shows that additional growth is uneconomic. When GDP is growing, while the ISEW or the GPI remains constant, it is not only economically irrational but ecologically irresponsible to continue GDP-growth.

3.3.5 Institutional Change for the Steady-State Economy

An economic crisis is today understood as any threat to economic growth. If an economy fails the perpetual growth promise, it will produce social instability. It is fairly clear and well-documented: no increase in GDP means no jobs, no revenues, the collapse of the pension system, hence the rise of radical ideologies and social conflict. Since the SSE would presumably maintain GDP constant, is the SSE then a threat to the social fabric? The answer offered by Daly is the following:

> The fact that an airplane falls to the ground if it tries to remain stationary in the air simply reflects the fact that airplanes are designed for forward motion. It certainly does not imply that a helicopter cannot remain stationary. A growth economy and a SSE are as different as an airplane and a helicopter. (Daly 1991: 126)

In this section, ten broad policy recommendations for institutional change required to achieve and eventually manage a SSE are discussed.[58] They are shown in Box 3.1.

Policy recommendations one and two are intended to restore the autonomy of the 'community of communities' (nation-states). Re-regulating international commerce means that we should move away from the ideology of global economic integration: free trade, free capital mobility (financial globalisation) and export-led growth, in short, the core constituents of what is called globalisation. Daly is not against international trade, international treaties, international alliances, and so on. However, as the word suggests *international* relations are between nations, and they should remain the basic unit. Global economic integration implies national economic *dis-integration*, the progressive erasure of national boundaries, in order to be reintegrated into the new whole: the globalised economy.

---

[58] Some of the policy recommendations discussed here can be fairly understood through Daly's theoretical tenets explained in the previous sections. There are other policy recommendations which would require extensive explanation. As extensive explanations are impossible in the limited scope of this chapter, the reader may consider consulting Daly and Farley (2011).

Box 3.1: Institutions and the Steady-State Economy

1. Re-regulate international commerce.
2. Downgrade the IMF, WB, and the WTO.
3. Move to 100% reserve requirements.
4. Free up the length of the working day, week, and year.
5. Limit the range of inequality regarding income distribution.
6. Reform national accounts.
7. Enclose the remaining commons of rival natural capital in public trusts.
8. Use cap-auction-trade systems for basic resources.
9. Use ecological tax reform.
10. Stabilise population.

Apart from the theoretical flaws upon which this policy is based, globalisation makes nation-states too dependent even for their basic survival, especially poor countries. It also pre-programmes international tensions and conflict. Poor countries should re-direct their efforts to build their agricultural capabilities for their domestic food demands rather than growing cash crops for unstable and highly speculative international markets. By and large, the development of domestic production for internal markets deserves priority. According to this policy of development, poor countries should use, for example, protective tariffs against subsidised agricultural products from rich countries.[59] Conversely, rich countries should also adopt protective tariffs in order to remain able to enforce rational national policies in the environmental and labour realm, that is, from standard-lowering-competition of poor countries with laxer environmental laws and lower wages. In organisational terms it means the downgrading of the IMF, WB and the newer WTO, perhaps reconsidering the original idea of Keynes at Bretton Woods. Keynes' original plan in Bretton Woods was to create an International Clearing Union, which would charge penalty rates on trade surpluses as well as on deficits in order to avoid imbalances of trade among their members.[60]

---

[59] The position of Daly also supported by a new generation of so-called 'post-autistic' development economists such as Ha-Joon Chang. He showed that today's developed countries, beginning with Britain, promoted their industrial basis and became rich through all sorts of protectionist measures, for example tariffs and subsidies, and later on 'kicked-away the ladder' for development in poor countries. Chang attempted to show the little empirical basis of the claim that development was achieved through free trade embodied in IMF and WB policies. Interestingly enough, the same argument, based on historical evidence was also formulated by Karl Polanyi (2001) in his critique of the classical liberal economists. He speculated on the dire consequences for Britain, had she ever followed the doctrines of Ricardo. I will come to Polanyi later. Chang's policy recommendation for development is roughly to repeat this pattern followed by rich countries in the past and maintained in many respects in the present (see Chang 2003, 2008).

[60] Keynes blamed impoverishment, wars and revolutions for trade's imbalances. The International Clearing Union would be a similar institutional arrangement which governs payments within

Policy recommendation 3 is primarily concerned with putting an end to the fractional reserve banking system (money creation) and implementing 100% requirements. The reasons are the following: first, and most evident, because money creation is one of the many institutional arrangements which fuels economic expansion and increases cyclical instability.[61] Money and debt can expand exponentially ('the magic of compound interest') while man-made capital cannot do so. According to Daly there also exists a conceptual confusion between capital and money: 'money fetishism'. The abstract symbol (money) came to dominate the concrete reality being symbolised (man-made capital). Daly treats it as a FMC (Daly 1996: 38). Second, we came to accept the idea of money creation as normal, yet 'the leading economists of the early twentieth century, Irving Fisher and Frank Knight, thought it was an abomination' (Daly and Farley 2011: 290). If money fetishism cannot be avoided, Daly prefers to conceptualise money as a public good (a non-rival 'resource'). It follows that seigniorage would be public revenue, instead of the money supply being privately loaned into existence at interest. Third, allowing private banks to become *too big to fail* has always been ill-advised on the same grounds of allowing industrial monopolies to emerge. Banks and other private organisations which are too big to fail are simply too big to exist.

Policy recommendation 4 is of overriding importance with reference to the intentions behind the growth policy, namely, to combat unemployment. The consumption-driven growth policy was the cure which Keynes proposed to tackle the disastrous consequences of mass unemployment after the great depression. Hence, the current disproportionate reliance on economic growth certainly has historical reasons which explain the 'glittering anomaly' noted previously. Nevertheless, other feasible economic policies also exist to combat unemployment in a SSE, the most obvious one is the shortening/sharing of working hours. Implementing this policy should be understood in rich countries more as a great benefit than a cost. It would allow for more options arising for leisure, such as hobbies, family, friendship, and community – in short, time for all those other activities which make a human life worth living. I will return to this issue in Sect. 3.5. It is worthwhile to underline, that this policy is probably the most amenable of *gradual* implementation and testing. It is a reminder for those social thinkers and politicians who are genuinely concerned with the possibility that

---

nations and would manage an international monetary unit (bancor). Clearance of balances between countries would be carried out by central banks through the accounts at the ICU. See for a recent discussion on the issue Piffaretti (2009).

[61] For a rigorous treatment of the issue see Biswanger (2009). Biswanger proposes also an interesting set of policy reforms to cut off the dependency of modern societies to grow. His monetary explanation of growth led him to propose a change in the stock and bond markets along with further changes in the institutional setting of joint-stock companies (corporations). Corporations are in his view the main drivers of economic growth nowadays. The support for corporations that for economic, political and ecological reasons should be directly challenged ought to be transferred to other legal forms of entrepreneurship less subjected to growth that are typical in small- and medium-sized firms. He also proposes to encourage the formation of cooperatives and foundations.

people would dedicate their increased leisure time to socially damaging activities. This policy also offers a possible solution in rich countries facing the problem of aging populations instead of the highly doubtful 'productive ageing'.[62]

Policy recommendation 5 is believed to have direct consequences for the general welfare of the community and is complementary to the former one. A minimum wage has popular support and already exists in many countries. What is missing in Daly's view is a debate and eventually an agreement upon a *maximum* wage. Recall that growth is celebrated as the main means with which to eliminate poverty, and yet, rich countries categorised as such many decades ago, are experiencing increasing levels of poverty. Growth is by no means an economic policy which replaces policies fostering equality, such as tax progression. True, complete equality would be unfair, but unlimited inequality is also unfair even if a country could even approximate the normative purpose of 'equality of opportunity'. Furthermore, in the middle run gross inequality is politically damaging for any society. We might lack a clear-cut scientific standard which tells us how much inequality is 'gross', yet the same clear-cut scientific standard is missing regarding of how much equality of opportunity really exists in a given society.

Reforming national accounts is the sixth policy recommendation. The main message of this policy is to separate GDP into a cost and a benefit account, and to then compare both accounts at the margins. It is what Daly and Cobb did with the ISEW, which also operationalised the central tenet that capital drawdown should not be counted as income. The remaining global commons of rival natural capital, such as the Amazon basin, should be priced and enclosed in public trusts. This is policy recommendation 7. At the same time, the non-rival commonwealth such as knowledge and information should be freed from patent monopolies. The guiding principle of this policy is to stop the treatment of the scarce as if it was non-scarce, and the non-scarce as if it was scarce. Intellectual progress is customarily a collective process. In academia and arts people have freely shared and built upon the ideas of other's for centuries. Great thinkers and artists have been driven by the habitual 'making a living', but also by curiosity, intellectual satisfaction and glory rather than by the profit-motive.[63] Copyrights and Patents which were initially awarded for 14 years have been extended under corporate lobbies up to 95 years

---

[62] This specific argument is advanced by Höpflinger (2010).

[63] In 2001, 41 pharmaceutical companies took the South African government to court for importing cheaper 'copy' drugs from countries like India and Thailand to deal with its severe HIV/AIDS problem that could not be properly tackled given the high costs of these drugs. After international social uproar that showed the companies in a bad light, they withdrew the lawsuit. The companies argued that, without enforceable patents, there would be no more incentive for innovation. The argument that seems compelling is in reality only the half-truth. Many researchers all over the world come up with new ideas all the time, many government research institutes and universities even explicitly refuse to take out patents on their inventions. At the height of the HIV/AIDS debate, 13 fellows of the highest scientific society of Britain, the Royal Society, stated the following: 'Patents are only one means for promoting discovery and invention. Scientific curiosity, coupled with the desire to benefit humanity, has been of far greater importance throughout history' (The Financial Times 2001 quoted by Chang 2008: 124).

and in so doing hampering further intellectual progress. On the other hand, since technologies change so fast, the over-extension of patents keep technologies out of the public domain until they are obsolete:

> The irony is that patent rights are protected in the name of the free market, yet patents simply create a type of monopoly – the antithesis of the free market. (Daly and Farley 2011: 177)

Policy recommendations 8 and 9 are closely related, however, Daly strongly prefers cap-auction-trade systems over ecological taxes. He gives two reasons: first it gives the correct order for institutional design: (1) environmental sustainability, (2) social justice and, (3) market efficiency. This order is superior to making environmental sustainability and social justice dependent on market efficiency, which is too often considered to be an end itself. The cap (or quota) effectively limits the scale of economic activity according to the resource limitations or natural sinks constraints, the auction captures scarcity rents for equitable redistribution, and trade allows for efficient allocation. This greatly resembles the concept of 'embeddedness' coined by Karl Polanyi in 1944 which he expected to be operative in industrial societies. I will address Polanyi's ideas in Sect. 3.4.1. In addition, it is worth noting that cap-auction-trade systems cut off the Jevons' paradox by starting with a quantitative limit which would raise relative resource prices but not quantities. Second, caps or quotas are effective in other contexts, that is, protecting ecosystems or in general renewable resources from liquidation, especially as long as the global financial system remains unstable and speculative. For instance, if a country depending on wood exports becomes the victim of an economic crisis, it might have to devaluate its currency, and consequently it may be forced to overexploit the forest beyond its sustained-yield.

Whenever the cap-auction-system is not necessary or difficult to enforce then an ecological tax reform could perform better. The principle underlying this proposal is to shift the tax base away from the value-added (labour and capital) on to which the value is added (entropic throughput). This procedure will internalise negative externalities and will raise revenue. Population control is a central measure for poor countries when the problem still exists. For the US, which is the exception of a rich country whose population is still growing and which is the focus of Daly, the policy is to achieve a balanced population so that births and immigrants are equal to deaths and out-migrants. Daly, following Mill's doctrines, has asserted long ago that the reason for the pro-population attitude of commercial elites is due to the effect on wages (Daly 1970). When a given population does not grow, commercial elites will tend to favour laxer migration policy or the moving of production to where labour is abundant and therefore cheaper. In present circumstances, it moves well-paid jobs and high environmental standards from the North to bad-paid jobs and lower environmental standards in the South. It is a transaction which makes the air cleaner in the North, dirtier in the South, but which effectively warms the atmosphere for both.

This short list of policy recommendations for institutional change leaves aside physical and political complexities which Daly is aware of.[64] It is also worth noting that some of these policy recommendations have been gradually implemented over the last few years, at different levels and in different regions, namely cap-auction-trade systems for GHG such as the EU Emission Trading Scheme and ecological tax reforms in Europe, both certainly not without dispute.[65] Other policy proposals have been made in the past, and partially implemented. Freeing up the length of working time in Western Europe was articulated by French Philosopher André Gorz in the 1970s and was a central condition for attaining the German 'qualitative growth' demanded by labour unions during the 1980s until they became weakened during the triumphing march of global laissez-faire policies (Loske 2011: 25, 27–28). Equally, the proposal of putting a part of the Amazon in an international public trust was launched by the Ecuadorian president in 2007: The Yasuní ITT initiative. Yasuní is a biosphere reserve with oil reserves in the ground. Enclosing the national park would tackle three policy goals at the same time: (1) the reduction of 407 million tons of carbon dioxide (Gobierno Nacional de la República de Ecuador 2009), (2) the protection of biodiversity and the Amazonian forest, and (3) the protection of indigenous communities' rights to live at Yasuní in 'voluntary isolation'. The initiative which was gradually gaining international support seems now to be in a deadlock after the German federal minister for Economic Cooperation and Development withdrew his initial commitment to co-finance the trust fund in September 2010.

There have also been similar policy proposals which are fairly old and highly controversial not only due to the theoretical background which supports them, but because of the scale of vested interests involved. For instance, the move to a 100% reserve requirement is one of the main proposals made by the ex-congressman Ron Paul in the US, who was running for president in 2008 and who is also in the presidential race of 2012. He wants to bring the gold standard back which would put an end to what he sees as an inflationary monetary policy of the Federal Reserve. Limiting the range of inequality distribution and reversing or at least slowing down

---

[64] Daly and Farley (2011: 414–417) advance six policy design principles: (1) Each independent policy goal requires and independent policy instrument, (2) Policies should strive to attain the necessary degree of macro-control with the minimum sacrifice of micro-level freedom and variability, (3) Policies should leave a margin of error when dealing with the biophysical environment, (4) Policies must recognise that we always start from historically given initial conditions, (5) Policies must be able to adapt to changed conditions, and (6) The domain of the policy-making unit must be congruent with the domain of the causes and effects of the problem with which the policy deals.

[65] The institutional design and potential effectiveness cap-and-trade designs depend on several factors such as the geographical scale of the pollutant, whether the polluter/polluted can be clearly identified, the physic-chemical characteristics of the pollutant, and whether sanctions are truly enforceable. On the shortcomings of the implementation of the EU Emission Trading Scheme see Gilbertson and Reyes (2009) and Clifton (2009). These analysts are claiming to better use eco-taxation. One of the cap-and-trade creators, economist Thomas Crocker argued in the US against it, given the little possibilities of enforceability (Hilsenrath 2009).

the pace of globalisation is a popular demand which took shape almost immediately when globalisation was pursued in the early 1990s by the IMF and the WB (Stiglitz 2002). The reform of national accounts, especially GDP has been a frequently discussed topic for almost 40 years, if one set the seminal paper *Is growth obsolete?* by Nordhaus and Tobin (1972) as the starting point of the debate and the final one, the report of the Commission on the Measurement of Economic Performance and Social Progress in France published in 2009 (Stiglitz et al. 2009). All in all, there have been similar policy proposals which have been partially implemented and which have been the subject of on-going political disputes. Nevertheless, with regards to all of these policy proposals the overall policy objective has stayed undisputed: economic growth. The general intellectual contribution of Daly is having coherently subsumed these policy proposals from the standpoint of Ecological Economics into an overall policy objective, the SSE.

3.4 The De-growth of Serge Latouche

3.4.1 Intellectual Foundations: Illich, Bookchin and Polanyi

Serge Latouche is the most visible French anthropological economist behind the contemporary promotion of de-growth in Western Europe.[66] Economically and politically speaking, his theories reach back to the French utopian socialists and the following socialist libertarian views – an ecumenical body of anti-authoritarian ideas – inspired in the writings of Jean Jacques Rousseau and Pierre Joseph Proudhon in its individualist version, and later in Michail Bakunin and Pjotr Kropotkin in its collectivist version. The latter social thinkers differ from popular Marxists by favouring and theorising on economic decentralisation and cooperation rather than the planned economy and a centralised state. Latouche takes the libertarian socialist ideas from Murray Bookchin's libertarian municipalism and Social Ecology, both conflated into *ecomunicipalism*.

---

[66] As in the case with Daly, the reduction of Latouche's intellectual foundations to, in this case, three social thinkers, is somehow arbitrary. Others social thinkers will be briefly mentioned in the following text. Additionally, two of the external reviewers have raised doubts as to whether Murray Bookchin is a key reference of Latouche and that in an earlier draft I neglected to mention Karl Polanyi. From my readings of Latouche's texts I concluded that Bookchin is, if not a key reference, at least one of increasing importance (see Latouche 2009: 44–47, 54, 77, 90). Other social thinkers who have influenced Latouche's thoughts, perhaps even more than Bookchin, are André Gorz and Cornelius Castoriadis. Concerning Polanyi, I did not find any important reference made in the texts consulted. A reason is perhaps that Polanyi is simply 'too known' to be explicitly mentioned, or that I may have missed Latouche's former texts when the influence was more evident. Following these recommendations, I corrected this mistake and succinctly describe Polanyi's main insights.

Libertarian municipalism challenges parliamentary democracy as a means for public representation and policy formulation. In its place, the citizens of the municipality (the town or the village) through direct democracy should formulate policy; therefore, decision-making is not a hierarchical activity left to professional politicians, bankers, or in general, technocrats, but to the municipal assemblies. It does not mean that expert's knowledge is discarded; it simply means they do not take decisions which have overarching impacts on the community. Furthermore, Bookchin claims that ethics based on the values sharing, cooperation, and solidarity can only be pursued through direct democracy and politics as was practiced in classical Athens:

> Direct democracy, the formulation of policies by directly democratic popular assemblies, and the administration of those policies by mandated coordinators who can easily be recalled if they fail to abide by the decision of the assembly's citizens. (Bookchin 2007: 48–49)

On a more aggregate level he sees a confederation of eco-communities instead of the state. For Bookchin the ecological crisis has its roots in the hierarchical mode in which society currently functions, and he extends this insight for the relationship to human-nature. In his view, the ecological crisis can neither be understood let alone solved without this understanding.

Another important part of Latouche's thought is the European cultural critique of modernity. Four years before Schumacher's *Small is Beautiful* was published, French philosopher Bernard Charbonneau published his *Le Jardin de Babylone* (1969) in which he deplored the 'gigantism' and the power of the 'technique' in the industrial world (Martínez-Alier et al. 2010: 1742). His reflections on the technique were further developed by French philosopher Jacques Ellul who pointed out its alienation effects, whereby humans became the instruments of their own instruments. According to both philosophers, escaping the dark-side of modernity requires cultural change, in which the values of productivity and individualism are replaced for quality of life, solidarity, frugality and voluntary simplicity (Martínez-Alier et al. 2010: 1742–1743). Another highly influential author of the cultural critique to modernity is the Austrian philosopher Ivan Illich. In the assessments of Latouche on the notion of development, he would prefer to see 'convivial societies' in rich and poor countries, rather than rich 'developing' the poor (Latouche 2001, 2003a). The notion of convivial societies is taken from Illich. Illich argued along the lines of Ellul, that machines were created under the hypothesis that they would replace slaves. As this hypothesis proved wrong it must be discarded: 'neither a dictatorial proletariat nor a leisure mass can escape the dominion of constantly expanding industrial tools' (Illich 1973: 10). One of the effects of the hegemony of the machines was the degrading of humans as mere consumers. From this perspective, it follows that this expansion must be limited and the positions of dominance must be inverted if the values of survival, justice and self-defined work are worthwhile to be fostered and protected. Conviviality is the opposite of industrial productivity, and it means:

Autonomous and creative intercourse among persons, and the intercourse of persons with
their environment; and this is in contrast with the conditioned response of persons to the
demands made upon them by others, and by a man-made environment. I consider conviviality to be individual freedom realized in personal interdependence and, as such, an
intrinsic ethical value. (Illich 1973: 11)

Illich believed that reversing the direction of dominance between machines and
humans would set in motion an evolution of new life styles and political systems.
Illich then moved on to outline his programme for a convivial reconstruction.

Although Latouche drew on both intellectual traditions of apparently similar
lineage, there exist irreconcilable tensions. Bookchin is a believer of reason and
Hegelian dialectics, therefore he disdains the anti-rational bias of post-modernism,
its anti-technological attitudes, and anti-civilisational tendencies of the central
European cultural critique which emerged in the 1960s. For instance, commenting
on the best known book of Ellul *The Technological Society* published in 1964 he
stated:

Ellul advanced the dour thesis that the world and our ways of thinking about it are patterned
on tools and machines (*la technique*). Lacking any social explanation of how this 'technological society' came about, Ellul's book concluded by offering no hope, still less any
approach for deeming humanity from its total absorption by *la technique*. (Bookchin 1995:
30. Italics in original text)

Although Illich later corrected this fatalism; he ended up with innocuous
recommendations for lifestyle changes which were rather conducive to inwardness,
narcissism and individual mysticism, therefore nipping in the bud any social
cooperation needed to produce any real social change. Bookchin disliked the
wide-spread anti-technological attitude of the time for two additional reasons:
first it veils the cause of social dislocation and ecological destruction, which are
in his view the hierarchical social relations of capitalism; and second, those thinkers
forget that the same technology which extraordinarily raised productivity could be
harnessed in a more 'rational' society as to meet unsatisfied needs and to free
humans of mindless toil for more creative and rewarding activities (Bookchin 1995:
29–30). Bookchin also explained that the minor changes which the cultural critique
of the late 1960s and 1970s produced were easily absorbed and channelled in the
economic and political market, whilst the structures they attempted to change
remained intact.

The economic and environmental vision of Latouche was further enlarged by his
experience as an anthropological economist in Africa being under the intellectual
influence of Polanyi's *Great Transformation* written in 1944. Polanyi's book
offered a vivid description of Britain's social dislocation during her early paths
towards industrialisation, and examined the idea of the self-regulating market
envisaged by political economists, developed to its fullest by David Ricardo.
Ricardo conceptualised humans ('labour') and the environment ('land') as
commodities to be exchanged in an ideal self-regulated market. Polanyi insisted,
had Britain and later other European powers ever followed Ricardo's doctrines,
they would have destroyed themselves – literally speaking. Thus, prior to the Great
War they did so in only limited time spans, given the emergence of mechanisms of

self-protection such as mandated improvements in working conditions, pension systems, embryonic environmental legislation, and the like, which laissez-faire ideologues condemned as market distortions. Strikingly, these mechanisms emerged uncoordinatedly in countries with exceptionally different cultural and political outlooks such as the liberal Victorian England and the strong Prussian state of Bismarck. Another uncoordinated self-protection mechanism which emerged was the export of social conflict. Polanyi explained the renaissance of colonialism, outmoded between 1770 and 1880, as an additional societal mechanism of self-protection against the attempts to forcefully implement free trade doctrines:

> The difference was merely that while the tropical population of the wretched colony was thrown into utter misery and degradation, often to the point of physical extinction, the Western country's refusal to trade was induced by a lesser peril but still sufficiently real to be avoided at almost all cost [...] to expect that a community would remain indifferent to the scourge of unemployment, the shifting of industries and occupations and to the moral and psychological torture accompanying them, merely because of economic effects, in the long run, might be negligible, was to assume absurdity. (Polanyi 2001: 224)

It is worth emphasising that Polanyi was neither vilifying Ricardo nor arguing against markets. Ricardo (and Malthus for that matter) honestly believed that he was discovering the 'laws' which British society should respect for her own long term benefit. The market, Polanyi explains, is an institution that has existed virtually since the Stone Age. He was merely warning against renewed attempts to subordinate the substance of society (humans and nature) to market 'laws', for they will necessarily culminate in catastrophe once again. In the 1930s, the laissez-faire movement and its counter movement found themselves in political stalemate, until fascism seized power and broke with laissez-faire, democracy and peace. The self-regulated market was a strong utopia which Polanyi hoped to see transcended after the Second World War – as indeed was greatly accomplished in central and North Europe the years thereafter. Given Polanyi's insights, Latouche saw the attempt to replicate Britain's pattern towards industrialisation in poor countries under the heading of 'development' and 'progress' as socially and environmentally ill-advised, including the persistent practice of exporting social conflict.

Latouche was also aware of Georgescu-Roegen's work (Latouche 2004a: 63). It was Jacques Grinevald and Ivo Res who introduced into the Francophone world Georgescu-Roegen's writings. The title of the French translation in 1979 of some of Georgescu-Roegen's writings was *'Demain la décroissance'*.[67] They translated the English verb 'decline' into the French substantive *la décroissance*, and that word was translated back to the English language as 'de-growth' (Grinevald 2008: 15). The back-and-forth translations that Georgescu-Roegen himself agreed upon, given his literacy in the French language and personal relations with French Philosopher Grinevald,[68] fully reflected his opinion that 'the necessary conclusion [...] is that

---

[67] De-growth for tomorrow.

[68] I am indebted to Prof. Martínez-Alier for this biographical note.

the most desirable state is not a stationary, but a declining one [. . .] Undoubtedly, the current growth must cease, nay be reversed' (Georgescu-Roegen (1975): 368–369. Italics in original text). In this passage he was arguing against the SSE proposed by his former pupil Daly. This debate will be further examined in the next section. Latouche also embraced LtG reasoning, especially the immense destructive forces of exponential growth expressed in the ever expanding ecological footprint and carbon emissions. For Latouche the major problem of rich countries was that of overconsumption, and for emerging and poor countries was that of aspiring the overconsumption of the rich, encouraged by the policies and cultural dominance of the North and the elite's corruption of both.

3.4.2 From 'Developmentalism' to the Virtuous Cycles of Rs

As previously noted, Latouche belongs to the short but growing list of social scientists and practitioners who have criticised the so-called 'developmentalist' project. Indeed, in their view, this project destroys viable societies by uniform development and the imposition of the utopic market-society. His critique can be fairly summarised with the following quote:

> As long as hungry Ethiopia and Somalia still have to export feedstuffs destined for pet
> animals in the North, and the meat we eat is raised on soya from the razed Amazon
> rainforest, our excessive consumption smothers any chance of real self-sufficiency in the
> South. (Latouche 2004b: 2)

Since the developmentalist project slipped in the sustainable development discourse, Latouche rejected sustainable development altogether. In his view, it was not only a contradiction in terms as Georgescu-Roegen and Daly previously claimed from an entropic point of view, but also a pain-relieving discourse in view of the harsh socio-environmental realities that economic growth delivers, which were further deepened by the progressive re-implementation of globalisation (Latouche 2003a, b). The political question is then: how to escape the iron cage of growth which is destroying both nature and humans?

Latouche's strategy began at the bottom, with localism as a response to development and globalisation. At this level a transition process or a 'virtuous cycle of quiet contraction' would be initiated (Latouche 2009: 33). The reason for starting with the local was simply because it was the only space of political action left by the overwhelming financial and corporate power of today's world which have severely limited the scope of action of politicians. Placing the emphasis on political action, the term de-growth was the political slogan intended to defeat current pro-growth ideologies. As advocators of economic growth share a religious belief in it: 'we should be talking at the theoretical level of 'a-growth', in the sense in which we speak of 'a-theism', rather than 'de-growth'' (Latouche 2009: 8).

An important step which was central to Latouche's thought was what he repeatedly called the 'decolonisation of the imaginary' from 'economicism' and the economy (Latouche 2003a, 2004a: 115). This means the pro-active liberation of the mind from economic thinking which is so hegemonic in social life,[69] and the pro-active liberation in the material sense which is creating new autonomous spaces of social interaction and production, in which frugality and voluntary simplicity can be practiced. The requirement is that of a cultural revolution across all levels which may reach politics. The cultural revolution in politics would reduce the need for politicking, and will likely re-establish the dignity of the political profession. The ultimate end is the convivial and sustainable society. The intermediate means are the serene contraction which is composed of eight interdependent and, so expected, self-reinforcing R-guiding concepts (Latouche 2009: 33–43).

Re-evaluate: The re-evaluation of social values which are admired but hardly practiced, namely altruism and cooperation instead of egoism and competition. Other values re-directing preferences ought to be re-evaluated: local over the global, autonomy over heteronomy, and appreciation of good craftsmanship over productive efficiency. A sense of justice, responsibility, and solidarity must be won back. An example of how the sense of justice has been so badly distorted by *economicist* thinking is the accepting of almost everything which creates employment (growth) as inherently good, such as exporting pollution into poor countries, 'land grabbing', exaggerated expenditures in the military and the like. According to Latouche, re-evaluation along with re-localisation are the most important Rs in strategic terms.

Re-conceptualise: This means to deconstruct and reconstruct the meanings of wealth, poverty, scarcity, and needs.

Restructure: When values are changing then the productive apparatus must be changed accordingly. As the restructuring is on-going, the question about going beyond capitalism will inevitably be raised.

Redistribute: Within and among the countries. Rich countries should restore or, depending on the specific situation of the country, improve a system of fair taxation and the gains of economic booms. Redistributing from the North to the South is confronted with the 'payability' problem of the immense ecological debt accumulated by the North. Nonetheless, the mechanism is not too much in giving away but in taking less. Ecological footprints are a good metric for determining each country's drawing rights, hence through the mediation of markets, an exchange of quotas and permits to consume could be made possible.

Re-localise: It deals not only with the re-localisation of productive activities, but with culture and politics. The strategic importance of re-localising is to show that the 'concrete utopia' is doable in political and economic terms. Of great importance is the existence of the collective project which is territorially rooted, for example the town, the village and so on, hence fostering the sense of belonging which will allow the protection of the common good and the emergence of other values. Latouche mentioned several examples of on-going projects with differing scales

---

[69] In the texts reviewed Latouche blames the economic discipline on the whole.

4

in Europe, such as the province of Milan or in the Tuscany region. In fact, there are hundreds of on-going local projects which have emerged since the localisation movement appeared.

Reduce: This means especially reducing consumption. Nonetheless, reduce is also directed at reducing health risks, working hours, and mass tourism. Less working hours and work sharing is one of the formulas against one of the main arguments to keep the growth machine. On the other hand, we must overcome the 'tragedy of productivism', that is, our addiction to work (Latouche 2009: 40). It makes us unable to rediscover the repressed dimensions of life, such as the pleasure to engage or develop our talents and to practice our hobbies, to play, to enjoy conversations or to simply enjoy being alive.

Re-use/recycle: It is about the reduction of waste, to fight in-built obsolescence and recycle waste which cannot be reused. Latouche mentioned examples of firms which through product design make almost full recycling possible.

Resist: It is said to be the central R of the Cultural Revolution expected to be triggered and carried out by the rest of the Rs. Resist is contained in the rest of the Rs.

Against the accusations of the potential intransigent characteristics of the eight - R-guiding concepts, Latouche defended himself by claiming that they are a response of the system excesses with all of its 'overs': over-development, over-production, over-abundance, over-extraction, over-fishing, over-grazing, over-consumption, over-supply, and so on.

3.5   Steady-State or De-growth?

The ideas of Daly and Latouche differ mainly in their grade of theoretical elaboration and completeness. This asymmetry might be explained by the dissimilar time span in which each of them has been involved in the economic growth debate. Daly has been writing on this issue for 40 years with remarkable scholarship and in a holistic fashion. In this time he has covered practically all of the topics related to the issue. Latouche, on the other hand, started to write about de-growth in the early 2000s although he has already been arguing against the notion of development for a long time. He also seems less interested than Daly in the growth-debate with economists, and more interested in broader political, social and cultural aspects. Although both social thinkers follow largely different intellectual traditions, similar policy proposals arise, albeit with different wording. This is perhaps a result of the exchange of ideas in the 1970s between the US and Europeans thinkers, and recently from social thinkers of southern countries such as India, Ecuador and Bolivia. Another potential explanation may be that *some* of these proposals seem to be sheer common sense.

To make this point, a short review of the most convergent policy proposal shared not only by both thinkers, but also by a number of their intellectual mentors and many others should be provided: gaining leisure by working less, what conversely might free up jobs for others in the community. Keynes (2009: 198) was already, one might retrospectively say, dreaming in 1930 that the main problem of the worker would be 'how to occupy the leisure, which science and compound interest will have won for him, to live wisely and agreeably and well' and 2 years later Bertrand Russell even felt the necessity to praise idleness. The very same architect of the 'German miracle', economist Ludwig Erhard anticipated in 1957 a 'correction in economic policy' which would become necessary once people started to ask about the value of accumulation and eventually conclude that more leisure is more valuable (Radkau 2010: 46). In the 1970s French philosopher André Gorz was demanding leisure for French workers, while disenchanted Austrian philosopher Ivan Illich was blaming 'the machine' for robbing it, and even for enslaving humans. On the other side of the Atlantic, Georgescu-Roegen felt that we needed to realise that an important prerequisite 'for a good life is a substantial amount of leisure spent in an intelligent manner' (1975: 378). Daly transformed it in a policy proposal that perfectly fits into his SSE, while Latouche wanted us to move away from the 'tragedy of productivism'. From these reflections, it seems clear that the underlying question has always been: what is after all the purpose of a society classed as materially rich if not liberating the majority of the population from the 'toil and trouble' of work? Not even some societal sectors in peripheral countries which have recently gained voice are particularly enthused at the thought of becoming 'developed' by means of owning all sorts of gadgetry while losing leisure – especially due the bleak perspectives of becoming 'developed' under the present global status-quo.

Pertaining to SD, both thinkers are troubled with the current meaning of development which so strongly implies export led-growth. Nonetheless, Daly sees the necessity to maintain the sustainability political forum. Sustainable development is dialectic more than an analytic concept of the sort of justice or love; therefore it is preferable to continue the political attempt to shape its meaning. There is after all no other option in the international realm. Beyond Daly's ideas of development, he formulated in the 1970s the 'impossibility theorem': US or Western European high-mass consumption style economy for a world of nearly four billion people (at that time) is impossible. Even more impossible is the prospect of an ever growing standard of consumption for an ever growing population. The physical limits of the earth will not support a world population in a 'developed' state (Daly 1991: 151). Latouche on the other hand, sees no reason why poor countries should follow the development path of rich countries even if globalisation was inexistent. He introduces four additional Rs: Renew, Rediscover, Reintroduce and Recuperate:

> Renew contact with the thread of a history that was interrupted by colonization, development and globalization. Rediscover [...] cultural identity. Reintroduce specific products that have been forgotten or abandoned [...]. Recuperate traditional technologies and skills. (Latouche 2009: 58)

Thus, Latouche formulates the Rs in terms of cultural emancipation.

The only sharp difference in their conceptions of development is population control policy. Latouche rejects it for political and pragmatic reasons. He maintains that the intention behind the population control consensus that emerged in the 1970s was based on hegemonic intentions (Latouche 2009: 25). Furthermore, as the population rapidly increased in poor countries because of high fertility rates over the last decades, it will equally rapidly decline with lower fertility rates in the next ones.[70] On the question of technological progress, Daly understands it as a fundamental component in the SSE but as highly destructive within the growth paradigm. Latouche emphasises the alienating characteristics of technology and believes in simpler and more manageable *tools*. By and large, Daly's SSE subsumes virtually all of Latouche's Rs in his theoretical framework, excepting his re-use/recycle of waste, especially recycling of waste. Daly does not reject the practice of waste recycling, but this activity, independent of how we re-define 'waste', does not defeat the strict linearity of the entropy law. Daly also seems to be more specific about the ideas on the role of the state and the market-economy than Latouche.

Recently and Kerschner (2010) also attempted to compare de-growth and the SSE, drawing not only on Latouche but on other de-growth advocates. He addressed the following specific issue: de-growth proponents reject Daly's SSE and blindly cling to Georgescu-Roegen's judgment that, as mentioned in the last section, de-growth is preferable over the steady-state. Georgescu-Roegen was himself indeed highly critical of the SSE developed by his former pupil (see Georgescu-Roegen 1975: 366–369, 1977; 1979: 102–105). Before examining Kerschner's arguments in depth, it is convenient to first consider Georgescu-Roegen's arguments against the SSE. They can be reduced to:

1. Growth, zero-growth (SSE) and even de-growth (declining) cannot exist forever in a finite environment.
2. The SSE offers no basis for determining even in principle the optimum levels of (a) population and (b) man-made capital.
3. The SEE does not offer a guide with which to determine the appropriate stock of capital for human's 'good life'.

Argument 1 arises simply by the strict application of the entropy law; therefore a SSE will be of finite duration.[71] Yet the same fate is also shared by a declining path.

---

[70] The United Nations' estimations on global population growth have been recently readjusted. The current world population (7 billion) is projected to reach 10.1 billion by 2100, reaching 9.3 billion by 2050 according to the median variant. This increase is projected to come from high-fertility countries (an average woman has more than 1.5 daughters), which are mostly located in sub-Saharan Africa, but also in nine countries in Asia, six in Oceania and four in Latin America (UN 2011). On the other hand, as made clear several times, Daly and Latouche agree that current global threats stem rather from consumption in rich and emerging countries. After all, population is just one variable which put pressure on the natural environment. Those poor countries with still high-fertility rates consume virtually nothing compared to the consumption levels in rich countries.

[71] Beckerman (1995) also pointed out that the LtG stabilisation path scenario was cut off in the year 2100, hence omitting the declining availability of resources that would, according to the same model, arrive beyond 2100.

Argument 2a is true, 2b is *somehow* true and 3 was true when Georgescu-Roegen was writing on the issue. Concerning 1 the discussion must ultimately rest on the *duration* and on the ethics implied, consciously or not, in any economic and ecological policy in the time span that humanity at large will live, which, needless to say, is highly speculative. However, for most people it might be unacceptable to ground public policy for 150 years, that is, roughly the life time of three generations, and for the majority it would be ludicrous to ground public policy for 5 billions years, which is the estimated time before the sun becomes a red giant which vaporises all life on earth, and the earth herself. With this said, one must also have a sense of proportion. As Daly observed more than 30 years ago:

> In the very long run of course nothing can remain constant, so our concept of an SSE must be a medium run concept in which stocks are constant over decades or generations, not millennia or eons. (1979: 80)

The question which then arises is: why did Georgescu-Roegen prefer declining economies if after all humanity is doomed to disappear? His answer was because 'the population is too large and part of it enjoys excessive comfort' (1975: 368). His persistence on *extravagant* wants (rich's problem) were more frequent than the population problem (poor's problem): 'If we understand well the problem, the best use of our iron resources is to produce plows or harrows as they are needed, not Rolls Royces [...]' (1971: 21). However, this did not come out of a capricious personal taste, but out of his proposed ethical principle: 'Love thy species as thyself' (1977: 270) that would require the overcoming of the 'dictatorship of the present over the future', and replacing the maximisation of present utility by the minimisation of 'future regrets' (1979: 102).[72] The latter requirement is, in his view, not served by discounting rates arbitrarily set by present people – usually economists – against the future generations who cannot bid for the choice of present resources. Once this principle is internalised, 'right' prices, production, distribution, and pollution will naturally emerge. Therefore, in spite of the critical comments on the SSE, he judged as an improvement the ultimate-means (low entropy matter-energy) – ultimate-ends (religion) spectrum which Daly (1979: 70) was elaborating:

> This paper thus strengthens the impression emerging from his previous writings that the essence of Daly's conception is not economic or demographic, but, rather, ethical – a great merit in a period in which economics has been reduced to a timeless kinematics. (Georgescu-Roegen 1979: 102)

Argument 2a is true, but this is also only the case when Georgescu-Roegen fills the void with a proposal. Indeed, the SEE offers no guidance as to determine the optimum level of population and the guidance Georgescu-Roegen offers is quite concrete. In his 'minimal bioeconomic program' consisting of eight points (1975: 374–379) he advises that world population should gradually be reduced to a level at

---

[72] He was by no means indifferent to the differential gradient between rich and poor nations that was in his view 'an evil itself' (Georgescu-Roegen 1975: 377).

which it can adequately be fed *only* through organic agriculture. He considered modern agriculture (and most modern technologies) as energy squanderers.[73]

Arguments 2b and 3 are thornier. Concerning the optimum level of man-made capital Georgescu-Roegen insists that natural resources must be governed by quantitative regulations/restrictions, strictly rejecting the aiding of market efficient allocation through Pigovian taxation/subsidies, for these measures would simply end up benefiting the already wealthy and the political protégés (1975: 377). Quantitative regulations/restrictions are also Daly's preferred political instrument, but he does not reject taxation/subsidies as shown in Sect. 3.3.5. In the case of quantitative regulations/restrictions, Daly's aim is to induce substitution through technological progress once restrictions are set. This is also stubbornly rejected by Georgescu-Roegen as it precisely implies that an economy cannot possibly be in a steady-state (argument 1) and that Daly's proposal would mean 'joining the club of the believers in exponential progress' (1979: 104).

Regarding the appropriate stock of man-made capital for human's good life (argument 3), this argument *was* true, as at that time there was no refined framework with which to handle the issue. Nevertheless, Daly's notion of optimal scale was polished and operationalised years later through his ISEW. Here, Daly allows for the freedom of the individual to decide about the 'good life', which will be eventually shaped by the forces of market once quantitative regulations are in place.[74] Additionally, Daly is also highly critical of *extravagant* wants stimulated by advertising and clings to the old economic principle of declining marginal utility: 'if nonsatiety were the natural state of human nature then aggressive want-stimulating advertising would not be necessary, nor would the barrage of novelty aimed at promoting dissatisfaction with last year's model' (Daly and Cobb 1994: 87–88). In other words, if want-stimulating advertising was absent then individual choices aiming for the good life would ultimately require far less throughput. Georgescu-Roegen also advised us to educate ourselves to despise fashion and to make durable goods even more durable and to design them so they are repairable. In spite of argument 3 and as noted before, he wanted us to understand the importance of leisure for the 'good life'. Similarly, with the resources freed by the prohibition of the production of all instruments of war, it would be possible to help poor nations to arrive as quickly as possible at 'a good (not luxurious) life' (1975: 378).

The last paragraphs thoroughly cover Georgescu-Roegen's criticisms of the SSE. Coming back to Kerschner's paper, he is basically troubled with argument 1, especially with the word 'annihilation' used by Georgescu-Roegen (1975: 367):

---

[73] By the same token, his advice was to master the harnessing of solar energy which is the most abundant source of energy (albeit flow-limited) and by so doing lowering the increasing rate of terrestrial entropy. Additionally, he saw it as the safest technology compared to other technologies such as nuclear power.

[74] As hopefully noticed, Daly is not a radical individualist.

The crucial error consists in not seeing that not only growth, but also a zero-growth state, nay, even a declining state which does not converge toward *annihilation*, cannot exist forever in a finite environment. (emphasis supplied)

The first problem Kerschner sees is that de-growth proponents have adopted Georgescu-Roegen's position against the SSE while conveniently omitting the word 'annihilation' when they cite him (ibid.: 548). The second problem is that Georgescu-Roegen's position is 'a path without a constructive goal for policy making [...]' (ibid.: 547). This assessment has, however, been previously softened when Kerschner speculated that Georgescu-Roegen was referring to the entropic death of the universe. My own assessment is that the issue is to some extent overestimated. Clearly, neither Georgescu-Roegen, nor de-growth proponents (at least Latouche) want us to de-grow humanity to death, therefore it is somehow immaterial whether they quote the annihilation-part of the sentence or not.[75] As mentioned before, Georgescu-Roegen was simply reminding us that humanity is mortal. Finally, and differing from Kerschner's view, Georgescu-Roegen did offer several policy options in his minimal bioeconomic programme. If 'constructive' meant 'politically possible' that would be a separate discussion which I will return to later. It is also useful to recall the time at which the debate was being conducted (Fig. 3.1).

A very interesting point raised by Kerschner is that Georgescu-Roegen appeared inconsistent in his critique against the SSE, and that he even implicitly supported it. Kerschner mentioned his organic agriculture/population proposal which would ultimately imply a stabilised population, as Daly's SSE requires. One may add the proposal of helping poor nations to arrive as quickly as possible at a good life, which would mean a movement towards a SSE – through growth! It is difficult to accept that Georgescu-Roegen's erudite mind was being inconsistent,[76] but one must agree with Kerschner's assessment. Judging from his proposals on population and aid for the poor, Georgescu-Roegen views favourably fit Kerschner's observation that both movements: growth for the poor, de-growth for the rich, and a stable population for all towards a (quasi) steady-state is required.[77]

---

[75] Admittedly, Kerschner has been far more involved in the de-growth discussions than I have been. This experience may have told him that the point ought to be made.

[76] Georgescu-Roegen was a trained mathematician, economist, and philosopher of science with ample knowledge of physics and biology. He was the social thinker who was once called by Paul Samuelson 'a scholar's scholar, and economist's economist' and included in economist historian Mark Blaug's book *Great Economists since Keynes* published in 1985 (Daly 2007: 125). In Blaug's view it was this erudition and complex style that led his colleagues to ignore him so persistently hitherto. Daly adds as a reason that he, the mathematician, was severely criticising the excessiveness of mathematics in economics, the very element of orthodox economists' proudness which confers the scientific status of the profession – or at least the appearance thereof (Daly 2007: 126).

[77] The term 'quasi' steady-state was used by Georgescu-Roegen (1975) when he was explaining that such societies indeed existed in the past but they were rather culturally and technologically stagnant.

Globally understood, both schools are indeed complementary as Kerschner concluded – in fact they must *logically* be so. In order to reinforce Kerschner's conclusion it should re-emphasised that Daly's SSE has been continuously refined and expanded in the last 30 years. He built upon Georgescu-Roegen's bio-economics (the lowest part of Daly's spectrum), improved the body of economic theory, integrated political/institutional insights and fully handled the 'ultimate end' (the highest part of his spectrum) in *For the Common Good* co-authored with theologian John Cobb and published in 1989. The book was available only a year later after the institutionalisation of the merge between ecology and economics through the inauguration of the journal Ecological Economics. This was a merge which Georgescu-Roegen already saw as inevitable in the 1970s – although he was not fully satisfied with it as his goal was rather to replace the, in his view, fatally flawed mainstream neo-classic economics with bio-economics and not to be relegated merely as a school of secondary importance (Levallois 2010). Finally, the metrics which came out in the late 1980s and have been refined and tested ever since, could replace GDP and thus must also be mentioned. Indeed, they are the kind of metrics which would tell us how much growth/de-growth is economic for the 'good life'.[78] After all, the SSE, as the name implies, is an (approximate) state, while de-growth, as well as its antithesis growth, are processes. In a nutshell, it would be incomprehensible if today's de-growth proponents, for whichever reasons, deliberately neglected 40 years of intellectual work which is in any case the outgrowth of Georgescu-Roegen's ideas, humanism as well as the sense of responsibility and urgency which flows from his intellectual legacy.

The preceding discussion brings me back to the comparison of Daly's SSE and specifically on Latouche's de-growth. As previously mentioned, Daly's SSE is theoretically more elaborated and comprehensive than Latouche's de-growth, yet apart from Latouche's own theoretical contributions, his significance lies undoubtedly in the resonance of de-growth as a political slogan which has been capable of re-launching an academic and public debate in Europe, and to some very limited extent in the US. In the academic domain three international conferences on de-growth have been held, the first in April 2008 in Paris, the second in March 2010 in Barcelona and the third in June 2011 in Berlin. A fourth will be held in Montréal in

---

[78] For sure an index does not do justice to the richness of the meaning of the 'good life' that varies in different cultural settings, but this fact does not make indexes superfluous. As previously note, Daly is not particularly motivated with the shaky ground upon which the notion 'development' rests, and even less is he an enthusiast of bringing it to the rest of the world. This is not only because of ecological but also because of cultural reasons. After all, he wants to transform economic thought so that it serves specific communities (see for example Daly and Cobb 1994: 133–137). On the other hand, while developing his ISEW with Cobb, he was keeping an eye on custom, or 'path-dependency' using a more fashionable term. If GDP is the national account in which statistical efforts have been invested in the last decades one has two options: (1) to disregard it and to force the introduction of several indexes – as it has been proposed several times, or (2) to replace it with a better index building on available information currently collected by statisticians, hence building upon the general obsession with a single index. Daly preferred the latter option.

2012. Additionally, a new academic journal based in France called *Entropia* was launched. A number of books on the issue have recently been published in other core European countries such as *Prosperity without Growth* by ecological economist Tim Jackson (2009) in Britain and the edited book *Postwachstumsgesellschaft*[79] by Irmid Siedl and Angelika Zahrnt (2010) in Germany. In the political domain Latouche explains that by now at least the French public is familiar with the slogan 'de-growth' (Latouche 2010: 201). In Germany, green politician Reinhard Loske (2011) has argued (again) for abandoning 'growthmania', thus proposing a set of political reforms which would enable this, some of them very similar to those proposed by Daly and Latouche. Furthermore, the European Commission published in September 2010 an article dealing with what was called *sustainable de-growth* through their news alert service (EC 2010). They drew on an article authored by ecological economists Martínez-Alier et al. (2010). Although the intensity of the debate on economic growth originated mainly in the US, it remains today a strong taboo not only for professional politicians but also for the public at large; nonetheless, some minor ripples from the growth discussion in Europe have spread back to that country (Schor 2010).

The de-growth slogan advanced by Latouche was also debated in the journal of Ecological Economics, particularly its feasibility for political implementation. Van den Bergh (2011) has asked what should de-grow: GDP?, consumption?, throughput? or work-time? These questions can be answered with what has been written so far. What is important to highlight is his assertion that GDP de-growth, consumption de-growth and 'radical' de-growth[80] are likely to meet strong resistance in democratic systems. He was certainly correct in his judgment that striving 'for political feasibility nationally and internationally is an important precondition for getting such as policy package implemented' (ibid.: 888). He then proposed what he sees, an effective policy package of five items, one of them was regulating commercial advertisement more stringently and the other one was taxing status goods. It seems difficult to realise how such policy proposals will not meet strong resistance in a democratic system. He also argued in favour of 'a-growth', that is, to encourage economists, politicians and media to 'ignore' GDP. In this case, it is also implausible that propagating an *attitude* towards GDP will automatically reduce the ecological impact of effectively growing economies structurally designed to do so. A comprehensive and appropriate response was put forth by Kallis (2011). The fact of the matter is that any policy package which challenges growth will receive strong opposition and it will be rated as politically 'impossible' – as has been the tenor of the last 40 years. On the other hand, a broader and dynamic understanding of democracy could be helpful. The preconditions which Van den Bergh accurately identified can only be created bottom-up, be it for enacting his or Daly's proposals, or for that matter any proposal aiming at de-growth towards a (quasi) steady-state.

---

[79] The post-growth society (traduced by the author).

[80] A notion too ill-defined as he himself admits.

It is the role that Latouche and others are playing, which incidentally constitutes another dimension which makes de-growth and the SSE complementary. Perhaps it is useful to bring to back to memory that in the West, the ideas upon which social institutions such as slavery and patriarchy were based went on *millennia* without being challenged – not a few decades as in the case of growth – and when challenges emerged they were held by some to be politically impossible. Finally, slavery was ended and patriarchy was undermined. In choosing between tackling a political 'impossibility' and a biophysical impossibility, reason tells us to judge the latter to be more impossible and to take our chances with the former.

3.6 Conclusions and Prospects

The aim of this chapter was to study the economic growth debate hitherto, and to review and compare two alternatives to it: the SSE of Herman Daly and de-growth of Serge Latouche. The growth debate emerged out of the convergence of several ecological and political factors in the late 1960s in rich countries. The position of economists became divided on the issue with the majority maintaining the growth commitment. It was, however, the *Limits to Growth* report published in 1972 which projected the debate well beyond academia. The debate remained strongly polarised until the Brundtland report was published in 1987 settling the issue at the international political level. The Brundtland report which recognised the natural environment in essential ways was however, a (inevitably) product of political compromise that as such, neglected many important issues, namely the phenomenon of social-engineered wants already well-documented at that time. It also ended up making recommendations such as improvements in energy/matter efficiency, while ignoring scale effects (Jevons' paradox) which were also widely known, albeit strongly disputed. In spite of these disputes, the world economy continued to expand as measured, for example, by the ecological footprint. Years later, laissez-faire doctrines took over the world with a new formula for growth which was expressed in the ecological domain by the radical optimism of economists such as Julian Simon. From the 1990s onwards the public focus shifted towards climate change which by the beginning of the 2000s evolved into a political stalemate. Climate change was given a boost through the Stern report in 2006, whose proposals became politically feasible only after the last economic crisis prompted a renovate interest in Keynesianism. The new circumstances allowed for the notions of the 'green economy' and 'green growth' to find its way into the official environmental discourse which have been commonly used in Europe. The Green Economy report launched by the UNEP in 2011 was more coherent than the Brundtland report and reflected the tendency of a gradual shift away from market-fundamentalism and the integration of many elements of Ecological Economics such as state investments in green research, ecological restoration, public goods and more generally, investments in the global commons such as the atmosphere. Nonetheless, the report failed to get to grips with the issue of scale which logically allows for the preservation of the growth commitment.

The two alternatives beyond the environmental official discourse remain the SSE of Herman Daly and the cultural change called upon by Serge Latouche to realise de-growth in affluent countries. Daly drew on the ideas of Mill and Georgescu-Roegen. For Mill the stationary-state was highly desirable because of ecological and social reasons. Georgescu-Roegen examined the implications of the first and the second law of thermodynamics in the economic process, and concluded that the growth policy had become untenable. Indeed, he even criticised the SSE which was being developed by his former student Daly. Daly proposed a SSE in which low-entropic throughput is minimised but the service maximised. He put forward economic (qualitative) development instead of economic (quantitative) growth, and cogently demonstrated that the latter can also be 'uneconomic'. A set of policy recommendations for institutional change consistent with the SSE was suggested covering virtually the entire spectrum of economic and environmental policy. It is useful to underline his policy of quantitative restrictions proposed in order to tackle the Jevons' paradox, a topic left inconclusive in the Green Economy report. He proposed quantitative limits selected according to the most stringent necessity (depletion or pollution) and letting production/consumption to adapt to the new prices. Serge Latouche built upon the cultural critique to modernity of central European thinkers such as Jacques Ellul and Ivan Illich. From the Economic Anthropology of Karl Polanyi, Latouche derived his critique against uniform patterns of development, and from Murray Bookchin's ecomunicipalism, he strengthened his cause for the local as a starting point. Latouche advocated for a cultural revolution which should expand gradually whilst being guided by a set of interrelating R-guiding concepts.

The set of policy recommendations which arose from both approaches were greatly similar, such as working-less and work sharing, but using different lines of reasoning and wording, given their, to a degree, dissimilar intellectual traditions. The only marked differences were: the waste recycling practices which Latouche advanced but that Daly saw with reservation given the entropy law; and the population control policy which Daly supported as a still legitimate means of development, but that Latouche rejected on political and pragmatic grounds. Excepting these differences Latouche's Rs could be subsumed in the detailed theoretical elaboration on which the SSE rests. The SSE and de-growth are not mutually exclusive approaches but necessarily complementary, unless we do not value human existence on the planet. At the bottom, the SSE is, as the name indicates a state, while de-growth indicates motion. The discussion will ultimately rest in:

1. The physical quantities which economies need (population and man-made capital) for the *good life* in the *long run*;
2. How to *decide* on them, that is, biophysical limits, Daly's metrics, and Georgescu-Roegen's organic agriculture/population proposal;

3. How to achieve them, that is, Latouche's cultural change and a dynamic understanding of democracy; and
4. How to *maintain* an approximate steady-state.

It is nearly impossible to add anything novel to the statements of Boulding, Mishan, Schumacher, Daly, Georgescu-Roegen, to mention only the most prominent scholars. Indeed they stated with unparalleled clarity, after having understood that consumption and production became *bad* things that as such they should be minimised instead of maximised in countries which had already achieved an unprecedented level of material comfort. Others such as Hardin placed greater emphasis on population growth, an emphasis which was frowned upon by many. The rich were blaming the weakest members of their societies and, at some point of world's poor for the calamities they saw looming. The honest mistrustful (and Hardin himself) too often missed the point that it was the *combination* of policies and not singled out policies which mattered. However, the conclusion remains fundamentally the same. By the present state of things, 'growthmania', world economic growth and population growth must cease, or even be globally reversed. With the latter objective some advancement has been made, while with the two formers virtually no strides have been taken in the arena and area which matters: the political arena of core countries – and newly, the emerging ones.

It has long since been well-understood that with the perpetual quest for economic growth instead of for example economic development – in Daly's sense – or what Europeans thinkers once referred to as 'qualitative growth', everything becomes more complex, vulnerable and, therefore intractable for human management. It has the effect of pushing societies to resort to doubtful plan B's such as geo-engineering proposals and the additional scaling-up of institutions. The increasing acceptance of geo-engineering proposals such as injecting sulphur into the atmosphere strongly correlates with the failure of getting a necessary international binding agreement on climate change. It is easy to note that this plan B fits perfectly well within the predominant cultural belief of humans dominating nature through technology, which conversely allows for the maintenance of the growth commitment. One can almost imagine installing a switch on the planet for when it gets too hot, similar to calibrating an air conditioning system; while running the risk of forgetting that climate scientists have not, and maybe never will, completely understand the wide array of dynamic interconnections between the climate and life-support systems; that we may run the risk of falling again into a progress trap, and that at this stage, we just begin to anticipate the potential consequences for international relations.

Scaling-up institutions which began with mandatory 'end-of-pipe' treatments of waste and pollution can be grasped as the reflexive societal response to tackle bigger ecological problems in an almost hopeless attempt to cope with increased entropy and overwhelmed ecosystems. Yet, scaling-up of institutions must be necessarily accompanied by scaling-up governance structures for the purposes of enforcement – the rub of the issue. Institutions devoid of feasible enforcing mechanisms will remain, at least at the international level, simply in good formalised intentions. At this juncture it should be acknowledged that this societal response, albeit necessary, further jeopardises parliamentary democracy and freedom, for the institutional scaling-up tends to shift decision-making away from the sub-institutional units of the nation-states. Additionally, the bargaining costs of co-shaping the content of these institutions will tend to increase proportionally. If bargaining costs tend to increase *pari passu* with scaling-up institutions, it implies that greater bargaining costs will likely be more easily borne by correspondingly bigger players, which include not only big states but most importantly nowadays, big private organisations. This trend conversely reinforces the trend already set by globalisation – not a natural law but a myth encroached by mere repetition, and in some instances certainly by deliberate cultivation. From this perspective, the aim of global de-growth, in terms of energy and material throughput; and *de-globalisation*, in terms of free trade and free capital mobility, are perhaps not sufficient but clearly necessary conditions for achieving environmental sustainability and for protecting, and in some cases even restoring freedom and democracy.

Contrastingly, arguments for freedom and democracy are raised against policy proposals aiming at de-growth/growth towards a steady-state. It is believed that by allowing too much intrusion of the state in the ecological realm, we will be on the *road of serfdom*, in which a tyranny may emerge in the form of an 'eco-dictatorship'. It would be foolish to deny this possibility. Although societies may have latent totalitarian forces waiting for their political window of opportunity to curtail freedom in the name of ecological salvation – or in the name of other societal goals for that matter – the arguments laid down above indicate that, the more accelerate entropy through unnecessary growth, the more likely are the chances opened to a potential 'eco-dictator'. Indeed, causal empiricism shows that a strict hierarchical control of throughput and therefore, of social life is often witnessed in places where resources are extremely scarce, for instance in small ships, space shuttles and the like.

On the other hand, those arguing against the intromission of the state for the reasons of preserving freedom will have a difficult time in arguing against some of most the famous philosophers of the subject, such as J.S. Mill. Furthermore, a classical liberal less known than Mill in Anglo-Saxon countries, but from who Mill took inspiration was Wilhelm von Humboldt. In his inquiry on the *Limits of State Action* written in the late eighteenth century, he stressed that theory must be guided by attempting to achieve the greatest freedom possible, while coercion must be guided by reality, hence:

> Either man or the situation is not yet adapted to receive freedom, so that freedom would destroy the very conditions without which not only freedom but even *existence itself* would be inconceivable [. . .]. (1993: 144–145. Emphasis supplied)

It can be argued that the meaning of freedom has progressed ever since, but if this progress is meant to be the purposeful conditioning of the human mind as to disregard the natural tendency of satiation in order to have the *freedom* of choice between hundreds of brands given growth-necessities, then this progress would appear to be rather a regress in the conceptualisation of freedom. Under this frame, 'consumer sovereignty' becomes a cynical notion.

Another argument often put forth against the policy proposals which emerge from de-growth and the SSE is that their proponents want industrial societies to go back to the caves, or rather to the trees. This argument overlooks that the challenge consists precisely of institutionally channelling technological progress, that is, innovation and efficiency, in a manner which leads us to a material steady-steady state. This order is necessary, for innovation and efficiency first will not yield frugality second – unless frugality is dismissed as a precondition to cope with ecological problems. Besides, there are already hundreds of local projects attempting to live up frugality in which high-tech is used, thus encouraging self-sufficiency in energy, that is, photovoltaic and small farms for bio-fuels; democratic participation and cooperation facilitated through social media; but also urban gardening, co-housing, local monetary policy, and so on. All of this requires technical knowledge in agronomy, architecture and economics. It must also be mentioned that these local projects are not only a product of the bucolic romanticism of the rich, as it is sometimes portrayed, but an act of reflexive self-protection and justice. It is an act of reflexive self-protection if it holds true that we are on the downside path of the Hubbert's curve – let alone the threats of climate change; and an act of justice if we resist to rationalise under the label of development the emerging trend of buying large tracts of land in poor countries ('land grabbing') for the purpose of securing future fuel for the globally increasing and constantly renewed automobile fleet. Those who value tremendously human ingenuity in the realm of technology, too often do not value human ingenuity in the social realm. True, 'social experiments' have desolately failed in the past, yet the same judgment can be made on certain technological experiments.

From the previous discussion, what are the emerging prospects for scholars, at least for those sharing the view that global growth must cease and converge towards a SSE? In recent years, there has been a renewed interest in re-evaluating GDP-growth. An example of this can be found in France where a commission led by Stiglitz published a report on the issue in 2009, and presently a similar commission is working on the same topic in Germany. Prior to these reports, there was an increasing number of publications dealing with the measurements of the many aspects of human happiness and welfare. These studies can be added to the vast of body on green indexes' research which emerged from the interrelated debates on sustainability and growth. Concerning de-growth and the SSE there is still room for research regarding potential combinations with previous indexes and for different regions. It is important, however, to highlight that although such indexes are undoubtedly needed, they must be complemented with the additional study and evaluation of alternative institutional arrangements. These alternatives may take the form of encouraging other judicial forms of companies such as cooperatives, familiar firms and foundations which, different from joint-stock companies, are more interested in a steady-income stream than in profits and expansion, as Binswanger (2009) explains. The assessment of these institutional forms which could make economies less dependent on economic growth with reference to factors such as employment is of vital importance. Otherwise, the discussion on metrics will remain in modern Platonism.

As previously mentioned, there are already hundreds of on-going local experiments consciously practicing frugality which may require closer study regarding for example, how they function and what is the potential for extending these models regionally and beyond. These 'experiments' are not only being pursued in local villages in rich countries using sophisticated tools, but also in poor countries – poor in income terms. In Latin America there are larger attempts to re-build sustainable societies, which are guided to some extent by autochthonous notions. They are, needless to say, highly controversial and even antagonised from inside and outside. For instance, a couple of years ago the constitutions of Bolivia and Ecuador introduced the indigenous notion '*sumak kawsay*' (good life) as an overriding societal goal instead of economic growth and development. Regardless of the difficulties of understanding this notion, it is enough to state that it gives nature or '*Pachamama*' (mother earth) an overriding place, in which human life and other sentient beings are contained. It follows that *Pachamama* cannot possibly be abused for insatiable human wants. Whilst being cautious with comparisons, it may resemble the line of argument of Polanyi with his term 'embeddedness' unlike the disembedded spheres or quasi-independent pillars of SD. If this comparison was allowed, it would support the theory that in the history of humanity nature once had a sacred place in culture, and that the deviation of this pattern is, by historical standards, rather novel. Anyhow, the study of these attempts, their on-going successes and failures open up the possibilities of research for cross-national comparison and broadly understood, on international research cooperation.

Retrospectively seen, it seemed naïve when ecologists and some economists in the 1970s assumed that the product of small scientific revolutions, evidence, logic, refined modelling and common sense, would be enough to induce decision-makers to actually make rational decisions, thus ignoring the inherent messiness of human affairs. Although disciplinary research has become more holistic in methodology and content, it still aims almost *solely* at the provision of advice to decision-makers. To tackle this deficit, a new concept has been attracting attention in recent years: transdisciplinarity. In 't Veld (Chap. 1 in this book) presents a concise definition:

> Transdisciplinarity is to be defined as the trajectory in a multi-actor environment from both sources: from a political agenda and existing expertise, to a robust, plausible perspective of action.

From this definition the notion 'political agenda' should be underscored. In line with what has been written thus far, the understanding of political agenda should not be restricted to the agendas that professional politicians at the regional or national level and at a given moment happen to have, especially because these are usually pro-growth agendas. This argument is also supported in the schematic representation of the 'knowledge democracy' also detailed by in 't Veld. The third order of the scheme connects transdisciplinarity with participatory democracy and bottom-up media. These connections support the cause for the local. From this perspective, the action of 'boundary workers' should also include, and maybe even rather focus upon the boundary-work between science and *community*. This is what is habitually referred to as education, bearing in mind that modern educators mostly recognise the reciprocal character of their activities, that is, in the act of educating, they are also educated.

This idea is far from exceptional. In recent history, in the realm of economics and in a core country, it was Milton Friedman who initially understood that the role of the scholar should not be restricted to talking or giving advice to professional politicians, but directly to communities by means of numerous conferences and videos; in a time when social media was inexistent.[81] The redirection of at least a portion of the academic resources and efforts spent on advising established decision-makers to educate and learn from non-partisan representatives of civil society is also necessary for the following reasons: the almost immediate effects of an economic crisis (no-growth or 'negative' growth) mobilise societies in a direction – whatever it may be – while most of ecological problems seem distant. These problems happen in slow motion, sometimes not even discovered given many non-linear processes and middle term uncertainties; and impacting first and predominantly powerless nations. These features allow for adaptation and oblivion. Moreover, a cornucopian promising Eden on earth by letting business go as usual, and the neo-classical economist insisting that the only need is to get the prices 'right' in order to internalise social costs, will win over the 'pessimist' preaching the old-fashioned frugality and prudence. Indeed, this will meagerly counteract the enormous advertising budgets and the large adherence of the 'top-down media' to the growth call.

The former reflections do not imply a replacement of disciplinary/interdisciplinarity science (in the sense discussed in this book). It would be a mistake to become too enamoured with the local and transdisciplinarity for the following reasons. In the social realm, the preference for the local is merely because it is hoped that the constituents of professional politicians may be able to find new democratic ways of compelling them to abandon 'growthmania' and to correspondingly make policy proposals for a SSE. In other words, for the social researcher, as a member of the community, the hope rests in the ability to co-trigger a wide reflexive process or, being momentarily Hegelian, to further advance the de-growth anti-thesis. However, any local, regional or even national attempts can be easily discouraged at the international level which feeds back to the national one given the present forces of competition under which the current world functions. This case is crystal clear in the failed ratification of the Kyoto protocol and the uncertainty of the process in the

---

[81] See for instance the internet presence on the popular video portal YouTube of Milton Friedman (1912–2006) compared with his contemporary fellow economist Nicholas Georgescu-Roegen (1906–1994). The former yields 5,740 hits, while the latter only 20 hits. From these 20 hits, Georgescu-Roegen does not even personally appear in any video. A search on Herman Daly (1938–) yields 60 hits, while Serge Latouche (1940–) gets 310 hits. Fortunately this trend may be changing. British ecological economist Tim Jackson, who recently published *Prosperity without Growth* (2009) and similar to Friedman back in the late 1970s, has been recently engaged in proselytising activities, gets 12,600 hits (Search conducted on August 22, 2010).

following years. The same problem could be predicted if a serious attempt was undertaken to tackle the Jevons' paradox in the way proposed by Daly. On the same grounds, in the realm of the social sciences, it would be a mistake to become too obsessed with transdisciplinarity. This is because, as usual, any given methodology must be subjected to the scope and nature of the research problem. At this level, it is disciplinary/interdisciplinary science which must tackle the most formidable question of policy: how to transcend the international growth-race?

Mill, the intellectual grand-father of the stationary-economy explained that although this state was necessary, for 'the safety of national independence it is essential that a country should not fall much behind its neighbors in these things' (2004 [1848]: 690), 'these things' being increased production and accumulation. Back in the 1970s, Daly and Dutch politician Sicco Mansholt saw as a potential and promising 'deal'. This deal was the negotiation of economic de-growth in affluent countries for population de-growth in poor countries. This door seems to be by now entirely closed. Would China, for instance, who has saved a great deal of GHGs through the one-child-policy, who invests vast amounts of capital in green sources of energy agree with a view to becoming 'frugal'? Would the Chinese re-vive the habit of bicycle transportation gradually lost in the last years and stop growing their car fleet while the most important *overgrown* countries do not even consider de-growing their economies arguably for the reasons given by Mill more than one and half centuries ago? From this angle, it is difficult not to succumb to real pessimism on the international political ability to reverse what is truly new under the sun: the disproportionate space taken by humankind within the natural world. Georgescu-Roegen (1975: 379) with his usual causticity once speculated:

> Will mankind listen to any program that implies a constriction of its addiction to exosomatic comfort? Perhaps, the destiny of man is to have a short, but fiery, exciting and extravagant life rather than a long, uneventful and vegetative existence. Let other species – the amoebas, for example – which have no spiritual ambitions inherit an earth still bathed in plenty of sunshine.

At least his dream of attempting to harness solar energy has recently found its way in international politics, and his recommended ethical principle of leaving as much as possible an intact planet, that is, its life-support functions and services for the future generations was also adopted by the sustainability discourse years later. I believe it is a good principle in spite of the difficulties in defining the time span meant by the 'future generations' and that it may invite present inaction. It is a good principle in the sense that it is the only thing that the present generations can indeed do for the future ones, as happiness, welfare or even dignity are not transferable. If the future generations made themselves miserable with a relatively intact planet, this would be a choice which present generations would hardly be able to influence.

Acknowledgements This contribution greatly benefited from the enormous knowledge of Professor Joan Martínez-Alier in the history of economic and ecological thought and from the enlightening comments of Professors Herman Daly and Robert Ayres. They have been direct protagonists of the growth debate since its very beginning, which also enabled them to enrich this contribution directly from their memories. I would also like to thank Dr. Giorgos Kallis for his clarifications concerning the approach followed by Professor Serge Latouche, as well as Prof. Dr. Angelika Zahrnt for her precise clarifications on the growth debate in Europe and Germany, in which she has been an important actor in co-shaping the ecological discourse. This contribution was at its very conception, enhanced by the challenging comments which I received from Professor Roeland Jaap in 't Veld and Dr. Louis Meuleman. General yet equally helpful comments came from the observations of my colleagues in the TransGov project and my more distant colleagues at the Institute for Advance Sustainability Studies. Finally, I owe infinite gratitude to Professor Klaus Töpfer as he allowed me the complete intellectual freedom necessary to purse this controversial topic. This contribution also profited from his vast experience in politics and education pertaining to economics and the environment in many places and positions around the world. Any errors are entirely mine.

References

Alcott B (2005) Jevons' paradox. Ecol Econ 54:9–21
Anon (1997) Plenty of gloom. The Economist. http://econ.st/rjx9Ep. Accessed 20 Feb 2011
Anon, (2011) Rio+20 Expectations Unclear as CSD 19 Ends on Sour Note[online]. International Centre for Trade and Sustainable Development. Available from: http://ictsd.org/i/news/biores/106698/. Accessed 24 May 2011. Emphasis supplied
Arndt HW (1978) The rise and fall of economic growth: a study in contemporary thought. University of Chicago Press, Chicago
Arrow K, Bolin B, Costanza R, Dasgupta P, Folke C, Holling CS et al (1995) Economic growth, carrying capacity, and the environment. Science 268:520–521
Assadourian E (2011) It's time for millennium consumption goals, Worldwatch Institute. http://bit.ly/gb2SZ5. Accessed 2 Feb 2011
Ayres R (2008) Sustainability economics: where do we stand? Ecol Econ 67:281–310
Ayres R, Kneese AV (1969) Production, consumption, and externalities. Am Econ Rev 59:282–297
Barnett H, Morse C (1963) Scarcity and growth: the economics of natural resource availability. Johns Hopkins Press, Baltimore
Baumol W, Oates W (1988) The theory of environmental policy, 2nd edn. Cambridge University Press, Cambridge
Beckerman W (1972) Economists, scientists, and environmental catastrophe. Oxford Econ Papers 24:327–344
Beckerman W (1974) In defence of economic growth. Jonathan Cape, London
Beckerman W (1995) Small is stupid. Gerald Duckworth, London
Biswanger M (2001) Technological progress and sustainable development: what about the rebound effect? Ecol Econ 36:119–132
Biswanger HC (2009) Vorwärts zur Mäßigung. Murmann, Hamburg
Bookchin M (1995) Social anarchism or lifestyle anarchism: an unbridgeable chasm. AK Press, Oakland
Bookchin M (2007) Social ecology and communalism. AK Press, Oakland
Boulding KE (1966) The economics of the coming spaceship earth. In: Jarret H (ed) Environmental quality in a growing economy. resources for the future. John Hopkins University, Baltimore, pp 3–14
Bromley D (1990) The ideology of efficiency: searching for a theory of policy analysis. J Environ Econ Manage 19:86–107

Brookes LA (1979) A low energy strategy for the UK by G Leach et al.; a review and reply. Atom 269:3–8

Catton WR (1980) Overshoot: the ecological basis of revolutionary change. Chicago: University of Illinois Press

Chang HJ (2003) Kicking away the ladder. Development strategy in historical perspective. Anthem Press, London

Chang HJ (2008) Bad Samaritans. The myth of free trade and the secret history of capitalism. Bloomsbury Press, New York

Clarke M, Sardar IM (2005) Diminishing and negative welfare returns of economic growth: an index of sustainable economic welfare (ISEW) for Thailand. Ecol Econ 54:81–93

Clifton S-J (2009) A dangerous obsession: the evidence against carbon trading and for real solutions to avoid a climate crunch. Friends of the Earth, Wales

Cole HDS, Freeman C, Jahoda M, Pavitt KLR (eds) (1973) Thinking about the future: a critique of the limits to growth. Chatto and Windus for Sussex University Press, London

Cooter R, Rappoport P (1983) Were the ordinalists wrong about welfare economics? New York University, New York

Daly HE (1970) The population question in northeast Brazil: its economic and ideological dimensions. Econ Dev Cult Change 18:536–574

Daly HE (1972) In defense of a steady-state economy. Am J Agric Econ 54:945–954

Daly HE (1979) Entropy, growth, and the political economy of scarcity. In: Smith VK (ed) Scarcity and growth reconsidered. John Hopkins University Press, Maryland, pp 67–94

Daly HE (1991) Steady-state economics, 2nd edn. Island Press, Washington

Daly HE (1996) Beyond growth. Beacon, Massachusetts

Daly HE, Cobb J (1994) For the common good: redirecting the economy toward community, the environment, and a sustainable future. Beacon, Massachusetts

Daly HE, Farley J (2011) Ecological economics: principles and applications. Island Press, Washington, DC

Dasgupta P, Mäler KG (1998) Analysis, facts, and prediction. Environ Dev Econ 3:504–511

DB (2008) Economic stimulus: the case for "green" infrastructure, energy security and "green" jobs. Deutsche Bank, New York. http://bit.ly/qzmjP6. Accessed 20 Mar 2011

Gobierno Nacional de la República de Ecuador (2009) Yasuní-ITT. Una iniciativa por la vida. http://yasuni-itt.gob.ec/. Accessed 13 Feb 2011

Dryzek JS (1997) The politics of the earth. Oxford University Press, New York

Du Pisani JA (2006) Sustainable development – historical roots of the concept. Environ Sci 3:83–96

EC (2010) Sustainable de-growth: an alternative to sustainable development? European Commission. Environment News Alert Service. http://bit.ly/rD5EMB. Accessed 10 Nov 2010

EEA (2001) Late lessons from early warning: the precautionary principle 1896–2000. European Environment Agency, Copenhagen

Escobar A (1992) Imagining a post-development era? Critical thought, development and social movements. Soc Text Third World Post-Colon Issues 31/32:20–56

Ewing B, Moore D, Goldfinger S, Oursler A, Reed A, Wackernagel M (2010) The ecological footprint atlas. Global Footprint Network, Oakland

Fromm E (2007) To have or to be? Continuum, New York

Galbraith JK (1958) The affluent society. Houghton Mifflin, New York

Georgescu-Roegen N (1971) The entropy law and the economic process. Harvard University Press, Massachusetts

Georgescu-Roegen N (1975) Energy and economic myths. South Econ J 41:347–381

Georgescu-Roegen N (1977) The steady state and ecological salvation. Bioscience 27:266–270

Georgescu-Roegen N (1979) Comments on the papers by Daly and Stiglitz. In: Smith VK (ed) Scarcity and growth reconsidered. John Hopkins University Press, Maryland, pp 95–105

Gilbertson T, Reyes O (2009) Carbon trading: how it works and why it fails. Dad Hammarskjöld Foundation, Uppsala

Goldenberg S (2009) Obama focuses on green economy in speech before Congress. The Guardian. http://bit.ly/9qi44Z. Accessed 4 Mar 2011

Grinevald J (2008) Introduction to Georgescu-Roegen and degrowth. In: Flipo F, Schneider F (eds) Proceedings of the first international conference on economic de-growth for ecological sustainability and social equity, Paris, pp 14–17. http://bit.ly/rhrS4I. Accessed 14 Oct 2009

Hammitt JK (1998) Environmental false alarm and policy implications. Environ Dev Econ 3:511–516

Hardin G (1974) Lifeboat ethics: the case against helping the poor. Psychol Today 8:38–43

Herring H (1999) Does energy efficiency save energy? The debate and its consequences. Appl Energy 63:209–226

Hickman L (2011) The population explosion. The Guardian. http://bit.ly/eB5VNX. Accessed 7 May 2011

Hicks JR (1966) Growth and anti-growth. Oxford Econ Papers 18:257–269

Hilsenrath J (2009) Cap-and-trade's unlikely critics: its creators [online]. The Wall Street Journal. http://on.wsj.com/yIF5L. Accessed 19 Mar 2011

Hirsch F (1977) Social limits to growth. Routledge & Kegan, London

Höpflinger F (2010) Alterssicherungssysteme: Doppelte Herausforderung von demografischer Alterung und Postwachstum. In: Seidl I, Zahrnt A (eds) Postwachstumsgesellschaft. Metropolis, Marburg, pp 53–63

Hopwood B, Mellor M, Brien GO (2005) Sustainable development: mapping different approaches. Sust Dev 13:38–52

Illich I (1973) Tools for conviviality. Marion Boyars, London

IPCC (2007a) Synthesis report. http://bit.ly/lZwL4. Accessed 5 Oct 2009

IPCC (2007b) Contribution of working Group II to the fourth assessment report of the intergovernmental panel on climate change. http://bit.ly/pfG6FP. Accessed 5 Oct 2009

Jackson T (2009) Prosperity without growth. Earthscan, London

Jenkins J, Nordhaus T, Shellenberger M (2011) Energy emergence: rebound and backfire as emergent phenomena. Breakthrough Institute, Oakland

Johnson J, Pecquet G, Taylor L (2007) Potential gains from trade in dirty industries: revisiting Lawrence Summers' memo. Cato J 27:397–410

Jonas H (1979) Das Prinzip der Verantwortung: Versuch einer Ethik für die technologische Zivilisation. Suhrkamp, Frankfurt am Main

Kallis G (2011) In defence of degrowth. Ecol Econ 70:873–880

Kapp KW (1950) The social costs of private enterprise. Harvard University Press, Cambridge

Kaysen C (1972) The computer that printed out W*O*L*F. Foreign Affairs 50:660–668

Kerschner C (2010) Economic de-growth vs. the steady-state economy. J Cleaner Prod 18:544–551

Keynes JM (2009) Essays in persuasion. Classic House Books, New York

Khazzoom DJ (1980) Economic implications of mandated efficiency standards for household appliances. Energy J 1:21–40

Kneese AV (1998) No time for complacency. Environ Dev Econ 3:516–520

Krebs JR (1998) Predicting the environment: time series versus process based model. Environ Dev Econ 3:521–523

Kriström B, Löfgren KG (1998) For whom the market tolls: one-armed economists and 'Plenty of gloomsters'. Environ Dev Econ 3:524–526

Latouche S (2001) Les mirages de l'occidentalisation du monde: en finir, une fois pour toutes, avec le développement. Le Monde Diplomatique, Mai 2001

Latouche S (2003a) Pour une société de décroissance. Le Monde Diplomatique, Novembre 2003. http://bit.ly/5ZOtmN. Accessed 10 May 2011

Latouche S (2003b) Sustainable development as a paradox. Symposium Baltic Sea 2003. Gdansk, Kaliningrad, Tallinn, Helsinki and Stockholm

Latouche S (2004) Survivre au développement. De la décolonisation de l'imaginaire économique à la construction d'une société alternative. Mille et Une Nuits, Paris

Latouche S (2004b) Why less should be so much more. Degrowth economics. Le Monde Diplomatique, Novembre 2004 http://bit.ly/4yOHfY. Accessed 10 May 2011

Latouche S (2009) Farewell to growth. Polity Press, Cambridge

Latouche S (2010) Inzwischen kennt die französische Öffentlichkeit den Begriff 'Décroissance'. Interview mit Serge Latouche. In: Seidl I, Zahrnt A (eds) Postwachstumsgesellschaft. Konzepte für die Zukunft. Metropolis, Marburg, pp 201–204

Lawn PA (2003) A theoretical foundation to support the index of sustainable economic welfare (ISEW), genuine progress indicator (GPI), and other related indexes. Ecol Econ 44:105–118

Lawn P, Clarke M (2010) The end of economic growth? A contracting threshold hypothesis. Ecol Econ 69:2213–2223

Levallois C (2010) Can de-growth be considered a policy option? A historical note on Nicholas Georgescu-Roegen and the Club of Rome. Ecol Econ 69:2271–2278

Levin SA (1998) Anticipating environmental disasters. Environ Dev Econ 3:527–529

Loske R (2011) Abschied vom Wachstumszwang. Konturen einer Politik der Mäßigung. Basilisken Presse, Rangsdorf

Malthus TR (1998 [1798]) An essay on the principle of population. Prometheus Books, New York

Maneschi A, Zamagni S (1997) Nicholas Georgescu-Roegen, 1906–1994. Econ J 107:695–707

Mansholt S (1972) Le Chemin du Bonheur. Le Nouvel Observateur No. 396. http://bit.ly/ouuMK4. Accessed 15 Aug 2011

Martínez-Alier J (1987) Ecological economics: energy, environment and society. Blackwell, Oxford

Martínez-Alier J (1995) The environment as a luxury good or 'to poor to be green'? Ecol Econ 13:1–10

Martínez-Alier J, Masjuan E (2008) Neo-Malthusianism in the early 20th century. Ecological economics encyclopaedia. http://bit.ly/lNZWUe. Accessed 20 July 2011

Martínez-Alier J, Pascual U, Vivien F-D, Zacca E (2010) Sustainable de-growth: mapping the context, criticisms and emergent paradigm. Ecol Econ 69:1741–1747

Max-Neef M (1995) Economic growth and quality of life: a threshold hypothesis. Ecol Econ 15:115–118

McNeill JR (2000) Something new under the sun: an environmental history of the twentieth-century world. Norton, New York

Meadows DH, Meadows DL, Randers J, Behrens WW (1972) Limits to growth: a report for the Club of Rome's project on the predicament of mankind. Universe Books, New York

Meadows DL, Randers J, Meadows DH (2004) Limits to growth: the 30-year update. Chelsea Green, Vermont

Mill JS (2004 [1848]) Principles of political economy. Prometheus Books, New York

Mishan EJ (1967) The costs of economic growth. Staples Press, London

Mishan EJ (1977) The economic growth debate. George Allen & Unwin, London

Munasinghe M, Stewart R (2005) Primer on climate change and sustainable development. Facts, policy analysis, and applications. Cambridge University Press, Cambridge

Nordhaus WD, Tobin J (1972) Is growth obsolete? In: Nordhaus WD, Tobin J (eds) Economic research: retrospect and prospect, vol 5, Economic growth. NBER, New York, pp 1–80

Norgaard RB (1990) Economic indicators of resource scarcity: a critical essay. J Environ Econ Manage 19:19–25

Nuttall N (2008) Global green new deal – environmentally – focused investment historic opportunity for 21st century prosperity and job generation. UN Environment Program. http://bit.ly/qnvbbl. Accessed 4 Mar 2011

Odum HT (1971) Environment, power and society. Wiley, New York

Otway H (1987) Experts, risk communication, and democracy. Risk Anal 7:125–129

Packard V (1960) The waste makers. Pocket Books, New York

Pearce D (2002) An intellectual history of environmental economics. Ann Rev Energy Environ 27:57–81

Perrings C (1998) Introduction: environmental scares – the Club of Rome debate revisited. Environ Dev Econ 3:491–492

Piffaretti NF (2009) Reshaping the international monetary architecture: lessons from Keynes' plan. Pol Res Work Pap Ser World Bank 5034:1–28

Polanyi K (2001) The great transformation: the political economy and economic origins of our time. Beacon, Massachusetts

Polimeni JM, Mayumi K, Giampietro M, Alcott B (2008) The Jevons Paradox and the myth of resource efficiency improvements. Earthscan, London

Polin B (1998) Environmental scares, science and media. Environ Dev Econ 3:500–503

Pollard S (1968) The idea of progress. C.A. Watts, London

Pollin R, Garrett-Peltier H, Heintz J, Scharber H (2008) Green recovery: a program to create good jobs and start building a low-carbon economy. Political Economy Research Institute. University of Massachusetts. http://bit.ly/P6O3X. Accessed 16 Nov 2009

Portney RR, Oates WE (1998) On environmental gloom and doom. Environ Dev Econ 3:529–532

Radkau J (2010) Wachstum oder Niedergang: ein Grundgesetz der Geschichte? In: Seidl I, Zahrnt A (eds) Postwachstumsgesellschaft. Konzepte für die Zukunft. Metropolis, Marburg, pp 37–49

Robins N, Clover R, Singh C (2009) A climate for recovery. The colour of stimulus goes green. HSBC Bank. Global Research. http://bit.ly/qNUMCP. Accessed 20 Aug 2010

Rockström J, Steffen W, Noone K, Persson A, Chapin FS III, Lambin et al (2009) A safe operating space for humanity. Science 461:472–475

Sachs W (ed) (1992) The development dictionary. Zed Books, London

Sachs W (1999) Planet dialectics. Explorations in environment and development. Zed Books, London

Schor J (2010) In der US-Amerikanischen Öffentlichkeit und Politik ist Wachstumskritik ein Tabu. In: Seidl I, Zahrnt A (eds) Postwachstumsgesellschaft. Konzepte für die Zukunft. Metropolis, Marburg, pp 214–218

Screpanti E, Zamagni S (2005) An outline of the history of economic thought, 2nd edn. Oxford University Press, Oxford

Seidl I, Zahrnt A (eds) (2010) Postwachstumsgesellschaft. Metropolis, Marburg

Sen AK (1999) Development as freedom. Anchor Books, New York

Simms A (2008) The poverty myth. New Sci 200:49

Simon J (1996) The ultimate resource 2. Princeton University Press, Princeton

Smith A (1991 [1776]) An inquiry into the nature and causes of the wealth of nations. Prometheus Books, New York

Sneddon CS (2000) Sustainability in ecological economics, ecology and livelihoods: a review. Progr Human Geogr 24:521–549

Solow R (1973) Is the end of the world at hand? Challenge 16:39–50

Solow R (1974) The economics of resources or the resource of economics. Am Econ Rev 64:1–14

Starr C (1969) Social benefit versus technological risk: what is our society willing to pay for safety? Science 165:1232–1238

Stern N (2007) The economics of climate change: the Stern review. Cambridge University Press, Cambridge

Stiglitz JE (2002) Globalization and its discontents. Penguin, London

Stiglitz JE, Sen A, Fitoussi JP (2009) Report of the commission on the measurement of economic performance and social progress. Commission on the measurement of economic performance and social progress. http://bit.ly/JTwmG. Accessed 15 Mar 2011

Strasser S (1999) Waste and want. A social history of trash. Metropolitan Books, New York

Topfer K, Bachmann G (2009) One man – one vote – one carbon footprint: knowledge for sustainable development. In: In 't Veld RJ (ed) Knowledge democracy. Consequences for science, politics, and media. Springer, Berlin, pp 49–61

Truman H (1949) Truman's inaugural address. Harry S. Truman. Library & Museum. http://bit.ly/cdPKqY. Accessed 10 Feb 2011

Turner GM (2008) A comparison of the limits to growth with 30 years of reality. Glob Environ Chang 18:397–411

UN (2011). World population to reach 10 billion by 2100 if fertility in all countries converges to replacement level. United Nations Press Release. http://bit.ly/qU2BH3. Accessed 10 May 2011

UNDP (2010) The real wealth of nations: pathways to human development. United Nations Development Programme, New York

UNEP (2011) Towards a green economy: pathways to sustainable development and poverty eradication. United Nations Environment Programme, New York

UNFCCC (1992) The United Nations framework convention on climate change. http://bit.ly/q0ZILz. Accessed 10 Jan 2011

UNSNA (2009) System of national accounts. United Nations, New York

Van den Bergh J (2011) Environment versus growth – a criticism of 'degrowth' and a plea for 'a-growth'. Ecol Econ 70:881–890

Vitousek PM, Ehrlich P, Ehrlich A, Matson P (1986) Human appropriation of the products of photosynthesis. Bioscience 36:368–373

Von Humboldt W (1993) The limits of state action. Liberty Fund, Indianapolis

Wackernagel M, Rees W (1996) Our ecological footprint. Reducing human impact on the earth. New Society Publishers, Gabriola Island

Wackernagel M, Rees W (1997) Perceptual and structural barriers to investing in natural capital: economics from an ecological footprint perspective. Ecol Econ 20:2–24

WCED (1987) Our common future. The world commission on environment and development. Oxford University Press, Oxford

Weizäcker EV, Hargroves K, Smith M, Desha C, Stasinopoulos P (2009) Factor five: transforming the global economy through 80% improvements in resource productivity. Earthscan, London

Whitehead AN (1978) Process and reality: an essay in cosmology. The Free Press, New York

Worster D (1993) The wealth of nature. Oxford University Press, Oxford

Wright R (2005) A short history of progress. Canongate Books, Edinburgh

Zongguo W, Kunmin Z, Bin D, Yadong L, Wei L (2007) Case study on the use of genuine progress indicator to measure urban economic welfare in China. Ecol Econ 63:463–475

Chapter 4
Development, Sustainability and International Politics

Jamel Napolitano

Abstract The chapter argues for a lecture of the notion of development as strongly linked to the uneven distribution of material and non-material sources of power among groups. It thus analyses the rise of a public environmentalist awareness in the late twentieth century as a challenge to the capitalist pattern of production and consumption. Finally, the chapter aims to shed some light on the process of mainstreaming these claims by subsuming them within the western model of societal transformation, under the new, catchy label of sustainable development.

Pressing for institutional solutions to environmental depletion has meant to further spread the sustainability goal worldwide. On the other hand, it has also implied a kind of betrayal of the truly transformative instances of many social movements and local communities, which were seeking for a revolutionary, rather than reformative, path to societal change.

4.1 Introduction

This chapters deals with the history of sustainable development by going back to the very notion of development.

As development has mostly been dealt with through international lenses, in spite of the particular local issues raised by processes of societal change, the international structure stands back as a framework able to co-explain the main processes which will be discussed. Once we are aware of the profound power and geopolitical inequalities among states and – more correctly – social groups worldwide, this framework in turn proves to lead to a critical understanding of the rise of the development notion, as well as of its continuous reviews and improvements.

J. Napolitano (✉)
Fondazione Istituto Carlo Cattaneo, Bologna, Italy
e-mail: jamel.napolitano@gmail.com

L. Meuleman (ed.), *Transgovernance*,
DOI 10.1007/978-3-642-28009-2_4, © The Author(s) 2013

Beginning from point four of the inaugural address by President Truman, the first section critically addresses the so-called *developmentalist* era. Since then there have been many culturally specific interpretations of social change and transformation. However, our deep conviction is that, at the apogee of American and – broader speaking – western pre-eminence, the scientific and cultural mainstream was scarcely pluralist. Rather, developmentalism *modernization-style* tended to overlook the history of colonial exploitation of most of the countries invited to replicate the western road to well-being. This storyline, in other words, took for granted a level playing field, thus missing the relational point and looking at any single unit in the 1950s as if it started from the same departing point as western countries in the nineteenth century.

Since the 1960s, however, and thanks to the ideological opposition between the first and the second world, third world countries have expressed their unavailability to be absorbed into one of the two geopolitical blocs. The rise of the non-alignment movement coincided with the rise of, and was in turn analytically fuelled by, lines of thought such as the *dependency school*. Against the modernization school – whose main points are mentioned in the first section – many Latin-American scholars have queried the atomistic understanding of development, outlining the mutual relationship between development and underdevelopment and coming to propose a *delinking* strategy for less wealthy countries. To be sure, dependency as well as other counter-theories had its own internal fallacies. However, it concurred to stimulate debates and initiatives focused on a fairer economic structure on international grounds.

Criticisms also paved the way for a new attention to non-material dimensions such as the cultural one. That cultural turn represented a first, highly valuable breach into the economicist wall of many developmentalist accounts. However, accounting for cultural particularisms has too often meant keeping the binary opposition between modern and backward societies, under the label of modern and backward cultures – with the inevitable and implicit assumption of the superiority of the formers which were, not by chance, the devisors of these asymmetric and mutually exclusive counter-concepts. Cultural intervening variables' misuse has thus led to the attempt to universalise a particular culture, exactly the western one, as the most appropriate to the goal of economic growth.

As all these competing scientific trends were built, the international hierarchy of power has experienced its own changes, the most dramatic one being the fall of the socialist bloc. Thus, after a couple of decades of unquestioned unipolarism, the most common description of the current distribution of power among nations is multipolarism. As shown by Sect. 4.3, economic figures confirm the rise of new economic giants on the international scene. However, outlining the new role of national powers such as China or India – and thus speculating on a new national leadership according to a strict hegemonic reasoning – does not seem enough if we are interested in picking out the new cultural and scientific trends underpinning the current structure of global governance.

Emphasising the soft sources of international power requires paying attention to the ideational grip of a set of ideas, beliefs, institutions and so on and their ability to not only gain the general consensus, but stimulate emulation. In spite of the longstanding appeal of many dimensions of American scientific and popular culture, the main promoters of the new ideational trends are far from representing a single nation's worldview. Accordingly, taking for granted the weight of political collective actors belonging to government levels different from the national one, one of the core goals of the chapter is to underline the influential role of a broader group of actors. Epistemic communities, entrepreneurs, media, lobbyists, civil servants and executives from multilateral organisations, activists from social movements, volunteers and practitioners from NGOs: they all participate in the process of shaping the changing rules of the game thanks to a faster scientific and lay knowledge production and dissemination. When their grip on processes of submission and selection of social problems, agenda building, decision-making, policy implementation and evaluation has a worldwide outcome – eventually in spite of the local feature of the issues addressed – we can talk about them as transnational elites.

With respect to the development discourse, the role of these elites over the last decades has been twofold. First of all, they have been able to save developmentalism from the impasse it precipitated because of the many theoretical criticisms and empirical failures (exemplified by the lost decade of development), by adjusting it consistently with sensitivities such as the environmental one. On the other hand, they have contributed to mitigating the most drastic demands expressed by niches, incorporating the topic of environmentalism without taking seriously into account the problems connected to the very topic of development.

This is where sustainable development comes from. It was born thanks to the popularisation of instances and claims originally disregarded by agencies and institutions in the development sector during the apogee of western cultural and scientific power. Actually, environmentalism could be looked at as a part of the broader anti-systemic movement, aiming at a radical change of the capitalist lifestyle. Then, it has been legitimised and, as usual, the institutionalisation of conflict has led to a noticeable reduction of its revolutionary contents. Sustainable development, as pursued by most of the institutions in charge of global governance, represents today a reformist strategy, in spite of a long-standing, radical view of it diffused especially at the base.

This is why, among the most genuine sustainable development promoters, its development element, with its intrinsic reference to economic growth, still represents the tricky ingredient of the recipe. The new wine appears to have a good potential for being a very good one, provided that we are wise enough to throw away the old bottle.

4.2   Setting the Development Goal

Since the nineteenth century, a divide was established between natural sciences and the humanities, especially within the English educational and research system, as synthesised by the title of Snow's 1959 lecture, *the two cultures* (Snow 1990). Between those two poles, a third autonomous field of research, social sciences, had emerged by the middle of the twentieth century. According to Weber, social sciences represent a kind of via *media* between the search for general laws of nomothetic sciences on the one hand, and the idiographic accounts of humanities, on the other. In the aftermath of the Second World War, within social sciences themselves, the line between sociology (tackling the issues of how people live and relate to each other), economics (focusing on wealth production and distribution), and politics (the art of governing the *res publica*), has been further fixed. Meanwhile, new fields of research had been institutionalised: anthropology, furnishing usable knowledge on 'others' traditions once the decolonisation had been launched; and psychology, addressing individual behaviour, emotions, shocks and so on.

Thus, different bodies of knowledge have tackled their own issues, mainly relying either on the nation-state or the individual agent as their basic units of analysis. 'The division of labor among the social sciences has been a practical necessity, but it has had the unfortunate side effect of overspecialization' (Hofstede 1995: 213). For instance, typical anthropological concerns such as cultural diversity were paid scarce attention by non-anthropologists during the post-war period.

The developmentalist discourse has risen exactly in the framework of that general scientific environment (McCarthy 2007). Since the 1950s, the goal of development became institutionalised on international grounds, put forward by the United States as a kind of promise of improved living-conditions (So 1990; Rist 1996; Di Meglio 1997; McMichael 2004).

Before that era, development – as well as the broader issues of change, transformation and transition – had been a controversial analytical dimension for the social sciences, be it for the feared, often unspoken link with societal and political revolutions, be it for the trend to rely on static analytical categories. Thenceforth, however, development has been understood as a desirable, cumulative and linear process that every country was supposed to experience in order to replicate the western path of economic growth grounded on English industrialisation and then on the mass production and consumption goal reached by the United States. In fact,

> Few realize that Americans in 1776 had the same income level as the average African today. Yet, like all the present-day developed nations, the United States was lucky enough to escape poverty before there were Developmentalists. [...] George Washington did not have to deal with aid partners, getting structurally adjusted by them, or preparing poverty-reductions strategy papers for them. (Easterly 2007: 35)

To be sure, the idea of a one-style-fits-all model for the enhancement of living conditions had been envisaged in the western political, social and economic agenda well before the 1949 inaugural address by President Truman. However, 'it is only from that moment on that development policy became a truly global endeavor in which the world was divided into two groups of countries or regions, the developed and the underdeveloped' (Lepenies 2008: 205), with the formers devoted to provide the latters with development assistance.

Once that the pre-modern constrains preventing the full deployment of the economic and political revolutionary processes – respectively led by the UK and France (Touraine 1994) and epitomised by the rise of a working class employed in the industrial sector and of national democracies led by elected officials – were overcome, the goal of western countries had become the accomplishment of economic growth, and then the building of representative democracies. Truman's speech, and especially its *point IV*, has thus only contributed to the universalisation of such aspirations on a world-wide scale, launching the development era and introducing the notion of underdevelopment and the unit of measurement of Gross National Product (Rist 1996). In spite of the fact that 'one can expect definitions of the quality of life concept to be culturally dependent' (Hofstede 1984: 389), the recipe for national development was tailored on the western path of economic, political and social change, and on western peoples' experiences and desires in terms of labour market structure, gender and family roles, religious beliefs and so on. Drawing upon older analytical oppositions such as those proposed by Maine, Tönnies, and Durkheim, the gap between modernity and backwardness became the catching all dichotomy of the post war political, scientific and economic jargons.

4.2.1 Western Social Sciences and Third World's Claims

During the Cold War, approximately two million people, many barely freed by the colonial subjugation, discovered their status of underdeveloped or, in the best case, developing countries: countries and peoples, namely, *to-be-developed*. Against the western and Socialist[1] worlds, the collective label for the to-be-developed peoples was Third World.[2] Developmentalism found a warm welcome in those target countries, which enabled the US to pursue its liberal order on an international basis. In fact,

> Rather than involving whole nations, this acceptance came from small indigenous groups who had been educated in Europe or had in some other ways come into contact with European ideas. (Tenbruck 1994: 199)

---

[1] The Soviet Union also tried to offer a socialist version of the formula towards the moral and material progress of backward societies, pointing, alike the US, to gain the loyalty of peoples and countries against the antagonist geopolitical bloc, consistently with the bipolar geopolitical frame (So 1990; Di Meglio 1997; McMichael 2004). However, according to some strands of literature, those two narratives shared many prescriptions for the developmental nation-state – first of all, industrialization – and aimed at the same goal: bridging the standard of living divide between rich and poor. For instance: 'the particular recommendations of the United States and the Soviet Union were not substantially different: strengthen the urban sector, expand education, engage in judicious protectionism, mechanize production, and coping the pattern of the leading state' (Wallerstein 2007: 56). That regimes as different as western liberal democracies and socialist states came up with a quite similar understanding of development, in spite of the competing visions of social, political and economic organizations they displayed, is hardly surprising. Indeed, at least since the eighteenth century, scholars as different in their own political and ideological persuasions such as 'Comte, Hegel, Marx, Spencer and others [had] described the inexorable, irreversible, stage by stage and unstoppable advance of humankind through successive stages towards a golden age on Earth' (Du Pisani 2006: 84).

[2] In 1952, Alfred Sauvy, paraphrasing the 1789 title by Sieyès (*Qu'est-ce que le tiers-état?*), introduced in an article called *Trois Mondes, Une Planète* the notion of Third World, thereafter become quite common to indicate both less developed countries and non-aligned ones.

Regardless, this was enough to guarantee the success of developmentalism on the side of target countries as well.

Thereafter, the process of knowledge production and dissemination in the just established scientific field of development studies endeavoured to flourish world-wide, consistently with the model agreed upon by studies of the history of science, which 'have shown that science is a cultural, social activity permeated with values and preferences' (Turnhout 2010: 26). Social sciences reconciled with the 'dangerous' topic of change and commenced to understand development through a normative approach. In other words, the transformation issue gained full legitimacy within theoretical and practical debates. Drawing upon the former idea of progress – and legitimising it definitively after centuries of diatribes between conservatives and progressives – development came to be known as a linear, cumulative and ameliorative trajectory towards modernity, consistently with the older functionalist and evolutionist approaches. This scientific, political and institutional view of the so-called *modernization* school implied the reference to a metaphor, projecting the main features of the development of natural organism – *directionality, continuity, cumulativeness, irreversibility* – onto the social world: this analytical artifice led to the naturalisation and universalisation of a particular history, the western one (Rist 1996).[3] Consistently with an ascending vision of the history which has seldom recognised other approaches to the temporal dimension as equally legitimate (Pomian 1979; Du Pisani 2006; Featherstone and Venn 2006; Ribeiro 2007), developing countries were supposed to pass through a number of historical steps until the full accomplishment of modernisation. 'The new assumption was that, if the countries of the South would only adopt the proper policies, they would 1 day, some time in the future, become as technologically modern as wealthy as the countries of the North' (Wallerstein 2005: 1264).

Of course, modernity referred to the widespread diffusion of the capitalist mode of production and consumption. It also implied the downplaying of those unequal power relations (Pieterse 1994) underpinning western economic path of development both with regard to the social imbalances inside the northern states themselves and the exploitative relationship between richer states and their peripheral colonies. This is why, among the many criticisms the development discourse has triggered, it has been defined as a *project* (McMichael 2004) or a *colonial discourse* (Escobar 1995). It has also been considered an *ideology* the same way as communism, for it favoured the attainment of collective outcomes and presented itself as a scientific theory framed by technicians, scientists, experts, planners and the like: 'it shares the

---

[3] 'This identification of modernity with the process of modernization, this absolute confidence in the 'progress of the human spirit', to quote the title of one of Condorcet's works, and in the necessity of destroying the old world was so total, so obvious to the majority of Westerners, that still today, at the end of a century defined by a great diversity of modes of modernization and resource development [...] the Western countries resist any analysis of their own specific mode of modernization, so convinced of their own incarnation of universal modernity itself' (Touraine 1994: 121).

common ideological characteristic of suggesting there is only one correct answer, and it tolerates little dissent' (Easterly 2007: 31).

Thus, during the golden age of developmentalism, the modernization view informed, first of all, the economicistic approach, epitomised by the evolutionistic work of Walt Rostow, who equated the stage of *mass consumption*, following the phases of *take-off* and *maturity*, with the final stage of the path nation-states follow to become developed. *The Stages of Economic Growth. A Non-Communist Manifesto*, was published in the early 1960s, at the very end of a more than 10-year leadership of the MIT Center for International Studies. Rostow's involvement in US foreign policy is not astounding; rather, it provides us with a clearer idea of the link between American geopolitical concerns during the Cold War and the zenith of the developmentalist discourse. American-style modernisation had to be realised even at the cost of an externally driven, bloody revolution (So 1990). This was the view taken within one of the most authoritative schools of economics and international politics of the time; a school which has traditionally been 'more loosely oriented to democratic values than that by sociologists of modernization or by comparative political scientists' (McCarthy 2007: 12).

Indeed, mirroring the disciplinary specialisation of that period, there was also a strong research line on political modernisation. Under the aegis of the Social Science Research Council's Committee on Comparative Politics, and the leadership of Gabriel Almond previously, and of Lucien Pye later, scholars such as Coleman carried out their inquiry in the field of political development 'pervaded by the dominant ethos of scientificity, with its emphasis on behavioralism, value-free inquiry, quantitative measurement, the discovery and testing of empirical laws [. . .]. And it generally underwrote the need for strong postcolonial states to direct the modernization process through central planning guided by scientifically trained experts' (ibid.: 11).

Finally, there was a stricter sociological approach to modernisation, too. Its main research centre was the Harvard Department of Social Relations, under the leadership of Talcott Parsons. Strongly relying on Darwinian naturalistic explanations and Weberian culturalistic legacies, Harvard University scholars such as Levy and Smelser focused on the gap between modern and backyard societies and, with David McClelland's works on the *achieving society*, were also able to propose a psychological reading of the process of modernisation. Briefly, the general thesis was that

> The development process [postcolonial societies] had already begun under colonial regimes could best be completing by their adopting Western attitudes, values, practices, and institutions   including market mechanisms and state bureaucracies, industrialisation and urbanization, secularization and rationalization, the rule of law and democratization, social mobility and mass education, and so forth. And all this could best be accomplished with the assistance of already developed societies and under the management of strong national states. (ibid.: 10)

On the domestic ground, the main agencies of these developmentalist strategies were the nation-states, the main unit of analysis in the field of social sciences. Besides the emphasis on economic growth – to be pursued through industrialisation – the second *universal ingredient* of the development project was thus the nation-state (McMichael 2004).[4] Nation-states were developmental states, strongly involved in the goal of economic growth – that is to say, of obtaining an increased per capita GNP, the traditional development measure (Easterly 2007) – and, to a lesser or to a greater extent, also concerned with citizens' wellbeing – consistently with the apogee of Keynesian welfare state (McCarthy 2007) and its implementation within national frameworks differing with regard to their own specific administrative, social, economic and religious traditions.

The geopolitical context of the golden age of development was also relatively stable:

> Cold war rivalry governed much of the political geography of the development project. (McMichael 2004: 48)

Among its main political, military and socio-economic effects, Cold War with its corollary of the balance of power between the US and the URSS and in the more general framework of decolonisation influenced first of all the developmentalist discourse. Moreover, it had a dramatic impact on the international relations between developed and developing countries as well as among the non-aligned countries themselves.

For instance, the bipolar context both stimulated and somehow frustrated political ventures such as the 1955 Bandung Conference hosted by President Sukarno and joined by many Asian and African countries – against the neo-imperialism and neo-colonialism of the two major superpowers; as well as the formal establishment of the non-aligned movement led by Indonesia, India, Egypt and Yugoslavia, and inspired by the principle of non-interference in international affairs. Since the 1960s, 'the Non-Aligned Movement shifted from primarily political preoccupations, such as the liberation of the remaining colonies, towards a focus upon economic underdevelopment as the root cause of their political impotence' (Worsley 1994: 85). From the economic point of view, at stake was the economic model of development pointed out by the existing multilateral institutional order and epitomised by the Bretton Woods system.

One of the first collective challenges against the international economic structure underpinning developmentalism was the establishment of the Group of 77, joined by Third World countries and attempting to obtain the reform of the international trade especially through the United Nations Conference on Trade and Development. If nothing, these claims had an institutional impact on the way development was understood by core international agencies: since the late 1960s, for instance, a new focus on equity was introduced within the developmental discourse, as demonstrated by the growing attention towards the matter of basic needs – a topic whose roots were definitely non-institutional. After a strong emphasis on economic growth as the way to improve material wellbeing, the traditionally economicistic analyses of development institutions were widened by a new

---

[4] 'A discipline which emerged in the early post-World War II period, [...] development studies always took for granted the context of national economies and nation-states' (Rapley 2008: 180).

attention to social, cultural and political dimensions. It was pursued, for instance, through the incorporation of the Human Development Index and the Human Freedom Index, whose establishment and diffusion owed quite a lot to the activities of the United Nations Development Program (McCarthy 2007). Unfortunately, the institutionalisation of the basic needs approach led to its adoption as a theoretical as well as a practical paradigm by many international aid agencies without triggering any serious reassessment of development projects.

A further expression of the issues collectively raised by many to-be-developed took the form of the 1974 proposal to the United Nations for a New Economic International Order. The initiative of the G-77 was strongly influenced by Third World representatives struggling for a united South and stressing in particular the aims of economic growth, the expansion of international trade and the increasing of aid – notions, according to Rist, even too consistent with the old order dominated by principles of capitalism and thus advantaging, at the best, national bourgeoisies of the Third World, rather than local populations and communities (Rist 1996).

4.2.2 The Humanistic Turn

As mentioned, criticisms raised against the old fashion approach to development, with its technocratic and economicistic bias, have brought back into the development discourse an increasing attention towards non-material dimensions of processes of societal transformation. Among the most important achievements for development studies addressing wicked problems such as the material gap between different areas of the globe and the more sustainable paths to transform this state of affairs, we should mention the introduction of an increased sensitivity towards cultural differences.

At the apogee of the development era, the concept of culture experienced many reformulations, criticisms and rethinks within the anthropological community itself (Wolf 1984), while other scientific fields have overlooked it completely. The result was that cultural diversity 'was neglected for a long time because it did not fit in the dominant paradigm of the post-war period: rational choice theory' (Meuleman and in 't Veld 2010: 276). As for the development field, the acknowledgement that, besides formal laws and institutions, market economies also need 'norms or social values that promote exchange, savings, and investment' – that is, a correlate set of cultural, non-written patterns of thinking and believing fitting with the *economic behavior* (Fukuyama 2001: 3130) – has been too often neglected. In the aftermath of the Second World War, development programmes aiming at the export of capitalist modes of production and consumption towards regions whose economies were rather regulated through different mechanisms had not paid attention to the *embeddedment* of economics within the social whole (Polanyi 2001).

Quite the opposite, nowadays cultural diversity[5] can be defined as a *global discourse* (Ribeiro 2007), informing a number of social sciences accounts but still treated with scepticism by many anthropologists, especially those concerned with cultural (Shweder 2001) and post-colonial studies (Fougère and Moulettes 2006).

Currently, there is a widespread awareness that 'different cultures have different need hierarchies' (Hofstede 1984: 396). For example, while tackling the issue of closing the material gap between rich and poor, we should be aware of how our developers approach might fail to fit needs and aspirations of to-be-developed. As far as quality of life is concerned, 'researchers approaching the issue in Third World countries have relied too much on definition of 'quality' derived from North American and, to a lesser extent, West European countries' (ibid.: 397). This top-down decision-making concerning both the identification of the goals and the one-style-fits-all model to accomplish them, is often condemned and viewed as hierarchical and unfair by the very people who are supposed both to cooperate in and to benefit by processes of development.

Moreover, if we adapt Hofstede's statements on the issue of the humanisation of work to that of development, we come up with further fruitful insights into the risks experienced by developers attempting to offer a high quality lifestyle in accordance with their own, particular value-standards (ibid.). This risk is still high when development projects involve local practitioners: even developers originally coming from non-western countries are often socialised to the same set of beliefs and principles as their colleagues and peers from North America or Europe, at least with regard to their own business.

> Many Third World social scientists have been educated in North America or Western Europe. It is difficult for them to free themselves from the ethnocentricity of the Western approaches. This ethnocentricity is never explicit but is hidden behind 'scientific' verbiage. (ibid.: 397)

In fact, since the end of nineteenth century, scientific and political paradigms inspired by the civilising project or the idea of a white man burden, were condemned due to their *developmental* or *evolutionary* approach to culture. However, as we have briefly mentioned, these criticisms are still being raised specifically against the use and the meaning of culture often relied upon within fields such as development studies. This happens because misuses of the notion of culture are common among many development specialists who still rely on the dichotomy between modern and backward society, blaming the latter for its cultural inability to fill the gap with the former. Thus, 'in development economics [. . .], the view that 'culture counts' or that 'culture matters' is now popular in part because it is a discrete way of telling 'underdeveloped' nations (either rightly or wrongly) that the 'Westernization' of their cultures is a necessary condition for economic growth' (Shweder 2001: 3155).

---

[5] For an extensive treatment of cultural diversity, see Meuleman, Chap. 3 of this volume.

For instance, since the 1980s and especially the 1990s, cultural factors have been evoked by agencies such as the World Bank and the IMF 'as key variables explaining successful transition strategies' towards the building of market economies (Fukuyama 2001: 3132). As mentioned, the introduction of the Human Development Index to measure standard of life improvements from a non-economic point of view 'was one of the most radical paradigm shifts in development policy ever' (Lepenies 2008: 207). This innovation, however, lost some of its revolutionary meaning once it was appropriated by the most powerful western development agencies and thus institutionalised from the theoretical as well as the practical point of view. Referring to intervening variables such as human and cultural ones might imply a kind of *blaming the victim* logic which does not take into account, for example, the possibility that development strategies might be useless or even harmful when pursued in some contexts. The risk, thus, is that of a paternalistic account along the lines of: we provided you with the right knowledge, institutions, resources, but you have not been able to take advantage of them due to your own cultural constraints which prevent you from appreciating the good quality of this external help.

The point is that cultural variables as evoked by some developmentalist narratives are often associated with the implicit universalisation of a particular culture. Indeed, there have been scholars such as Geertz, addressing relativism by establishing a connection between it and the value system. Furthermore, as for the anthropological community, the joint influence of history and materialism has led Wolf to claim that culture is 'ideology-in-the-making' (Wolf 1984: 399). The *uni*disciplinary world-systems approach, in turn, asserts that 'the very construction of cultures becomes a battleground' as it is a value- and interest-driven process, rather than a neutral one (Wallerstein 1994: 39). In Europe, Bourdieu has stated that the classical humanistic notion of culture refers to 'the beliefs and behavior of the 'dominant class''. According to him, this 'culture' is just a 'culture' amongst many others, but it is imposed as the only legitimate one by school, universities, and other cultural institutions' (Harouel 2001: 3182–3).[6]

As we are about to see, the universalisation of a particularism reflects existent power relationships at the international level. The ideal of material progress, a typical trait of western culture, has been 'exported' specifically under the scientific and practical umbrella of development thanks to the hierarchical distribution of power among developed and underdeveloped states. Developmentalism modernization-style has indeed been sold as a good recipe for every single country, consistently with American capability to project its own way of life and to stimulate consensual emulation processes at least until the end of the twentieth century.

---

[6] Situating culture in the frame of power (and economic) relations, however, is not a specific feature of Marxist analyses of processes of culture production. Rather, in the 1950 and 1960s, it also characterised functionalist approaches such as the well-known works by Talcott Parsons (Paterson 2001).

4.3   A New National Hegemony?

By the end of the 1980s, state-led developmentalism was dismissed due to market-driven criticisms aimed at the failure of previous Keynesian recipes and the corruption they had fostered among most ruling groups. Furthermore, theoretical and empirical claims concerned with the worldwide diffusion of western liberal values and practices as both the most desirable and realist scenario for the twenty-first century experienced a further dissemination since the disappearance of the Soviet Union, which brought an additional flow of Western economic and political principles and left the United States as the *lonely superpower* (Huntington 1999).[7]

Thus, at the apogee of the *Washington Consensus*, structural adjustment was at the core of most development programmes. However, development, understood as *participation in the world market* and based on comparative advantage (McMichael 2004), could not represent the suitable catching-up strategy to improve the destiny of postcolonial states

> For the global economic playing field is by no means level. Its general contours were laid out by the modern history of colonialism. [...] Moreover, the rules of the 'free market' game are, as usual, heavily skewed in favor of the most powerful players, who dominate international associations, agencies, and agreements, from the IMF and World Bank to the G-7 and World Trade Organization. (McCarthy 2007: 16)

Meanwhile, in spite of triumphalist western accounts of the years following the end of the Cold War, the indisputability of American leadership over the rest of the world proved to be quite brief. Rather, current years are marked by the decline of unipolarism and the rise of other state and non-state actors powerful enough to impact many areas of global governance. While, with regard to new powerful nation-states, traditional power measures such as GDP still make some sense, the increased involvement of non-state actors in the current process of reshaping the rules for global governance requires a new attention to non-material sources of power.

It is true that, after the fall of the Soviet bloc, western liberal values, whose bishop was obviously the United States, seemed to be finally free to spread across the world. However, after the initial enthusiasm, it is becoming even clearer that the US is losing its primacy over the rest of the world from an economic and political point of view.

Among OECD countries, the growth of Gross Domestic Product is currently slackening (World Bank 2010). The estimated US GDP growth was -2.4% in 2009, while, according to the World Bank, the Euro area is performing even worse. We should notice that, around this time, several Asian countries were experiencing a steady economic growth before September 2008 and were still weathering the financial and economic crisis better than other economies. For instance, China and India were growing at rates of 9.5% and 8.2%, respectively. Similarly, while European recovery appeared the slowest (with an estimated GDP growth of 0.7%

---

[7] See also Fukuyama (1992).

in 2010 and 1.3% in 2011), and while the US is expected to grow approximately by 3% during the period 2010–2012, both China and India are expected to achieve a GDP growth higher than 8.0% in 2011 (ibid.).

These figures are hardly surprising. Rather, they perfectly mirror longstanding Western concerns regarding the economic boom of Asian countries: Japan first, the Asian tigers next, and finally China or even India.

Among International Politics analysts, these arguments date back to the 1970s, when several scholars stressed the relative decline in the overwhelming primacy once enjoyed by the United States, and anticipated that the days of American leadership were over. Indeed, the latest debates have focused on the supposed hegemonic decline of the United States (due to its loss of economic pre-eminence and/or ideological attraction), the identification of rising competitors (e.g. Japan, Russia, China, India and even the EU), and the projections of upcoming international scenarios – a new hegemony, a balance of power or a condominium of great powers (Kupchan 2002; Sur 2002; Foot 2006; Hurrel 2006).

Hence, global leadership appears today much fragmented with regard to both the material and the non-material dimensions of power. Indeed, beyond the traditional measures of power, it is even more noticeable that the shift towards multipolarism is well felt also within extra-material dimensions. Accordingly, besides the relative distribution of economic and military power in the international structure, there is a further point to make about the purported decline of the American ability to lead the rest of the world. It concerns the so called *soft power*, the broad cultural appeal that a powerful actor exercises over the others and through which it either gains a hegemonic position within the international structure or, at least, strongly impacts the rules of global governance.

> A country may achieve the outcomes it prefers in world politics because other countries want to follow it or have agreed to a system that produces such effects. In this sense, it is just as important to set the agenda and structure the situations in world politics as it is to get others to change in particular situations. This aspect of power – that is, getting others to want what you want – might be called indirect or co-optive power behavior. It [. . .] can rest on *the attraction of one's ideas or on the ability to set the political agenda* in a way that shapes the preferences that others express. (Nye 1990: 31, emphasis added)

Nye's now classical notion of soft power resembles, somehow, the notion of world hegemony. The latter, indeed, when unconstrained by a positivist operationalisation of power admitting only material, measurable dimensions such as economic and military strength, is made up by qualitative elements, too.[8] Hegemony, then, 'refers to the attainment of 'common sense' status by some set of ideas and institutions'. Furthermore, it implies the 'rule of a class or class alliance through a combination of consent and coercion, the capacity for a ruling bloc to set the agenda

---

[8] Among the many works making the point of the qualitative dimension of power from an IR point of view, (see: Cox 1983; Keohane 1984; Rapkin 1990; Wallerstein 1991; The Forum 1994; Rupert 1995; Robinson 1996; Taylor 1996; Modelski 1999; Brzezinski 2004; Fontana 2006; Lentner 2006).

for various institutions and actors without constantly resorting to force' (Sherman 1999: 87).

That the US has relied upon immaterial sources of power until now, is a matter of fact. What is less obvious is whether it will preserve its soft power in the near future. In recent years, American international behaviour has led to strongly criticised foreign policy decisions and to a reduced multilateral commitment in many issue areas,[9] such as the environment. Consistently, several analyses – some more, some less normative – have proliferated, concerning the weight of soft power and the need for multilateralism and eventually for a policy of burden sharing.[10] Even a former National Security Advisor, Zbigniew Brzezinski, has called for a more universalistic model of American leadership, for

> To be viewed as legitimate, that leadership has to reflect comprehensive global interests; to be effective, it must be backed by allies with similar popular convictions and societal values. (Brzezinski 2004: 87)

To sum up, the appeal of the American dream today appears quite doubtful, as well as its ability to be considered the best model to emulate and thus to gain the consensual loyalty of the so-called *followers*. However, even more controversial is the issue of the purported challengers' capability to not become hegemons themselves but, at least, to take the lead of global governance by means of the universalisation of their own pattern for action and thought.

China does not seem able to wield a widespread cultural and ideological attraction. First of all, it is not a democracy, which strongly invalidates its chances of being welcomed as a leading power by other countries and to project its domestic structure internationally as an appealing one. China, moreover, lacks any of the welfare measures which represent the foundations of citizenship within Western political cultures. Although social protections are more and more under attack even in European countries, a rearrangement of liberal social democracies consistent with Chinese political and economic architecture does not seem plausible. Finally, with regards to the material sources of international influence, we should mention that China's economic growth is strongly dependent on exports, as it still lacks a secure domestic consumption market (IMF 2009) until the full consolidation of its own middle-class and in spite of its demographic weight; and that its military capabilities, growing as they may be, still remain weak with respect to US military primacy (Weber 2005).

---

[9] The traditional anti-Americanism of a few European elites has thus turned into overt popular anti-Americanism (Markovits 2007) – especially during the Bush administrations (Parsi 2006) – with both a European and an extra-Western, and much bloodier, declination (Martinelli 2004). Public disappointment is echoed in academic literature, too: in most recent years, a number of scholars have expressed their concern about an imperial turn in US foreign policy (Jervis 2003; Golub 2004).

[10] Among others, (Calleo 1987; Kennedy 1987; Mastanduno 1997; Posen 2003).

India, for its part, has passed from the discouraging prospect of the *Hindu rate of growth*, to the 1980s *Hindu rate of reform*, which has driven the country to its current status of rising economic power (Boillot 2006). Unlike China, India does not have to face international legitimacy dilemmas such as a very reproachful, traditional neglect for addressing human rights issues; nor must it demonstrate to other democracies that its economic development has been matched by a consistent political development, since it is already a democracy. However, besides noting its limited military capabilities, it is also questionable whether India will take the lead for global governance because of its scant achievements with regard to the fight against poverty and its progress towards human development.

On the contrary, Europe might succeed in inheriting US strength, and in matching it with a greater concern for matters such as social justice and environment. However, aside from the issue of their material sources of power, Europeans seem unable to mount a cultural and moral leadership whose influence might supersede weakened US soft power. Quite the opposite, the EU has too often demonstrated its receptiveness of the American market discipline, as with respect to the debate about US-style labour flexibility as well as European rigidities and high unemployment rates. Currently, the EU risks missing the opportunity to fill the intellectual and scientific vacuum which would pave the way for the diffusion of fresh policy beliefs for the purpose of economic recovery and the establishment of a new framework for governance. This happens in spite of the link between the economic crisis and the mainstream approaches towards managing of economic and financial matters – approaches which were inspired by the US, before becoming a shared set of formal models and policy orientations with universal scope. Finally, the EU suffers because of the well-known problem of democratic deficit; it lacks a unitary political dimension, as well as a common defence policy; furthermore, as we have seen, prospects for economic recovery of the Euro area are not very bright.

Therefore, we are left with the puzzle that while the centre of economic power is moving away from Washington, it does not allow us to expect the advent of a new hegemonic nation-state able to lead the international system by means of a cultural and normative framework. Rather, the analysis should now shift towards the rise of *non-state actors* as agents able to impact the system of ideas, beliefs and biases in many areas of global governance – and thus to impact, even indirectly – decision-making processes with global reach. In many issue areas, theoretical accounts as well as practical exercises of global governance are further fuelling a longstanding dissatisfaction with methodological nationalism (Long Martello and Jasanoff 2004). Global governance, indeed, increasingly claims for the acknowledgement of the many different actors involved, often informally, in a policy making process which has worldwide impact (Cerny 2001). There are, first of all, non-state actors representing either the sub- or the supra-national level to account for. Secondly, and especially when wicked problems are on the table, policy making involves actors from sectors other than politics – such as scientists, entrepreneurs, stakeholders, activists and so on.

4.3.1 Epistemic Communities and Global Knowledge

Post-war American scientific prominence in the social sciences had stunted a genuine interest towards non-positivist analysis among IR scholars, and topics such as non-state actors and discursive power were regularly overlooked. However, since the 1970s, this trend is reversing with respect to both methodological nationalism and utilitarianism (Ruggie 1998).

Nowadays, a growing number of global politics specialists assert that methodological nationalism provides an inadequate analytical framework for examining the contemporary reshufflings of power among national and transnational actors. In fact, they argue that current power relations encompass more territorial levels, as demonstrated by the flourishing debate on multilevel governance (Pattberg 2006; Risse 2007).[11] Furthermore, the dissatisfaction with the rational assumptions underlying the once preeminent approaches to the study of IR, has produced an increasing interest towards ideas and beliefs (Yee 1996).

These new scientific sensitivities reveal an interesting feature of contemporary research: the trend to overcome disciplinary boundaries of the past. Thus, after the overspecialisation of the two more autonomous subfields of political studies, policy analysis and international relations, we can now notice a fruitful mutual exchange due to some interesting overlaps between their objects of research. Most important, current scientific trends mirror the unsuitability of analyses of global governance as exercised only within formal settings and by national actors. They, quite the opposite, pave the way for a genuine reconsideration of who are the main actors impacting the related processes of knowledge production and decision-making.

Hence, current political studies show an increasing interest in the role of non-state actors such as *policy networks* working from outside formal political structures (Capano and Giuliani 2005). Aside from the great differences among the possible operationalisation of the network, there seems to be the opportunity to identify a 'minimal or lowest common denominator definition' of it. Indeed, Tanja Börzel suggests that policy networks refer to 'a set of relatively stable relationships which are of non-hierarchical and interdependent nature linking a variety of actors,

---

[11] Recently, and with special regard to the topic of environmental politics, a number of research themes can be identified, as pointed out by Zürn. They all admit that 'international institutions do matter, world politics is much more than intergovernmental politics and includes a wider range of actors than states, and world politics is not only about power and material interests but is also about nonmaterial interests, ideas, knowledge, and discourses' (Zürn 1998: 619).

who share common interests acknowledging that co-operation is the best way to achieve common goals' (Börzel 1997: 1).[12]

Policy analysis and international relations share a great concern for a specific kind of network: epistemic communities. These networks are made up of experts and technicians relying on scientific approaches and often referring to similar interpretative and causal framework. They represent 'a principal channel through which consensual knowledge about causal connections is applied to policy formation and policy coordination. [...] As a consequence collective patterns of behavior reflect the dominant ideas' circulating, often supra-nationally, among epistemic communities (Haas 2001: 11579).

Epistemic communities, thus, hold a relevant quota of soft power, for they are able to shape the political agenda through the scientific knowledge produced in many issue areas. Knowledge-based networks of scholars are directly involved in the production and dissemination of scientific trends ranging from dominant economic doctrines to legitimised knowledge and narratives concerning, for instance, human rights, social justice and the environment. In turn, and especially when wicked problems are on the table, politicians may draw upon these scientific findings, provided that they are consistent with their own systems of ideas and the available policy choices.

> Members of transnational epistemic communities can influence state interests [...]. The decision makers in one state may, in turn, influence the interests and behavior of other states, [...] informed by the causal beliefs and policy preferences of the epistemic community. Similarly, epistemic communities may contribute to the creation and maintenance of social institutions that guide *international* behavior. (Haas 1992: 4, emphasis added)

Among the most evident feature of that knowledge production and dissemination process, there is its clear supra-national reach. Current literature on transnational networks 'concerns the weight of ideas, the significance of communication along transnational lines, and the capacity of nongovernmental groups to influence outcomes in international politics' (Zürn 1998: 620). Today, working on the impact exercised on policy making by *transnational networks of knowledge-based experts* in fields such as the environment represents one of the most important contributions made by the constructivist approach to the IR research community (Ruggie 1998).

Hence, scientific knowledge production can be described as an interactive process, based on continuative exchanges between scientific communities dispersed worldwide and yet linked together by similar research interests. The way knowledge is produced, the actors participating in this process, and the geographical

---

[12] Building upon the former concepts of *subsystem, subgovernment, iron triangle*, Anglo-American literature has been developing the notion of network since mid twentieth century (Jordan 1990) to better take into account how actors other than parliaments, governments, bureaucracies and political parties participate into the process of policy making. 'With the state no longer being the sole entity capable of organizing society, there is a dispersion of expertise and competence, a multiplication of channels for mediation and agreement, and the involvement of different levels of decision-making from the local to the supranational' (Coleman 2001: 11608).

spaces they come from and represent, are all changing from a specialised, hierarchical approach to a more transdisciplinary one – which is among the most pressuring concern of those scholars addressing the topic of knowledge democracy and societal transformations (in 't Veld 2010). Most innovative, non-mainstream approaches are thus making the case for the important role played today by values- and interests- driven actors, linked together transnationally, often socialised to a scientific approach that, if it is not the same, is nonetheless based on the same scientific criteria and able to influence policy making at many government levels.

This does not mean, however, that scientific research, be it produced or not in the attempt to furnish politics with usable knowledge, is free from value biases and pressures exercised by core power groups. Featherstone and Venn, for instance, consider *debilitating* 'the hold that western knowledge has on experts internationally, globalized in the form of the social engineering advocated by international NGOs like the World Bank and WTO and disseminated through countless courses in universities across the world, where the knowledge is taught as authoritative and universally valid' (Featherstone and Venn 2006: 3).

Actually, we should be aware of the mutual influence between epistemic communities and economic and political vested interests. We should also treasure classical insights from the sociology of science by Merton, who has underlined how, 'even in those countries in which the principle of 'freedom of science' is accepted, states and political decision makers clearly have an influence on the formation of epistemic communities' (Zürn 1998: 645). Indeed, as claimed by Turnhout,

> Science and policy are not separate domains but continuously influence and shape each other in dialectical processes of coproduction. [...] Difficulties in the relationship between production and use of knowledge are not due to a lack of information and communication. [...] Scientific controversies are often characterised by competing knowledge coalitions that use and reject knowledge based on vested interests. (Turnhout 2010: 26)

Thus, the innovative character of studies on transnational epistemic communities notwithstanding, this strand of literature is under attack due to 'its uncritical, almost blind confidence in the role of science, which is furthermore detached from the social context and relations of power in which it is embedded' (Epstein 2004: 49). For instance,

> Policy-making on complex issue like sustainable development is [...] usually a relatively fuzzy process in which many actors in the 'policy arena' are involved and influence each other. The production of knowledge to support policy-making is also not a neutral process, but is value-laden and influenced by actors in 'knowledge arenas'. Therefore, a strict separation between science ('the world of measuring') and the policy arena ('the world of weighing') is not possible. (Meuleman and in 't Veld 2010: 267)

Not by chance, the role played by politicians, businessmen and scientists within such a mainstream temple of knowledge as the MIT has been extensively underlined by Taylor in his critical review of the global discourse on environmental problems.[13] The scholar also states that too often these storylines neglect the issue of how social, political and economic inequalities impact negatively on the way that a truly sustainable transformation is pursued (Taylor 1997). Addressing the topic of societal transformation then requires the acknowledgement that the content of legitimate discourses and worldviews, as well as the process of knowledge production itself, are strongly influenced by geopolitical and power inequalities, as demonstrated by the contents, the methods and the prescriptions elaborated in many western think tanks, research centres and institutions.

While analysing the *globalization of culture and knowledge*, for instance, Featherstone and Venn suggest 'to give greater consideration to our participation in the globalization of western-centric knowledge' (Featherstone and Venn 2006: 1). Odora Hoppers, in turn, draws our attention 'to the non-neutrality of knowledge, especially given the unequal power to pre-empt the construction of meanings and to determine and control the rules governing speech and practise'. In her analysis of the validity of the centre-periphery dichotomical opposition, she claims for the 'acknowledgement of the continuing impact of global geo-politics and power relations on the legitimation of science' (Odora Hoppers 2000: 285).

Thus, in analysing the process of knowledge production and dissemination, we cannot overlook the point that even the most informal and avant-garde scientific, political and media agencies focusing on the ways sustainable transformation can be pursued, are affected by specific power relations and must always receive a validation feedback from the outside – usually from authoritative sources holding the power to decide what kind of knowledge is legitimate enough to circulate and which is not.

Moreover, since scientists 'interact closely in a global context' (Bunders et al. 2010: 126) and tend to adopt the same set of principles and the same approach to scientific research worldwide, especially when transdisciplinary research is concerned, the point of a *globalised knowledge* has been raised (Hulme 2010). According to Hulme, globalised knowledge 'erases geographical and cultural differences [...]. Rather than the view from nowhere, global kinds of knowledge claim to offer the view from everywhere' (ibid.: 559). Taylor, for example, strongly criticises the technocratic and moral approach of global environmentalism for it seldom recognises local differences due to peculiar historical paths. Furthermore, he states that a globalised understanding of sustainability tends to ignore trans-local dynamics accounting for how each local community derives its specificities from the continuous interaction between its own social, economic, political and cultural features, on the one hand, and external constrains and opportunities originating from other territorial scales worldwide (Taylor 1997).

These simplifications turn into a very critical issue while we aim to build a fairer governance structure supporting the transformation towards a more sustainable society. On a practical ground, it has been emphasised that sustainable development

---

[13] (See also Long Martello and Jasanoff 2004).

'as practised in the developing world is largely informed by Western notions and is often funded in accordance with the agenda of multilateral, bilateral, non-governmental and philanthropic donor agencies from the developed countries. This is viewed as problematic because it creates new dependencies for the developing world and raises concerns about whose agenda is being served' (Nurse 2006: 36). Accordingly, we agree on the advisability of questioning the assumption of networks' neutrality with respect to vested interests – thus standing back from Haas' claim regarding the *neutrality* of epistemic communities (Haas 2001). Consistently with the suggestion that knowledge is *situated*, it has indeed been argued that 'which issues are defined as meriting the world's attention has everything to do with who has the power and resources, including scientific ones, to press for them' (Long Martello and Jasanoff 2004: 5). Then it seems fruitful to enlarge the analysis of the most influential actors able to reshape the rules for global governance by taking into account, besides scientists and decision-makers, a larger group of people informally able to co-lead decision-making and policy implementation processes with a worldwide impact.

Actors currently involved in the interlinked processes of knowledge production and policy making come from political parties, lobbies, giant corporations, multilateral organisations, rating agencies, media, NGOs, universities, research centres and think tanks – as suggested by scholars focusing on many different social sciences topics (Haas 1992; Sklair 2000; Campbell 2002; Friedrichs 2002; van Elteren 2003; Brzezinski 2004; Buchanan and Keohane 2006). They tend to share similar higher education patterns and have an outward-oriented approach; in other words, they usually belong to the same, particular cultural framework. For instance, in spite of their legal citizenship and of their own business, we can expect most of them to have higher education levels – often from well rated, Anglo-American style colleges attracting a cosmopolitan attendance – and a good record of work experiences in many parts of the world. Besides the consistency among their formal CVs, they also tend to rely on a high, shared social capital even from a more informal point of view.[14]

Hence, consistently with the transdisciplinarity through which multilevel governance of wicked problems is exercised, there is, beyond politicians and experts, a wider range of actors to look at in order to investigate who are the most powerful figures reshaping the rules of the game in fields such as development and sustainability in this current era of power reshuffling. For instance, when we look at a specific working environment such as international development, we are not

---

[14] For example, it has interestingly been noted how scientists, politicians, lobbyists participating in the British great season of policy change at the beginning of the twentieth century joined 'the same clubs, associations and other social venues' (Campbell 2002: 31). Informal social links also play a role in the reproduction of a working environment. For instance, 'within the development field, personal relations are critical in such relevant moments as recruitment of new staff members and promotion of like-minded political allies. [...] Networks usually congeal into cliques' (Ribeiro 2002: 173). Grant Jordan (1990), too, recalls the image of *revolving door* with reference to human resource exchanges within stable networks.

surprised by the many, powerful profiles involved. The development sector, defined by Ribeiro as a *power field* because of the different power positions occupied by insiders and outsiders, is said to be made up of

> Local elites and leaders of social movements [...]; officials and politicians at all levels of government; personnel of national, international and transnational corporations [...]; and staff of international development organizations [...]. Institutions are also important members of this field: they include various types of government organizations, non-governmental organizations (NGOs), churches, unions, multilateral agencies, industrial entities and financial corporations. (Ribeiro 2002: 169–170)

Focusing on the broad spectrum of actors involved either in knowledge production, decision-making or policy implementation – and stating that those processes are intimately related – we could conceive them collectively as a kind of elite which, well beyond scientists and politicians, also includes, for example, influential members of the media and business sectors.

In broad terms, elites have been sociologically understood as

> Small groups of people who exert substantial power and influence over the public and over political outcomes. This power is based on the possession and control of various resources, including economic ones [...], control of organizations, political supports, symbolic means [...], and personal resources. (Etzioni-Halevy 2001: 4420)

Growing globalisation has paved the way for the advent of *transnational* elites, because of the increased weight of multilateral organisations; the legitimacy progressively gained by several NGOs with a global range; and the proliferation of many other political, economic and scientific *fora*. Transnational elites embody the ideational and practical stances of public and private institutions, usually having their physical headquarters in the *global cities* (Sassen 1991).[15] They are involved in governance processes whose reach is a multilevel one.

Most of these institutions and organisations date back to the period of unquestioned American leadership over Western political and economic systems and still maintain the ideational and practical orientation of that epoch. However, they have also been experiencing a visible de-territorialisation, which means greater reception of non-US concerns and autonomy from their former, single mentor. This change mirrors, first of all, the reduced international clout of the US, which justifies a multipolar description of the current international structure of power. Secondly, the increasing visibility of global actors tabling the needs and wills from local levels confirms the urgency to revisit the analytical assumption of methodological nationalism.

Studies on the superseding of nation-states as the unique and most appropriate level of analysis are all but new. After the introduction of the notion of *transnational society* by Raymond Aron (Aron 2003) in the 1960s, scholars such as Nye and Keohane have pointed out *transnational relations*, whose key feature is the involvement of non-governmental actors. They stated that 'any unit of action that

---

[15] (See also Martinelli 2005).

attempts to exercise influence across state boundaries and possess significant resources in a given issue area is an actor in world politics' (Nye and Keohane 1971b: 733). The analysis of transnational relations raises, among other things, the *attitude* issue. Attitudes are beliefs, norms, 'opinions and perceptions of reality of elites and nonelites within national societies' (Nye and Keohane 1971a: 337). According to Nye and Keohane, attitudes are also shaped by non-state actors, and the process of new attitudes fostering is an asymmetrical one because only the most affluent and powerful segments of world population 'are able to take full advantage of [this] network of intersocietal linkages' (ibid.: 345).

Then, transnational elites are, even indirectly, involved in many processes connected to global governance thanks to their participation in the stages of agenda setting, decision-making and policy implementation and evaluation. As we are about to see, they are also the main agents able to legitimise and disseminate *world culture*.

4.3.2   Transnational Elites and World Culture

Actors such as experts and scientists specialised in the same field of knowledge and collectively understood as an epistemic community with a supranational reach, 'are often responsible for generating the very ideas that constitute the *world culture*' (Campbell 2002: 30, emphasis added).

Transcending specific scientific fields, this world culture impacts the systems of ideas and beliefs of many people, thus showing both its popular declinations – for instance, McDonaldisation – and higher expressions, as we are about to see.

> World culture refers to the cultural complex of foundational assumptions, forms of knowledge, and prescriptions for action that underlie globalized flows, organizations and institutions. It encompasses webs of significance that span the globe, conceptions of world society and world order, and models and methods of organizing social life. (Boli and Lechner 2001: 6261)

Recently, the idea of a world culture has been circulating insistently among social scientists. For instance, since the last decades of the twentieth century, and consistently with the weakening of both rational choice theory and hierarchical-bureaucratic approaches, political scientists have devoted much attention to 'how ideas, that is, theories, conceptual models, norms, world views, frames, principled beliefs, and the like, affect policy making' (Campbell 2002: 21). Surel has adopted an encompassing label, *cognitive and normative frameworks*, in order to address 'coherent systems of normative and cognitive elements which define, in a given field, 'world views', mechanisms of identity formation, principles of action, as well as methodological prescriptions and practices of actors subscribing the same frame' (Surel 2000: 496). Thus, one of the outcomes of belonging to the same frame is that individuals share a *collective consciousness*, a subjective sense of belonging, producing a specific identity' (ibid.: 500).

The broader definition of culture, as well as its understanding in terms of dynamic learning processes, leads scholars such as Featherstone (1994, 2006) to make the case for the *globalization of culture*. Pieterse, in turn, has outlined how this global culture must be looked at with reference to a process of *hybridization* and *creolization*. Outlining the continuous, relational process of mutual cultural exchange and learning would allow us to overcome the bias concerning the uniformity of culture. He also points out how even western culture has been made up during the centuries trough the interaction with, and the absorption of, other cultural forms and practices with no regard for formal political and geographical boundaries (Pieterse 1994).[16]

The consolidation of a global culture, moreover, should not be conceived as referring to a simple dichotomic framework – an either-or logic between diversity and homogeneity (Featherstone 1994) or local and global.[17] In addition, it should not refer to simple Americanisation and Westernisation. Rather, one of the main features differentiating today's global culture from ancient and modern processes of cultural colonisation lies in the current lack of one or more centres from which cultural elements irradiate (Appadurai 1994) – a validation of our hypothesis regarding the transnational combination of elements from many different geographical scales. Hannerz, for instance, places the origins of world culture in the 'increased interconnectedness of varied local cultures, as well as [in] the development of cultures without a clear anchorage in any one territory' (Hannerz 1994: 236). Smith, in turn, claims that 'global culture would operate at several levels simultaneously: as a cornucopia of standardized commodities, as a patchwork of denationalized ethnic or folk motifs, as a series of generalized 'human value and interests', as a uniform 'scientific' discourse of meaning and, finally as the interdependent system of communications which forms the material base for all the other components and levels' (Smith 1994, 176).[18]

The points to be made here refer, firstly, to the cultural homogeneity of groups cross-cutting formal national, regional or continental borders; and, secondly, to the power differentials allowing for the primacy of a few cultural traits over others.

As pointed out by Hannerz, we might recognise cultures transcending arbitrary territorial boundaries such as nations and regions and carried, rather, 'as collective structures of meaning by networks more extended in space, transnational or even global' (Hannerz 1994: 239). These systems of beliefs, as this scholar goes on to say, 'tend to be more or less clearcut occupational cultures (and are often tied to transnational job market)' (ibid.: 243). Hence, we can imagine most liberal professionals involved in different fields such as politics, media, business, academy

---

[16] On the same topic of cultural contamination, (see Gruzinski 1988).

[17] See Chap. 2 by in 't Veld, this volume.

[18] Appadurai, too, goes on the difference between the globalization and the homogenization of culture stating that 'globalization involves the use of a variety of instruments of homogenization (armaments, advertising techniques, language hegemonies, clothing styles and the like), which are absorbed into local political and cultural economies' (Appadurai 1994: 307).

but all employed in the development sector, as belonging to a common cultural and scientific framework. These transnational networks encompassing the realms of politics, media and science, as well as the private business sector (Graz 2003), share the same world culture in its highest declinations.

Then, we should stress the ideational dimension of power dynamics, which has been too often underestimated by structural approaches (Golub 2004; Nabers 2008). In fact, leverage on knowledge and information is a major source for the exercise of power (Risse 2002) and, not by chance, the *high* world culture shared by most of the actors forming the transnational elite is strongly influenced by initially Western values and beliefs, and its grounding lies in a positive bias towards market economy. Supporting the capitalist organisation of economic relations, however, also involves a consistent vision of socio-political arrangements (Ikenberry 1992), allowing broad grounds for social, human and environmental concerns. Indeed, 'western-like aspirations include the desire for liberal democracy, free enterprise, private property, autonomy, individualism, equality, and the protection of 'natural' or universal 'rights'' (Shweder 2001: 3156). As Blyth points out, 'economic ideas can create the basis of a mutual identity between differently located economic and political agents' (Blyth 1997: 246).

As mentioned, looking at world culture as a simple by-product of American grip over the rest of the world would seem quite naïve. Instead, with the current multi-layered distribution of power, transnational elites project their influence at all the levels of governance because of their grasp on a number of national and sub-national politicians, policy advisers, lobbyists and intellectuals. Indeed, as Overbeek states, domestic regimes and 'internal structures of states are adjusted so that each can best transform the global consensus into national policy and practise' (Overbeek 2004: 11). From their seats within public and private institutions, transnational elites work as 'progenitor[s] of ideas, which they successfully spread through bringing together senior civil servants, business executives, and technical specialists in working groups that give real substance to the concept of epistemic community' (ibid.: 14). Local populations, in turn, are socialised to a set of values, beliefs and practices delivered as universalistic in spite of their particularistic origins.[19] They range from consumerism to individualism; from faith in democratic regimes to implementation of neoliberal recipes; from a notion of globalisation as a self-generating process to the idea of multilevel governance as a regime enhancing local populations' self-reliance while addressing the most alarming global issues such as environment depletion, global warming and energy shortage.

It can be said with certainty that there is nothing intrinsically wrong with these values and prescriptions. However, we have previously urged giving greater attention to intangible sources of power and to non-state actors as agents of power; similarly, we shall now stress how systems of formal and informal rules are all but

---

[19] (See Späth 2002; Dingwerth and Pattberg 2007).

neutral. Quite the opposite, they constitute hegemonic discourses framed and delivered by the most powerful actors.

Summing up, the notion of culture is, firstly, becoming even broader to include knowledge, beliefs, ideas, biases. In this regard, it is flowing into research areas traditionally less sensitive to the findings of the anthropological scientific community – as proved by the constructivist turn leading many post-positivist research programmes. Secondly, the notion of world culture enables us to escape the fixed borders of nation-states, by explicitly referring to many coexisting and overlapping scales, consistently with the transnational character of both the *cognitive paradigms* and the *normative frameworks* constraining the options perceived as either useful or legitimate by ruling groups and other elites (Campbell 2002). Finally, as we will see within the specific field of development, a prominent trait of this world culture is represented by a wider inclusion of different narratives within the main story line. Nowadays, 'the voices and the views of the Third World are increasingly prominent in world-cultural development' (Boli and Lechner 2001: 6262), thus provoking increased conflicts and fragmentation between the competing and concurrent processes of *cultural homogenization* and *cultural differentiation* (Appadurai 1994).

Developmentalism was one of the cultural and scientific product of American and – more generally – Western soft power, as shown by its broad application in domains such as international power politics, national policy making, scientific research, campaigns of NGOs, individual humanitarian concerns and the like. Since addressing the changing nature and scope of hegemonic discourses and agents responsible for their formulation is one of the major challenges for social sciences, we will describe how – among the non-economic claims underpinning the current world culture – we can recognise wicked problems such as environmentalism.

4.4 Whither Governance for Sustainable Transformation?

Since the 1970s, a vitriolic discontent with developmentalism has been circulating among most Third Word populations.

> The development process itself had displaced them from traditional lands and ways of life, but without corresponding opportunities for absorption into the modern cash economy. Dispossession, marginalization, hyper-urbanization, and the explosion of precarious settlements and informal economies became symbols of a development enterprise that had gone tragically wrong, betraying its most fundamental promises. (Carruthers 2001: 96)

While the days of developmentalism seemed to be over, today the development machine is alive and well. How was its survival possible?

Today, the topic of international development goes hand in hand, on the theoretical as well as on the empirical ground, with the notion of global governance. The latter is said to be 'based on shared expectations, as well as on intentionally designed institutions and mechanisms' (Benedict 2001: 6237). Global governance has a Western or, better, Anglo-American root (Friedrichs 2002; Martinelli 2005), is strictly connected to the process of globalisation (Friedrichs 2002; Pattberg 2006) and is carried out according to the assumption that 'human rights, monetary affairs or security are to be governed by a global elite, because otherwise the realm of chaos and violence [...] takes place' (Späth 2002: 1–2). Global governance includes many levels for governmental functions, consistent with the current pre-eminence of actors belonging to agencies which cut across state boundaries, with the changing role of nation-states, and with the increasing regionalisation connected to a multipolar structure (Pattberg 2006).[20]

Global governance can be understood as a common framework of principles, rules and laws necessary to tackle decision-making in several issue areas which are upheld by a diverse set of institutions at the sub-national, national and supra-national levels (Benedict 2001). In spite of the claims for truly multilevel processes of decision-making empowering local communities as depositaries of lay knowledge and practices, this set of guidelines is mostly set by transnational elites. To be sure, this is not always a unidirectional, top-down process. However, the concrete opportunity for common people to effectively lobby top managers, chief executives, leading politicians and intellectuals of the OECD countries remains scarce (Risse 2002), the connections between several NGOs and grass-roots movements notwithstanding. This is why the global character of governance has been questioned (Dingwerth and Pattberg 2007) or why it is said to create 'new borders of inside/outside' (Späth 2002: 1). Nevertheless, similar to developmentalism, global governance needs and actually has an ideological appeal for many people. It entails a vertical process of interiorisation of the transnational elites' policy beliefs by local officials and intellectuals, thus representing a consensual tool for the management of global affairs.

Accordingly, the current rules for management of matters perceived as having a worldwide impact are mostly set by restricted inner circles whose membership is far from mirroring old binary differentiations such as developed and developing/underdeveloped states or western and non-western countries. This means that the view understanding global governance as a subtle synonymous for an enduring American leadership or, even worse, a new empire, is a very naïve one.

In the aftermath of the Second World War, the US deployed its hegemonic project by establishing organisations and agencies with a considerable supra-national reach – such as the Bretton Woods and the UN institutions, as well as the original nuclei of the OECD and WTO. At the zenith of the *Pax Americana*, these institutions acted consistently with their major mentor. However, in the course of time, they gradually started losing their territorial connotation and attracted agents and goals from other emerging powerful players in the global

---

[20] On the functionalist research program, pointing to neutral or, in the worst case, technocratic expertise replacing several political tasks and responsibilities, (see Dingwerth and Pattberg 2007). For an evaluative notion of global governance as a favourable instrument for the empowerment of global civil society by the means of a multilevel, non-hierarchical and democratic exercise of governmental functions, (see Scheuerman 2007).

arena, thus incorporating competing demands and claims as the US economic pre-eminence was declining in relative terms. For instance, the older IMF and World Bank experienced an enhancement of their commitment towards and compliance with economic and political concerns of transnational elites: there was a normative turn towards neo-liberal programmes in the economic dimension (Harvey 2005) and democracy promotion in the political one (Robinson 1996). At the same time, new institutions for the management of world politics and economy – such as the G7 and the Trilateral Commission, later replaced by the World Economic Forum – were established. Finally, interventions of and prescriptions by NGOs gained greater legitimacy, thus concurring with the present awareness for social, political or environmental needs of several non-OECD populations.

To be sure, the Washington Consensus is not being seriously undermined, even in the context of US economic decline relative to other powers. However, shared goals, values and beliefs establishing common standards have evolved over the last decades. Formal and informal rules have been adapted to one of the most important structural changes, the end of the Cold War, as well as to the advent of new great powers with their own international ambitions. Moreover, non-state actors such as multinational corporations, financial agencies and international organisations have consolidated their own roles in the process: while they had been mostly set since late 1940s, we should actually appreciate that they have evolved through the course of time. One of the core changes we notice with regard to leverage positions in the context of globalisation concerns the increased assimilation of non-US officials, purposes and values within global governance institutions. Currently, transnational elites seem to be involved in a wide range of issue areas and frequently pursue their own agenda through international organisations, independently from and some-times even contrary to the declared policies of national governments. Indeed, their American trademark notwithstanding, global governance institutions are increas-ingly straying beyond US control because of both the institutionalisation processes and the rise of new agents of power, be they state or non-state actors.

This dramatic political and economic turmoil challenging the international structure of power is allowing an increased space for new criticisms within the social sciences themselves. This is not a peculiar feature of the end of the twentieth century. Rather, the literature linking previous hegemonic transitions and cultural change suggests that a critical review of 'the foundations of knowledge has characterized each transition' (Sherman 1999: 110).

As for development studies, we experienced the birth and the consolidation of the modernization theory – with its assumption of a step-by-step transformation of backward/underdeveloped societies until the goal of reproducing the western pat-tern – in the 1950s. Not by chance, it was precisely during the post-war era that the western pre-eminence and especially American primacy over the rest of the world reached its zenith. The dissolution of the Soviet bloc, leaving former socialist states with no developing model, enabled the diffusion of post-Keynesian doctrines based on economic liberalisation and shrunk government. The modernization school meanwhile had been by and large questioned by other schools of thought – such as the *dependencia*, the *école de la régulation*, and the *world-systems analysis* – since the 1970s. Following this, *post* theories cast many doubts on the notion of modernity as one of the most catching all metaphors underpinning developmentalist theoretical as well as empirical narratives.

Current diffusion of post-positivism and post-modernism signals that 'a new battle of the books has been engaged' (Sherman 1999: 111). This dispute on tools and methodologies is consistent with the weakening of the old positivist monopoly related to the Western, hegemonic worldview. Among the many outcomes of the current intellectual and scientific turmoil – named by Cerny (2001) as a *small kuhnian revolution* – we should notice an increased consideration of ideological and cultural factors underpinning scientific knowledge production and dissemination, as well as the most popular systems of ideas and beliefs. A growing number of social analysts are paying greater attention to how cultural elements might mediate or even influence trends and phenomena into the social, political, economic as well as scientific arena. Since the 1990s, the post-development approach, for example, has made a strong case for the interaction between power and knowledge within the development field, claiming that 'the knowledge deployed in development is a product of epistemic perspectives of the 'West'' (Jakimow 2008: 312).

Furthermore, growing criticisms against modernization and developmentalism, have led to greater attention being paid not only to the human dimensions of development but to the environmental ones as well. Sustainable development

> ... became part of the critique of neo-liberal development models [...]. In this sense the sustainable development paradigm should be viewed as [...] part of the growth of new social movements and the rising wave of discontent with conventional development theory and practice. (Nurse 2006: 35)

As pointed out by Du Pisani, environmental damage, natural resources exploitation and population growth were concerns already raised in many classical books. However, they have become truly popular issues only in the second half of the twentieth century, when 'the Enlightenment promise of the linear and continuous improvement of the human condition had proved to be a Myth of Progress, because it was based on human hopes and aspirations rather than human potentialities and limitations' (Du Pisani 2006: 89).

Hence, since the beginning of the 1960s, western people's consciences were shocked by the publication of *Silent Spring*, which cast a shadow on that phase of economic boom. In 1968, Garret Hardin tabled the Malthusian issue of the exponential growth of the population size. The *tragedy of the commons*, the paper said, is related to the inevitable destruction of those common-pool resources by the users: 'a finite world can support only a finite population; therefore, population growth must eventually equal zero' (Hardin 1968: 1243).

During this initial phase, the United States took the lead of the rising green politics. Consistently with its international primacy, it demonstrated its potential for innovation in that new field of policy, thus stimulating, in turn, the distinguishing emulative effect that great powers are able to set in motion with regard to their own innovations. The US:

... was one of the first leading industrialized nations to develop comprehensive environmental legislation and regulatory institutions. [...] Much of this state activity was underpinned by the world's most dynamic environmental movement, which came into existence in the mid-1960s. US environmental groups ranging from the more traditional bodies [...] to modern environmental nongovernmental organizations [...] worked to create broadly based domestic support for a more ambitious environmental policy at home and abroad. US scientists and activists came to play a leading role in the global environmental movement that began to emerge in the 1970s. (Falkner 2005: 590)

Over time, however, the US shifted towards a reduced commitment in the environmentalist field, acting sometimes even as a veto power. In spite of that, green initiatives were in the meantime being emulated by other countries. Many cultural meanings of development, questioning the often unsustainable western equation between it and economic growth, had begun to circulate outside few, narrow and heterodox strands of literature and social movements. Green sensibilities flowed into the official discourse proposed via science, media, politics and business; at the same time, an environmentalist awareness grew among middle and especially upper classes all over the world. Among the core agents of change – whose common trait is the global reach – a critical role for the aim of a *new sustainability paradigm* might be assigned to the 'wide public awareness of the need for change and the spread of values that underscore quality of life, human solidarity and environmental sustainability' (Global Scenario Group 2002: X)

At the end of the day, in spite of many dramatic changes experienced by the global structure of power and of the rise of new intellectual trends, the development discourse had found a way to keep afloat by co-opting in its rhetoric pre-existing environmental concerns: 'Green thinking about sustainability, a radical position 15 or so years ago, has long been institutionalized as 'sustainable development''' (Pieterse 1998: 350). Public environmental awareness, defined by Levy (1997) as a *challenge to hegemony*, has thus been co-opted into the hegemonic discourse itself.

4.4.1 The Institutional Discovery of Environmentalism

In 1968, the UN General Assembly launched the project of the Human Environment Conference. Under the leadership of Maurice Strong, it was held 4 years later, representing the first international acknowledgement of the need to address environmental problems – mainly, pollution and acid rains. Kanie and Haas, thus, link the date of the Stockholm conference to the beginning of the 'institutionalization of international environmental policy-making', whose narrow focus was, at that time, 'on the conservation and management of natural resources' (Kanie and Haas 2004: 1).

The same year, the newly established Group of Rome was laying the foundations of a holistic understanding of the links between phenomena such as industrial activities, natural resources deterioration and environmental exploitation. One of the core findings of their work regards the clear acknowledgement of earth's limitedness. Thereafter, 'their 'limits to growth' arguments were successfully used, on occasions, to challenge the dominant Enlightenment ideal of progress, which could only ultimately be sustained by pursuing industrial and technological growth wherever and whenever, at all costs' (Doyle 1998: 772).[21]

However, the political and economic international shocks of the 1970s opened a window of opportunity for once isolated environmental warnings to reach the general public, the broader scientific community as well as the more open-minded figures of politics and business. It was during this hard decade that 'proto-sustainability gained real social momentum via populist Green movements in America and Europe when global catastrophe seemed to be imminent' (Petrucci 2002: 104).

The 1975 *Dag Hammarskjöld Report on Development and International Cooperation* seems to represent one of the more challenging documents of the decade. Perhaps this is the reason why it has been left mainly unmentioned; on the contrary, it deserves more than a brief mention here.

What now was prepared for a Special Session of the UN General Assembly and wishes for *another development* – a need-oriented, endogenous, self-reliant and environment-friendly development, that is, a qualitative one – to overcome the crisis of contemporary development, whose little successes had been achieved only with regard to 'the privileged minorities who remain in most parts of the Third World [...]. For them the 'gap' has been bridged' (*Dag* Hammarskjöld *Report* 1975: 37).

The report, thus, takes its cue from the recognition of a critical situation, to be looked at as a whole made up of 'a few dominating countries and the majority of dominated countries', tied up by unfair, exploitative economic links (ibid.: 5).[22] Analysing the potential for structural transformation, it clearly states that the most critical point does not relate to resources' limits, but to their asymmetrical and unjust distribution, an obvious but too often downplayed outcome of economic, political and cultural power differences at the international level.[23] Given the 'diversity of starting points' (ibid.: 35) among the nations, the idea of a one-style-fits-all model is rejected; rather, 'the plurality of roads to development answers to

---

[21] For an extensive treatment of the debate raised by The Limits to Growth (and then by Our Common Future), see Chap. 4 by Perez-Carmona, this volume.

[22] 'The crises are the result of a system of exploitation which profits a power structure based largely in the industrialized world, although not without annexes in the Third World: ruling 'élites' of most countries are both accomplices and rivals at the same time' (*Dag* Hammarskjöld *Report* 1975: 5).

[23] 'Sometimes, transgression of the limits results directly from a system of un equal economic relations: peasants deprived of accessed to fertile soils monopolized by large land-owners or by foreign companies have no other resource but the cultivation of marginal zones, contributing to erosion, deforestation and soil exhaustion, while consumption by the rich, modelled on that of the industrialized societies, adds the pollution of wealth to that of misery. An unequal distribution of wealth threatens the outer limits from both sides at once' (*Dag* Hammarskjöld *Report* 1975: 37).

the specificity of cultural or natural situations' (ibid.: 7), which should be opportunely enhanced through processes of multilevel democratisation and decentralisation.

Arguing for 'radical changes in development policies and in international relations' (ibid.: 105), this report came up with policy proposals that would have produced, besides a number of green side-effects, an increased government involvement in the production and management of goods. As for the most affluent regions of the world, the pillars of the sustainable transformation envisioned by the authors refer to ceilings on, and price control of, meat and oil consumption; rationalisation of living units to be built as greenhouses; a less consumerist approach to consumer goods and the selling, on a non-profit basis, of high quality basic commodities; the abolition of private cars, to be replaced with public transportation in city centres and motor-cars rented by public owned companies for long drives. With regard to Third World countries,

> At the socio-economic level the reform implies ownership or control by the producers [...] of the means of production [...]. Commercial and financial structures must equally be changed in such a manner as to prevent the appropriation of the economic surplus by a minority. At the political level, the reform of structures means the democratization of power. [...] This is only possible through a thoroughgoing decentralization [...]. In other words, each local community should be able, on the basis of self-reliance and eco-development, to manage its own affairs and to enter into relations on equal footing with others. (ibid.: 38–39)

Quite the opposite, a first glance at 1980s economic theories prevailing among the main international institutions traditionally in charge of the delivering and administration of magic recipes to developing countries, would let to conclude that that was the decade of structural adjustment.

However, going beyond that still economistic understanding of development drawn upon by agencies such as the World Bank and especially the IMF, we notice that something was changing in the consciences of the more enlightened sections of science, politics, media and business.

In 1986, after the Chernobyl disaster had shocked the world, *Our Common Future* was published, stemming from the work of the World Commission on Environment and Development chaired by Gro Harlem Brundtland. The report focused on the link between social and economic development, on the one hand, and human environment and natural resources, on the other – thus building on the suggestions and findings of the 1972 Stockholm Conference. The Brundtland Commission questioned the old assumption that 'economic objectives, such as poverty alleviation and economic growth, should take precedence over environmental concerns', thus paving the way for current 'integrative and holistic management approaches' (Jabareen 2008: 185).

The Brundtland Report admitted the existence of natural limits; nonetheless, it also envisioned the chance to overcome them thanks to technical improvements and economic growth. Overall, northern lifestyle was not disputed: those affluent countries should pursue the target of a 3–4% economic growth, thus helping both the general economic activities worldwide and the recovery of poorer countries.

According to Rist, the Commission succeeded in outlining the imbalances menacing human beings. However, it missed topics such as mutual exchanges between societies and environment, and the cultural and historical dimensions of growth. This meant that the Brundtland Report was unable to come up with serious proposals for the solutions of said dilemmas (Rist 1996).

Almost 20-years on, an evaluation of the impact of Our Common Future states that critics were right in raising the problems of uneven power relations and, especially, of the 'fundamental contradictions between the renewed call for economic growth in developing countries and enhanced levels of ecological conservation' (Sneddon et al. 2006: 254). However, the non-mutual exclusion between economic growth and nature respect and preservation had been definitively legitimised at the international level thanks to one of the most mediatised event of the 1990s, the Earth Summit (Carruthers 2001; Bernstein 2002).

The 1992 United Nations Conference on Environment and Development welcomed almost 30,000 people, coming from national governments, NGOs, and the business sector.[24] On the UN side, besides the third generation rights and principles enumerated in the Rio *Declaration on Environment and Development*, the endorsement of the programme contained in the final text of *Agenda 21* must be mentioned. Instead of the strict separation between environmental, social and economic dimensions, 'it proposed integrated systems of management to ensure that environmental, social and economic factors are considered together in a framework for SD' (Jabareen 2008: 186). At the local, national and international level, the implementation of the programme for the century to come was supposed to strongly rely on initiatives and ameliorations achieved by science, technology, education and economy. Accordingly, Agenda 21 launched an innovative vision of transdisciplinary, multilevel governance for sustainable development, referred to as the *procedural* component of sustainable development by Kanie (2007).[25]

Summing up, by the end of the century, green concerns experienced a broadening of scale, from the local level to the global one (Levy 1997; Carruthers 2001). Sustainability 'has become the central adage of environmental policies around the

---

[24] Among the documents produced by representatives of the civil society gathering around the Global Forum at Rio de Janeiro, there was *Changing Course: A Global Business Perspective on Development*, a report written by the Business Council on Sustainability. It exemplifies, with its technocratic and mainstream understanding of sustainable development as a combination of economic growth and environmental preservation, the new interest towards environmentalism among multinational corporations. For a network analysis of the *corporate/policy interlocks* with regard to the WBCSD as one of the most influent transnational policy groups, (see Carrol and Carson 2003). From a more general point of view, an increased involvement of private firms as sustainability partners has been wished for in Daily and Walker (2000) and critically analysed in Levy (1997).

[25] Kanie notices the request for a 'broader participation in decision making. Sustainable development is no longer the pure domain of national sovereignty. Agenda 21 calls for multiple stakeholder participation, or 'major groups', at multiple levels of international discussions, including NGOs, scientists, business/industry, farmers, workers/trade unions, local authorities, as well as indigenous people, women, and youth and children' (Kanie 2007: 70).

globe, and the environmental discourse has been globalized and transcended national boundaries' (Jabareen 2008: 187), thus being subsumed by the exercise of global governance. Moreover, the echo of *Our Common Future* has gone down well with the specialised inner circles of development, and started reaching western middle and especially upper classes consciences, thus affecting their sensitivities and belief systems. Indeed,

> Since the UN Summit 1992 in Rio de Janeiro the agenda of sustainable development is programmatically linked to the inclusive and consensus-orientated decision-making that gets people involved as actors rather than only as voters, and that gets sustainability thinking *mainstreamed* in parliaments, the private sector, and science and humanities. (Töpfer and Bachmann 2010: 58–9, emphasis added)

4.4.2 Mainstreaming Sustainability

> Accompanied by liberal democracy and free markets, sustainable development is now a pillar of contemporary universalism, embraced from industrialized north, to the less-developed south, to the post-communist east. (Carruthers 2001: 93)

The new century approach to development, with its joint interest in the material well-being of the poorer, the 'traditional' cultural systems of non-western people and the preservation of natural environment and resources, strongly requires a change of perspective. After the rigid disciplinary specialisation dominating exactly when the developmentalist story-line was set, today coping with development studies requires genuine but challenging exchanges and comparisons between as many fields of knowledge as possible. Besides this interdisciplinary enhancement on the scientific side, we can also notice a trandisciplinary shift with reference to the increasing institutionalisation of partnerships between science, politics, business and so on (Bunders et al. 2010). Furthermore, scholars and practitioners have begun to enrich their analysis with factors and variables once neglected such as non-material dimensions and non-state actors. Sustainable development is certainly one of these wicked problems requiring the empowerment of actors belonging to different circles and geographical levels as well as the promotion of different cultural settings and belief structures. A genuine governance towards sustainable transformation requires that local contexts have always to be allowed for, since any specific place has its own characteristics and 'what is thought of as 'sustainable' is often dependent on assumptions and values' (Töpfer and Bachmann 2010: 60).[26]

---

[26] Indeed, for example, during a 1996 conference joined by the Environment and Developing Areas Research Groups of the IBG 'speakers analysed how far researchers can collect information about environmental change and physical processes in a manner which allows researchers to be aware also of their own social and cultural settings'. Moreover, they underlined the need to reform the knowledge production processes in the environmental field in order to 'enable new agendas to emerge, that might support previously unrepresented groups' (Batterbury et al. 1997: 126).

As mentioned, Western developers have discovered a new interest in local traditions and cultures underpinning lifestyles whose environmental impact seems less dangerous than the western one. Besides the realms of politics and science, media too have engaged in the processes of knowledge production for sustainable development within knowledge democracies, thus affecting the following process of decision making. Finally, environmental concerns are being incorporated into the vision and the strategy of many private, profit seeking firms, too. To sum up, as noticed by Du Pisani, even before the 1970s economic downturn,

> Ecological disasters received much media publicity. Films, TV programmes and pop music popularized the idea of an imminent ecological crisis. Earth Day was celebrated for the first time in 1970. The Green Movement took off, the first environmental non-governmental organizations (ENGOs), Greenpeace and Friends of the Earth, were established, environmental groups became more outspoken, ecologism became an ideology of some importance and green political parties started making an impact. (Du Pisani 2006: 89)

Hence, our thesis is that the acceptance of the environmental issue among decision-makers has followed the outside initiative model diffused, according to Cobb, Ross and Ross (1976), into egalitarian contexts. Strongly felt, at the beginning, among a few sectors of the civil society – such as activists and grass-roots movements – collectively referred to as anti-systemic movement, environmentalism has entered the public and, finally, the formal agenda. 'The language of sustainability was once a discourse of resistance, fusing radical environmental consciousness with a critical rethinking of a failed development enterprise. It provoked challenging questions about scarcity and limits, affluence and poverty, global inequality, and the environmental viability of westernization'. It has passed, however, from opposition to orthodoxy, argues Carruthers (2001: 93).

Of course, the rise of the environmentalist issue at the top of the policy agenda mirrored the difficult circumstances of the times and the absence of good alternatives: 'there were few ideational competitors. Resource management bodies had traditionally been staffed by neoclassical economists and resource managers, who had been discredited by broadly publicized environmental disasters and the energy crisis of the 1970s as well as the limits to growth debate, [. . .] and attendant popular fears of widespread resource depletion' (Haas 2001: 11584).

Anyhow, as far as our knowledge democracies are concerned, we might maybe see the glass at least as half full compared to the post II World War times. It is true that 'poorer and more peripheral societies are less able to bring their cultural models to the world-cultural table, but many participants in the global arena from richer societies have become strong advocates of the poor and peripheral' (Boli and Lechner 2001: 6264). Indeed, transnational elites involved into the development business are more aware than before of the need to take into account different development paths and to discard the previous dominating focus on material and economic factors. 'New formulations – grassroots development, pro-peasant development, eco-development, bottom-up development, people-centered development, and so forth – opened up myriad paths in the quest to conceive an alternative, ecologically sustainable, socially-just development trajectory for the South' (Carruthers 2001: 96). Development discourses have also incorporated a noticeable concern for matters of inter-generational equity and justice. As synthesised within the three Ps strategy, calling for economic, social and environmental responsibilities, development processes and transformations must pursue the joint goals of Profit, People and Planet, with a careful evaluation of long-term outcomes produced by policy implementation. Consistently, 1990s scientific literature 'presented evidence to show how environmental problems in developing countries are not the result of short-term impacts of rising population or economic growth, but instead the result of complex long-term human-environment interactions' (Batterbury et al. 1997: 127).

Global governance for a more environmentally friendly management of the twentieth century changes at the economic, political and social levels seems to have more chance of success than in the previous epochs of great transformations. The broad goal of sustainability has been adopted worldwide, thus facilitating the embracing of green policy alternatives that would have found many vetoes only a couple of decades ago. Haas can thus evoke a 'consensual wisdom within the international community of environmental policy analysts'. They share, indeed, a simultaneous concern for environmental degradation, economic growth and the material gap between the richest and the poorest segments of world population. The *new policy doctrine* associated to sustainability 'argues that most social ills are nondecomposable, and that environmental degradation cannot be addressed without confronting the human activities that give rise to it. Thus sustainable development dramatically expanded the international agenda by arguing that these issues needed to be simultaneously addressed, and that policies should seek to focus on the interactive effects between them' (Haas 2004: 570). This picture, however, also has a negative side.

> There is also the view that mainstream notions of sustainable development co-opt rather than challenge, for example, neo-liberal economic hegemony because it shares a similar foundational premise as hegemonic development approaches in that it still prioritizes capital accumulation, for example, concepts like growth and efficiency remain part of the sustainable development discourse. [...] Mainstream notions of sustainable development fall within the narrow confines of modernization theories of development which prioritizes an image and vision of development scripted in the tenets of Western technological civilization that is often promoted as the 'universal' and the 'obvious'. What it does is to legitimize so-called modern Western values and to delegitimize alternative value systems thereby constructing a global cultural asymmetry between the 'West' and the 'Rest.' (Nurse 2006: 35)[27]

And what is more, even after sustainability gained its current status of buzzword in the 1990s, 'the Northern way of life – with all its internal contradictions and stresses – remained on a non-negotiable track' (Petrucci 2002: 105). On the other hand, most of the local roads to sustainability sometimes compete with western theoretical and empirical understanding of (sustainable) development itself. For instance, local communities whose lifestyles are at odds with the tenets of individualism, hierarchy and commodification are often the real inhabitants of

---

[27] For a radicalisation of the opposition between the *West* and the *Rest*, (see Huntington 1997).

The environmental and distributive dilemma raised by reformist environmentalism willing to green the economy and not the whole societal organisation is not accidental. Quite the opposite, this state of affairs is a direct consequence of the diffusion of environmental awareness of the last decades.

Greening the developmentalist discourse has meant, first of all, a public attention towards environmental damages. However, as we have seen, it has done so by incorporating into that approach concerns, originally expressed by anti-systemic movements, that do not go as far as to challenge the common view of the growth imperative. Jabareen, using a conceptual analysis methodology, finds that linking the notion of sustainability to that of development has meant a change of focus from environment to capitalist economy. This *ethical paradox* implies that sustainable development 'is accordingly deemed able to cope with the ecological crisis without affecting the existing economic relationships of power. Capitalism and ecology are no longer contradictory when brought together under the banner of SD' (Jabareen 2008: 181–2).

The so-called *alternative soul* of the post development approach to global inequalities has thus lost its more critical features while being incorporated within a unique, mainstream approach to development. The old opposition between alternative and mainstream development has been replaced by a weaker opposition within the mainstream itself. Since the 1990s, 'several features of alternative development – the commitment to participation, sustainability, equity – are being shared (and unevenly practiced), not merely in the world of NGOs but from UN agencies all the way to the World Bank' (Pieterse 1998: 370). The continuum within current development discourse, then, runs from the human and social approach to the recipe of structural adjustment. 'Institutionally this rift runs between the UN agencies and the IMF, with the World Bank – precariously – straddled somewhere in the middle' (ibid.: 360).

In order to tackle the environmental degradation, our current approach to policy-making requires, instead, a more radical 'approach to governance – a paradigm shift in the way that governance is carried out and decisions are made and implemented' (Töpfer 2004: 2). In other words, there is still a long way to go if we want to adjust current cultural understanding and technical practice of governance to the goal of sustainable transformation.

4.5   Summary

In the aftermath of the Second World War, within the international framework of bipolarism, the goal of development as the main strategy to be pursued by new independent states was set. In fact, at the beginning of the twenty-first century, this aim has definitively been transformed into the ambition of Sustainable Development, a more encompassing process of societal transformation to be experienced by every human community in the new framework of a multipolar and chaotic international setting.

What has happened over the past half century? This chapter has attempted to explore the process of mainstreaming sustainable development, critically examining its roots and thus looking at its progenitor, the developmentalist approach. Situating that discourse in the broader picture of power changes at the international level, we appreciate how we have moved from a very restricted understanding of development to a multidimensional, qualitative concept.

The original developmentalist programme has been analysed with special regard to the modernization school, a twentieth century American version of evolutionism, which has underpinned this discourse both in academic theories and empirical practices in a specific international environment, the Cold War.

Currently, we tend to describe, in spite of the post 1989 claims of enduring unipolarity, the current structure of power as a multipolar one, especially from the economic point of view. However, this does not mean that we are about to see the rise of a new hegemonic country able to produce an agreed scientific worldview and to shape popular sets of ideas and beliefs about how the world should work and actually works. In other words, economic multipolarism is not matched by the rise of a powerful nation-state whose scientific mainstream in the development sector has been universalised as the dominant paradigm and whose lifestyle, broadly understood, has stimulated consensual emulation abroad. Soft power, quite the opposite, appears much more fragmented than in the past.

Today, recognising the main decision-makers is a hard task, consistently with the increasing overlapping of both institutional and informal power *loci* at the international level. In many issue areas, such as those related to environment, theoretical works on, and practical exercise of, global governance are stimulating a review of the old assumption of methodological nationalism. Besides the flourishing of sub- and supra-national government levels, many other actors from any geographical scale are engaged in the current process of laying the foundations for the governance of issues perceived as global, even in spite of their possible local origins. The increasing role of organisations, agencies and institutions, far from embodying a simple American worldview, reflects both the multipolar character of the international structure and the opportunity to replace a hegemonic understanding of social change with a new attention to non-state actors.

This chapter has thus stressed specifically the role of non-state actors, able to participate in processes such as selection of social problems and agenda-building, decision-making, policy implementation and evaluation. They belong to sectors such as politics, scientific research, media, and private business. They also impact the system of knowledge, ideas and beliefs in many issue areas of global governance thanks to the peculiar features of knowledge democracies. As the fall of a dominant mainstream has stimulated the rise of smaller, less powerful and yet influential storylines, it is our conviction that the analysis of governance towards processes of societal transformations must be enriched by paying greater attention to those non-state actors.

Specifically, we have worked on the role of epistemic communities, stressing their increasingly trans-national reach, their ability to influence decision-making as well as their non-neutrality with regard to topics such as power and geopolitical strategy to see them, sooner or later, finally taken into account by the public as well as by decision-makers.

Acknowledgements Apart from the whole project team, I would like to mention the friendly, challenging and suggestive help I received from the Institute for Advanced Sustainability Studies staff and, in particular, from Prof. Klaus Töpfer, Judith Enders, Alexander Perez Carmona and Moritz Remig. This chapter, indeed, has been mostly written during an exciting research experience at the IASS, which has represented a stimulating working environment to carry out the TransGov Project. I would also like to thank Dr. Philipp Lepenies, who made many highly valuable comments on the chapter, as well as the scholars who have had a considerable role in my academic training, Prof. Filippo Andreatta and Prof. Mauro Di Meglio. Moreover, I am grateful to Giulio, Diana, Martha, Camilla; to Filippo; and finally to Viola, born while I was approaching the conclusive steps of my inquiry. This chapter is definitely dedicated to her. Obviously, I remain the only one to blame for all inaccuracies and errors.

References

Appadurai A (1994) Disjuncture and difference in the global cultural economy. In: Featherstone M (ed) Global culture: nationalism, globalization and modernity. Sage, London, pp 295–310
Aron R (2003) Peace and war: a theory of international relations. Transaction, London
World Bank (2010) http://bit.ly/NOmmY. Accessed Nov 2010
Batterbury S, Forsyth T, Thompson K (1997) Environmental transformations in developing countries: hybrid research and democratic policy. Geogr J 163(2):126–132
Benedict K (2001) Global governance. In: Smelser N, Baltes P (eds) International encyclopedia of the social and behavioral sciences. Elsevier, Oxford, pp 6232–6237
Bernstein S (2002) Liberal environmentalism and global environmental governance. Global Environ Polit 2(3):1–16
Blyth M (1997) Review: "any more bright ideas?" The ideational turn of comparative political economy. Comp Polit 29(2):229–250
Boillot J (2006) L'économie de l'Inde. La Découverte, Paris
Boli J, Lechner FJ (2001) Globalization and world culture. In: Smelser N, Baltes P (eds) International encyclopedia of the social and behavioral sciences. Elsevier, Oxford, pp 6261–6266
Börzel T (1997) What's so special about policy networks? An exploration of the concept and its usefulness in studying European governance. European integration online papers. http://bit.ly/s4SLD9. Accessed June 2011
Brzezinski Z (2004) The grand chessboard: American primacy and its geostrategic imperatives. Basic Books, New York
Buchanan A, Keohane R (2006) The legitimacy of global governance institutions. Ethics Int Aff 20(4):405–437
Bunders JFG, Broerse JEW, Keil F, Pohl C, Scholz RW, Zweekhorst MBM (2010) How can transdisciplinary research contribute to knowledge democracy? In: In 't Veld RJ (ed) Knowledge democracy: consequences for science, politics, and media. Springer, Heidelberg, pp 125–150
Calleo D (1987) Beyond American hegemony. Basic Books, New York
Campbell JL (2002) Ideas, politics and public policy. Ann Rev Sociol 28:21–38
Capano G, Giuliani M (2005) Dizionario di politiche pubbliche. Carocci, Roma

Carroll W, Carson C (2003) The network of global corporations and elite policy groups: a structure for transnational capitalist class formation? Global Netw 3(1):29–57

Carruthers D (2001) From opposition to orthodoxy: the remaking of sustainable development. J Third World Stud XVIII(2):93–112

Cerny P (2001) From "iron triangles" to "golden pentagles"? Globalizing the policy process. Global Governance 7(4):397–410

Cobb R, Ross J, Ross MH (1976) Agenda building as a comparative political process. Am Polit Sci Rev 70(1):126–138

Coleman WD (2001) Policy networks. In: Smelser N, Baltes P (eds) International encyclopedia of the social and behavioral sciences. Elsevier, Oxford, pp 11608–11613

Cox R (1983) Gramsci, hegemony and international relations: an essay in method. Millennium J Int Stud 12(2):162–175

Daily G, Walker B (2000) Seeking the great transformation. Nature 403:243–245

Daly HE (2010) Sustainable growth: an impossibility theorem. In: Dawson J, Jackson R, Norberg-Hodge H (eds) Gaian economics. Permanent Publications, Hampshire, pp 11–16

Di Meglio M (1997) Lo sviluppo senza fondamenti. Asterios, Trieste

Dingwerth K, Pattberg P (2007) How global and why governance? In: Paper Presented at the sixth Pan-European international relations conference. Turin, 12–15 Sept, pp 1–8

Doyle T (1998) Sustainable development and agenda 21: the secular bible of global free markets and pluralist democracy. Third World Q 19(4):771–786

Du Pisani J (2006) Sustainable development – historical roots of the concept. Environ Sci 3(2):83–96

Easterly W (2007) The ideology of development. Foreign Pol 161:31–35

Epstein C (2004) Knowledge and power in global environmental activism. Int J Peace Stud 10 (1):47–67

Escobar A (1995) Encountering development: the making and unmaking of the third world. Princeton University Press, Princeton

Etzioni-Halevy E (2001) Elites. Sociological aspects. In: Smelser N, Baltes P (eds) International encyclopedia of the social and behavioral sciences. Elsevier, Oxford, pp 4420–4424

Falkner R (2005) American hegemony and the global environment. Int Stud Rev 7:585–599

Featherstone M (1994) Global culture: an introduction. In: Featherstone M (ed) Global culture: nationalism, globalization and modernity. Sage, London, pp 1–14

Featherstone M (2006) Genealogies of the global. Theory Cult Soc 23(2–3):387–419

Featherstone M, Venn C (2006) Problematizing global knowledge and the new encyclopaedia project. Theory Cult Soc 23(2–3):1–20

Fontana B (2006) State and society: the concept of hegemony in gramsci. In: Haugaard M, Lentner H (eds) Hegemony and power. Lexington Books, Lanham, pp 23–44

Foot R (2006) Chinese strategies in a us-hegemonic global order: accommodating and hedging. Int Aff 82(1):77–94

Forum T (1994) Hegemony and social change. Mershon Int Stud Rev 38:361–376

Fougère M, Moulettes A (2006) Development and modernity in hofstede´s culture's consequences: a postcolonial reading. Lund institute of economic research: working paper series, 2, pp 1–20

Frazier JG (1997) Sustainable development: modern elixir or sack dress? Environ Conserv 24(2):182–193

Friedrichs J (2002) Global governance as the hegemonic project of liberal global civil society. Paper presented at the CPOGG Workshop. Amerang, 1–3 Nov, pp 1–22

Fukuyama F (1992) The end of history and the last man. Avon Books, New York

Fukuyama F (2001) Culture and economic development: cultural concerns. In: Smelser N, Baltes P (eds) International encyclopedia of the social and behavioral sciences. Elsevier, Oxford, pp 3130–3134

Global Scenario Group (2002) Great transition: the promise and lure of the times ahead. Stockholm Environment Institute, Boston

Golub P (2004) Imperial politics, imperial will and the crisis of us hegemony. Rev Int Polit Econ 11(4):763–786

Goodin RE (2001) Politics of environmentalism. In: Smelser N, Baltes P (eds) International encyclopedia of the social and behavioral sciences. Elsevier, Oxford, pp 4685–4687

Graz JC (2003) How powerful are transnational elite clubs? The social myth of the world economic forum. New Polit Econ 8(3):321–340

Gruzinski S (1988) La colonisation de l'imaginaire: sociétés indigénes et occidentalisation dans le Mexique espagnol XVIe-XVIIIe siécle. Gallimard, Paris

Haas PM (1992) Introduction: epistemic communities and international policy coordination. Int Organ 46(1):1–35

Haas PM (2001) Policy knowledge: epistemic communities. In: Smelser N, Baltes P (eds) International encyclopedia of the social and behavioral sciences. Elsevier, Oxford, pp 11578–11586

Haas PM (2004) When does power listen to truth? A constructivist approach to the policy process. J Eur Public Pol 11(4):569–592

Dag Hammarskjöld Report (1975) What now. http://bit.ly/u4q7CJ. Accessed May 2011

Hannerz U (1994) Cosmopolitans and locals in world culture. In: Featherstone M (ed) Global culture: nationalism, globalization and modernity. Sage, London, pp 237–251

Hardin G (1968) The tragedy of the commons. Science 162:1243–1248

Harouel JL (2001) Culture, sociology of. In: Smelser N, Baltes P (eds) International encyclopedia of the social and behavioral sciences. Elsevier, Oxford, pp 3179–3184

Harvey D (2005) A brief history of neoliberalism. Oxford University Press, Oxford

Hilgartner S, Bosk C (1988) The rise and the fall of social problems: a public arena model. Am J Sociol 94(1):53–78

Hofstede G (1984) The cultural relativity of the quality of life concept. Acad Manage Rev 9:389–398

Hofstede G (1995) Multilevel research of human systems: flowers, bouquets and gardens. Hum Syst Manage 14:207–217

Hulme M (2010) Problems with making and governing global kinds of knowledge. Glob Environ Chang 20:558–564

Huntington S (1997) The clash of civilizations and the remaking of world order. Touchstone, New York

Huntington S (1999) The lonely superpower. Foreign Aff 78(2):35–49

Hurrel A (2006) Hegemony, liberalism and global order: what space for would-be great powers. Int Aff 82(1):1–19

Ikenberry J (1992) A world economy restored. Int Organ 46(1):289–321

IMF (2009) http://bit.ly/uNrXKi. Accessed Nov 2010

In 't Veld RJ (2010) Knowledge democracy: consequences for science, politics, and media. Springer, Heidelberg

Jabareen Y (2008) A new conceptual framework for sustainable development. Environ Dev Sustain 10:179–192

Jakimow T (2008) Answering the critics: the potential and limitations of the knowledge agenda as a practical response to post-development critiques. Prog Dev Stud 8(4):311–323

Jervis R (2003) The compulsive empire. Foreign Pol 137:82–87

Jordan G (1990) Sub-governments, policy communities and networks. J Theory Polit 2(3):319–338

Kanie N (2007) Governance with multilateral environmental agreements: a healthy or ill-equipped fragmentation? In: Swart L, Perry E (eds) Global environmental governance: perspectives on the current debate. Center for UN Reform Education, New York, pp 67–86

Kanie N, Haas P (2004) Introduction. In: Kanie N, Haas P (eds) Emerging forces in environmental governance. United Nations University Press, Tokyo, pp 1–12

Kennedy P (1987) The rise and fall of the great powers. Randam House, New York

Keohane R (1984) After hegemony. Princeton University Press, Princeton

Kupchan C (2002) The end of the American era. Knopf, New York

Lentner HH (2006) Hegemony and power in international politics. In: Haugaard M, Lentner H (eds) Hegemony and power. Lexington Books, Lanham, pp 89–108

Lepenies P (2008) An inquiry into the roots of the modern concept of development. Contrib Hist Concepts 4:202–225

Levy D (1997) Environmental management as political sustainability. Organ Environ 2(10):126–147

Long Martello M, Jasanoff S (2004) Introduction: globalization and environmental governance. In: Jasanoff S, Long Martello M (eds) Earthly politics: local and global in environmental governance. MIT Press, Massachusetts, pp 1–30

Markovits A (2007) Uncouth nation. Princeton University Press, Princeton

Martinelli A (2004) La democrazia globale. EGEA, Milano

Martinelli A (2005) From world system to world society? J World-Syst Res XI(3):241–260

Mastanduno M (1997) Preserving the unipolar moment. Int Secur 21(4):49–88

McCarthy TA (2007) From modernism to messianism: liberal developmentalism and American exceptionalism. Constellations 14(1):3–30

McMichael P (2004) Development and social change: a global perspective. Pine Forge Press, Thousand Oaks

Meuleman L, In 't Veld RJ (2010) Sustainable development and the governance of long-term decisions. In: In 't Veld RJ (ed) Knowledge democracy: consequences for science, politics, and media. Springer, Heidelberg, pp 255–283

Modelski G (1999) From leadership to organization: the evolution of global politics. In: Bornschier V, Chase-Dunn C (eds) The future of global conflict. Sage, London, pp 11–39

Nabers D (2008) China, Japan and the quest for leadership in East Asia. GIGA Working Papers, 67, pp 1–30

Nurse K (2006) Culture as the fourth pillar of sustainable development. Prepared for Commonwealth Secretariat, Malborough Housee, Pall Mall, London

Nye J (1990) Bound to lead: the changing nature of American power. Basic Books, New York

Nye J, Keohane R (1971a) Transnational relations and world politics: an introduction. Int Organ 25(3):329–349

Nye J, Keohane R (1971b) Transnational relations and world politics: a conclusion. Int Organ 25(3):721–748

Odora Hoppers C (2000) The centre-periphery in knowledge production in the twenty-first century. Compare 30(3):283–291

Overbeek H (2004) Global governance, class, hegemony. Working Papers Political Science Vrije Universiteit Amsterdam, 1, pp 1–19

Parsi VE (2006) L'alleanza inevitabile. EGEA, Milano

Paterson RA (2001) Culture, production of. In: Smelser N, Baltes P (eds) International encyclopedia of the social and behavioral sciences. Elsevier, Oxford, pp 3167–3173

Pattberg P (2006) Global governance: reconstructing a contested social science concept. GARNET working paper, 4, pp 1–22

Petrucci M (2002) Sustainability – long view or long world? Soc Justice 29(1/2):103–115

Pieterse JN (1994) Globalization as hybridization. Int Sociol 9(2):161–184

Pieterse JN (1998) My paradigm or yours? Alternative development, post-development, reflexive development. Dev Change 29:343–373

Polanyi K (2001) The great transformation. Beacon, Boston

Pomian K (1979) The secular evolution of the concept of cycles. Review II(4):563–646

Posen B (2003) Command of the commons: The Military Foundation of U.S. hegemony. Int Secur 28(1):4–46

Rapkin D (1990) The contested concept of hegemonic leadership. In: Rapkin D (ed) World leadership and hegemony. Lynne Rienner, Boulder, pp 1–19

Rapley J (2008) End of development or age of development? Prog Dev Stud 8(2):177–182

Ribeiro GL (2002) Power, networks and ideology in the field of development. In: Fukuda-Parr S, Lopes C, Malik K (eds) Capacity for development. Earthscan Publications, London, pp 169–184

Part 2
Sustainability Governance:
Topical Themes

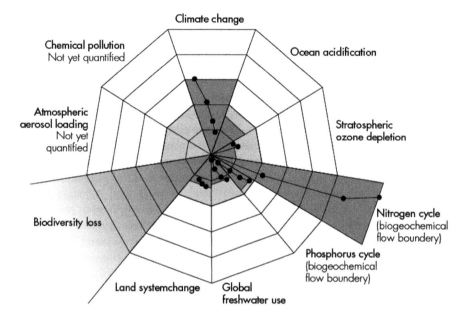

Fig. 5.1 Rockstrom et al. (2009a)

not sufficient to analyze and 'manage' these boundaries in isolation; they must be addressed in an integrative manner. Since the global freshwater availability is limited – a challenge that is severely aggravated by the fact that water resources are very unequally distributed globally – other sectors relying on freshwater such as food or biomass production have to take the 'freshwater boundary' into account. If an increased use of biomass for fuel production is one strategy to mitigate climate change, for example, the integrative nature of the boundary concept could present a useful tool for making decisions in an integrative way. That is, if strategies related to climate change have a negative impact on other boundaries such as freshwater and land, they should be applied with care and they have to be taken in full awareness of the *choices* to be made.

Furthermore, the actual proximity to a boundary may provide a rough indication of whether certain options are still at our disposal. If we are already well within the 'dangerous zone', i.e. beyond the point that should present a boundary for a given sub-system, human activities should not add further pressure to this area. For example, if the challenge of *food security* is continuously rising as currently anticipated (Ingram et al. 2010; Brown 2011: 175–191), necessary resources such as freshwater, land/soils or phosphorous may not be available at an equal scale for other services.

As a consequence of our emerging understanding of such interdependencies, the planetary boundaries concept is used to call for major governance and/or institutional reforms already within its initial or 'proof of concept' stage. In this respect, the (supposed) fragmented or (often) non-legally binding character of the global environmental governance system in place is often presented as a major weakness and is diagnosed as a 'patient' who needs to be cured (Hoff 2009; Walker et al. 2009; Steffen et al. 2011).

5.3 What Is (Not) Addressed

Because this article focuses on the governance implications of the planetary boundaries concept, it will not thoroughly discuss whether these nine candidates present the right set of boundaries, or if the boundaries in Fig. 5.1 are set correctly or arbitrarily. The issue of setting the boundaries at all, however, should be emphasised as an *act of governance* itself. This shifts the focus of analysis related to the planetary boundaries concept away from identifying the 'right' point where the boundaries have to be located towards governance concerns.

The correct identification of the individual boundaries may be a major reason for criticism within the academic community, but it may turn out that the fundamental challenge of very precisely setting the boundaries does not present a major problem of the concept. If we understand the planetary boundaries as 'boundary objects' – as introduced and elaborated upon by Stefan Jungcurt in this volume (Jungcurt 2012) – this may offer a different way of thinking about the concept. It would frame the nine boundaries as knowledge-claims about the earth system, which are both robust and flexible enough to meaningfully capture phenomena of a 'Planet under Pressure'.[7] In doing so, these boundaries would then not be necessarily described *exclusively* by scientific knowledge-claims. The authors of the planetary boundaries concept themselves call their proposals where the individual boundaries should be located in a trivialised manner 'a first guess' (Steffen et al. 2011: 5). This indicates that also the second and third attempt of specifying the boundaries will not result in very precise answers.

On the one hand, such considerations seem to run counter to the strong notion on 'solid science', which should underline the planetary boundaries concept. On the other hand, however, the nine boundaries identified are introduced as 'broad and vague concepts', which may be the only applicable way (in 't Veld 2012: 43–58). Furthermore, the authors of the planetary boundaries concept clearly acknowledge the following distinction. While system thresholds are 'absolute', i.e. set by the inner logic or functional conditions of the earth system, a 'boundary' is based on a 'normative judgement, determining a safe distance of how societies choose to deal with risk and uncertainty'. Boundaries are 'human-determined values of the control variable set at a 'safe' distance from a dangerous level' (Rockstrom et al. 2009a: 3).

---

[7] For 'Planet under Pressure' see Steffen et al. (2004). This is also the title of a major conference of the earth system science research community, London, 2012, which will, among other things, elaborate further on this concept: http://www.planetunderpressure2012.net/

decide and regulate who should export and who should import food.[15] On the other hand however, the concrete governance proposals made by these two authors in their concluding paragraphs are much less far reaching. They either refer in a general sense to Elinor Ostrom's concept of polycentricity, as many contributions to governance challenges do these days (Hulme 2010a; Underdal 2010; Steffen et al. 2011), or they suggest governance measures such as strengthening assessments or advisory councils, which are providing 'softer' or simply different functions than strict regulation.

One approach of conceptionalising various institutional functions was put forward by Young (1999, 2010) that is worth highlighting in this context. In his analysis, Young made clear that beyond 'classic' *regulatory* functions, institutions can perform *procedural* (e.g. providing a forum for negotiations and discussions on a regular basis), *generative* (e.g. reframing a problem such as the protection of nature towards sustainable use of ecosystem services) or *programmatic* functions (e.g. action programmes based on international agreed upon goals and targets).[16] The message from such a line of analysis for those who seek to establish new institutions based on the planetary boundaries concept is twofold.

First, to establish a legally-binding global regulatory regime or an international organisation with far-reaching authority to manage the 'myriad trade-offs' is neither politically easy to implement – again, nature will not care much about this – nor necessarily the best option (Meuleman 2012). Second, even if regularly arrangements for governing the planetary boundaries will not emerge in the near future, other governance functions should not be neglected. We may appreciate the *generative potential* of the concept that helps understanding the integrative nature of the earth system in the era of the 'Anthropocene', but we may be careful with calls for a global regulatory 'referee' who governs the boundaries in a top-down manner.

Regardless what the above quote of the authors of the concept means in the end, the proponents of the planetary boundaries conclude along similar lines. They present the following main functions that should be delivered by a governance arrangement which is informed by the concept of planetary boundaries: (a) early-warning systems, (b) dealing with uncertainties, (c) multi-level governance and (d) capacity to assimilate new information (Steffen et al. 2011: 5). Three out of these four functions clearly address the generative aspect of governance as just introduced. How could the concept be put in practice?

---

[15] Given that water is well within its boundary, according to the concept of planetary boundaries, a narrow-minded application of the concept could lead, hypothetically speaking, to the conclusion that the water community would (always) lose in making choices about the myriad trade-offs. For a political science analysis of these kinds of trade-offs see Zelli (2008) who presents a framework for understanding the battle among different issues in their quest for money, attention and definition power.

[16] See also Young (2008c), where Young states that governance research is still not well-equipped to understand the role globally agreed upon goals and targets play *as acts of governance*.

First, concerning its generative governance function, any of these boundaries is backed already by a more or less well organised scientific process in order to generate the information and knowledge needed for governing these issue areas sustainably. These research and assessment processes range from scientific and technical bodies as part of an established regime (Ozone) to intergovernmental panels or platforms for climate change (IPCC) and biodiversity (IPBES) to broader status reports such as the World Water Development Report (WWDR) to status reports on (agricultural) lands to more bottom-up driven scientific networks such as work on ocean acidification, the 'International Nitrogen Initiative'[17] or the 'Global Phosphorous Research Initiative'.[18] What a planetary boundaries perspective could contribute is the consideration of understanding different issue areas as an integrated 'system'. Such an integrative approach is already well-established within earth system science and it is partially and in an ad hoc manner already practiced in international environmental governance.[19] To think about a better institutionalisation of integrative assessment processes seems to be a candidate for a *still small but transformative change* in the context of Rio20+ and its focus on institutional reform for sustainable development governance. The planetary boundaries concept could help in this respect.

Second, how to govern the boundaries in relation to each other is not only an issue of creating integrative and transdisciplinary assessment processes. It also raises the question of how to set up multi-level and multi-sectoral governance arrangements within the spectrum of fully integrated approaches on the one hand and fully specialised approaches (one problem, one institution) on the other.[20] As the planetary boundaries concept points out, in addressing the challenges related to one boundary, the consequences of such action for other boundaries have to be taken into account. This has also been partially institutionalised within the UN system, for example, by inter-agency mechanism such as UN-Water or the Joint Liaison Group of the three Rio Conventions (Simon 2010).[21]

For example, the inter-agency mechanism UN-Water, for instance, was (re-) established in 2003 just after the Johannesburg Summit. It is worth highlighting that UN-Water could build on a relatively strong multi-sectoral approach in this policy

---

[17] http://initrogen.org/

[18] http://phosphorusfutures.net/

[19] One concerted political attempt was presented by the 'Bonn 2011 Conference The Water, Energy and Food Security Nexus. Solutions for the Green Economy' of the German Government, http://www.water-energy-food.org/de/

[20] The special issue on governance and resilience edited by Duit et al. (2010) dwells upon this as one of its major analytical problems to resolve. It is not surprising that the focus on resilience, a holistic concept itself, is both amble to capture and forced to 'solve' this classic dilemma of organization theory and practice (integration vs. specialization) as it is a litmus test for putting resilience into practice.

[21] The three Rio Conventions are the UNFCCC (United Nations Framework Convention on Climate Change), the UNCBD (United Nations Convention on Biological Diversity) and the UNCCD (United Nations Convention to Combat Desertification).

axis lists two options to frame the challenge, i.e. whether the concept will remain an environmental concept or can and should it be transformed into a governance approach relevant to sustainable development.

A first message is: Already today, every cell is covered, either by governance responses in reality or by proposals made in the past. In case the planetary boundaries concept is 'only' an environmental one, UNEP is the main focus of attention for reform proposals. Concerning assessment and monitoring of the state of the planetary boundaries, an upgraded Global Environmental Outlook (GEO) process seems to be the obvious candidate. GEO may have to be transformed marginally or substantially in the future, for example, to play at least partially a coordinative role vis-à-vis other assessment processes. The novelty would mainly be to develop a better integrative approach to the topics addressed by GEO and the process leading towards its results in order to understand planetary processes as a complex, coupled system.

Cross-sectoral coordination is the second important candidate for reform. Under a 'light' and obviously less challenging scenario, such coordination could be performed by the existing Environmental Management Group (EMG), an UNEP-led inter-agency mechanism of the UN system. In case the proposal for a UN Environment Organisation will be adapted, this coordination would get a stronger hierarchical notion and may be able to better integrate – most likely without fully merging – exciting MEAs. This could bring, among other things, some of the planetary boundaries already under one 'roof'. However, since many important processes would still be handled largely outside UNEP, such as land or water, an UNEO would also have to pay close attention to UN system-wide coordination.

It is time, 20 years after Rio, to re-consider the mandate and the functions provided by the UN Commission of Sustainable Development (UNCSD) as well – the often forgotten fourth institutional innovation of the Rio Summit (Beisheim et al. 2011). The fact that two of its 'policy sessions' collapsed in the past few years adds further reasons to this necessity for reform. UNCSD could be rather strong on the side of knowledge-production and facilitation of using this knowledge, for example in the form of its deliberations and negotiations of non-legally binding recommendations (Kaasa 2007). Since its reinvention of 2003, it has been a recurring pattern that the review sessions of UNCSD, which are not under pressure of policy sessions to produce a negotiated consensus outcome document, were perceived as constructive and even innovative learning platforms. They provided some space for 'outside the box thinking' and for thinking beyond the lowest common denominator (Kaasa 2007; Karlsson-Vinkhuyzen 2010; Schmidt 2011).

As an established *intergovernmental* process, and maybe in different incarnation as a Sustainable Development Council after Rio 2012,[22] such a body may be a surprisingly good candidate to take care of the assessment and monitoring-related

---

[22] Beisheim et al. (2011) present different formats for such a Sustainable Development Council that go well beyond the focus on how to institutionalise insights from the planetary boundaries concept.

governance functions identified for the planetary boundaries concept. This cell in Fig. 5.3 is thus highlighted, where one may find the best fit or form in the current system for the most promising functions provided by the planetary boundaries concept.

The fourth and final cell focuses on cross-sectoral coordination of the different domains of sustainable development. This could either be done as an inter-agency coordination, which would further highlight the role of the Chief Executive Board (CEB) within the UN system. As its coordinative character may not be sufficient, the proposed umbrella arrangement may be the right level of formalising this.[23] Again, along the lines of a 'both-and-thinking', an upgrade of UNEP would then not be an alternative but the other side of the same coin, i.e. a reform of UNEP and a reform of UN CSD/the establishment of a (new) umbrella arrangement. Both actors could pool resources and competences in governing the planetary boundaries. There is no need to perceive effective integration as a matter of fully merging existing structures and processes. Instead, taking into account a comprehensive picture and applying an integrative governance approach as required by the planetary boundaries concept may in the end only be achieved by actors who are connected to each other by being built on 'multiple engagements' in different institutional arrangements, as Roel in 't Veld (2012) has argued. Such a 'configuration' of a common agenda leaves ample room for diverse approaches.

5.7 The Costs of Inaction

As our world will reach more than nine billion people by the middle of this century, it may be pushed towards its limits and hard regulatory measures may be up for debate pretty soon. Inaction, i.e. if we do not achieve a transition towards sustainability, will aggravate this situation further. If one reflects on the current ability and willingness of key actors (nation state and non-nation state actors alike) to set up binding and effective regulatory arrangements at all levels of governance, it seems to be 'wise' to try our utmost to stay within what we currently perceive as 'our boundaries'. If we manage a transition towards sustainability, we may still have some room for 'maneuvering' and achieving desired outcomes even with less 'firm' governance responses. If the 'safe operating space' continues shrinking as we get closer to the individual boundaries, somewhat harder measures may be needed to remain stable within a 'limited terrain'. In this respect, this article has explored the question: can the planetary boundaries function as useful 'warning signs'? The answer is: *yes; but.* Klaus Töpfer continuously stressed during the research process of the TransGov project that keeping 'alternative pathways' open is key, if sustainable development for an open society should remain a meaningful concept. To cut

---

[23] The proposal of a World Environment and Development Organization as discussed by Bierman and Simonis (1998) is not discussed in detail here.

off alternatives for us and for future generations is the real price we may have to pay for our currently unsustainable practices.

Acknowledgements This paper benefitted from the reviews done by Professor Ernst U. von Weizsaecker and Dr. Louis Lebel, and comments by Professor Oran R. Young. The full responsibility for remaining inconsistencies and possible shortcomings stays with me as author.

References

Bachmann G (2012) Emergency response – clustering change. In: Meuleman L (ed) Transgovernance: advancing sustainability governance. Springer, Heidelberg, pp 301–326

Baumgartner T (2011) UN-water and its role in global water governance. Global water news. GWSP focus 2011: global water governance, pp 5–6

Beck U (2006) Reflexive governance: politics in the global risk society. In: Voß J-P, Bauknecht D, Kemp R (eds) Reflexive governance for sustainable development. Edward Elgar, Cheltenham, pp 31–56

Beck U (2010) Climate for change, or how to create a green modernity? Theory Cult Soc 27(2–3):254–266

Beisheim M, Lode B, Simon N (2011) Ein Rat für Nachhaltige Entwicklung. Vor der Rio-Konferenz 2012: Optionen zur Reform der VN-Nachhaltigkeitsinsitutionen, SWP-Aktuell 2011/A 46, Berlin

Berkhout F (2010) Reconstructing boundaries and reason in the climate debate. Glob Environ Chang 20(4):565–569

Berkhout F, Angel D, Wieczorek A (2009) Asian development pathways and sustainable socio-technical regimes. Technol Forecast Soc Chang 76(2):218–228

Biermann F (2010) Beyond the intergovernmental regime: recent trends in global carbon governance. Curr Opin Environ Sustain 2:284–288

Biermann F, Bauer S (eds) (2005) A world environment organization. Solution or threat for effective international environmental governance? Ashgate, Aldershot/Burlington

Biermann F, Simonis UE (1998) Eine Weltorganisation für Umwelt und Entwicklung. Funktionen, Chancen, Probleme SEF policy paper no. 9, Bonn

Breitmeier H, Young OR, Zürn M (2006) Analyzing international environmental regimes: from case study to database. MIT Press, Cambridge, MA

Brown LR (2011) Plan B 3.0. mobilization to save civilization. W.W. Norton, New York/London

Conca K (2006) Governing water: contentious transnational politics and global institution building. MIT Press, Cambridge, MA

Dessai S, Hulme M, Lempert R, Pielke R (2009) Climate prediction: a limit to adaptation? In: Adger N, Lorenzoni I, O'Brien K (eds) Adapting to climate change: thresholds, values, governance. MIT Press, Cambridge, MA, pp 64–78

Duit A, Galaz V, Eckerberg K, Ebbesson J (2010) Governance, complexity, and resilience. Glob Environ Chang 20(3):363–368

FAO (2011) The status of the world's land and water resources for food and agriculture: managing systems at risk. FAO, Rome

Folke C (2006) Resilience. The emergence of a perspective for social-ecological system analysis. Glob Environ Chang 16(3):253–267

Geden O (2010) Was kommt nach dem Zwei-Grad-Ziel? Die EU-Klimapolitik sollte für flexible Orientierungsmarken eintreten. SWP-Aktuell 2010/A 55, Berlin

Gehring T, Oberthür S (2008) Interplay: exploring institutional interaction. In: Young OR, King L, Schroeder H (eds) Institutions and environmental change: principle findings, applications, and research frontiers. MIT Press, Cambridge, MA, pp 187–224

Grin J, Rotmans J, Schot JW (2010) Transitions to sustainable development: Routledge studies in sustainability transitions. Routledge, New York

Gupta J (2008) Global change: analyzing scale and scaling in environmental governance. In: Young OR, King L, Schroeder H (eds) Institutions and environmental change: principle findings, applications, and research frontiers. MIT Press, Cambridge, MA, pp 225–276

Hoekstra A (2006) The global dimension of water governance: nine reasons for global arrangements in order to cope with local water problems. The value of water research report series by UNESCO-IHE, Delft

Hoff H (2009) Global water resources and their management. Curr Opin Environ Sustain 1(2):141–147

Hulme M (2010a) Problems with making and governing global kinds of knowledge. Glob Environ Chang 20(4):558–564

Hulme M (2010b) Cosmopolitan climates: hybridity, foresight and meaning. Theory Cult Soc 27(2–3):267–276

In 't Veld RJ (2012) Transgovernance: the quest for governance of sustainable development. IASS, Potsdam

Ingram J, Ericksen P, Liverman D (2010) Food security and global environmental change. Earthscan, London

Jaeger CC, Jaeger J (2010) Warum zwei Grad? Aus Polit Zeitges (32–33):8–15, Bonn

Jasanoff S (2010) A new climate for society. Theory Cult Soc 27(2–3):233–253

Jungcurt S (2012) Taking boundary work seriously: towards a systemic approach to the analysis of interactions between knowledge production and decision-making on sustainable development. In: Meuleman L (ed) Transgovernance: advancing sustainability governance. Springer, Heidelberg, pp 332–351

Kaasa SM (2007) The UN Commission on sustainable development: which mechanisms explain its accomplishments? Glob Environ Polit 7(3):107–129

Karlsson-Vinkhuyzen S (2010) The United Nations and global energy governance: past challenges, future choices. Glob Chang Peace Secur 22(2):175–195

Lenton TM, Held H, Kriegler E, Hall JW, Lucht W, Rahmsdorf S, Schellnhuber HJ (2007) Tipping elements in the Earth's climate system. Proc Natl Acad Sci 105:1786–1793

Meuleman L (2012) Cultural diversity and sustainability metagovernance. In: Meuleman L (ed) Transgovernance: advancing sustainability governance. Springer, Heidelberg, pp 60–118

Neumayer E (2001) How regime theory and the economic theory of international environmental cooperation can learn from each other. Glob Environ Polit 1:122–147

O'Brien K, St Clair AL, Kristoffersen B, O'Brien K, St Clair AL (2010) Towards a new science on climate change. In: Kristoffersen B (ed) Climate change, ethics and human security. Cambridge University Press, Cambridge, pp 215–227

Oberthür S, Stokke OS (2011) Managing institutional complexity. Regime interplay and global environmental change. MIT Press, Cambridge, MA

Pahl-Wostl C, Gupta J, Petry D (2008) Governance and the global water system: towards a theoretical exploration. Glob Gov 14:419–436

Pattberg P, Stripple J (2008) Beyond the public and private divide: remapping transnational climate governance in the 21st century. Int Environ Agreements Polit Law Econ 8(4):367–388

Rechkemmer A (2004) Postmodern global governance: The United Nations convention to combat desertification. Nomos, Baden-Baden

Rockström J, Steffen W, Noone K, Persson Å, Chapin S, Lambin E et al (2009a) Planetary boundaries. Exploring the safe operating space for humanity. Ecol Soc 14/2:32. http://bit.ly/6Z3Wmg. Accessed 20 Oct 2011. Also published as Feature in (2009) Nature 461:472–475

A disaster is often caused by nature (blizzard, hurricane, floodings, earthquake, tsunami, drought), but can also have human origins (e.g. chemical spills, nuclear incident, climate impact irregularities and climate engineering), and in particular the emergency implications of a disaster are increasingly determined by social and economic factors and the vulnerability of human settlements. The emergency itself is of unforeseen and often of a sudden nature. The extent to what the emergency has been unforseeable in substance and in process is subject to scientific and social debate, but is not substantial for the definition.

If successfully carried out, emergency response delivers 'hard' solutions while the 'sustainable development' delivers 'soft' aspiration and bearing points in a broader sense. Both are connected. This connection is neither merely circumstantial nor is it hard wired. It is rather a soft binding or coupling, making the connection by learning and accepting (or opening) new choices and new lines of responsibility through knowledge based informed debates. If however this reflexivity cannot be achieved, action might fail both regarding the aspect of the practical problem on the ground as well as the wider aspect of building societal values towards a sustainable future.

Sustainable Development is a societal aspiration which invites and involves people to share their values and to empower their abilities and competences for a better future. However, it can never be as concrete as the striking problems which it sets out to remedy. There is a two anchor process of change: the first is driven by the urgency of today's problems and those of the foreseeable future, whilst the other is enforced by the aspiration of people and their thinking with regards to fundamental issues of human life, nature and prosperity.

Emergencies have the potential to play a major role in change processes. This assumption sounds strange, at first glance.[1] Change processes often rely on design whilst emergencies do not. Emergencies happen haphazardly and at random, not by design. Change is perceived as a rather long term lineup of a multitude of steps whilst emergencies involve individuals. Change and transition are often perceived as a function of time. They assume at least some kind of stretched time line or even linearity. Emergencies on the other hand, with their emphasis on the 'now and here' are quite the opposite. In terms of governance issues, change is seen as something which one can manage (you can manage what you can measure, or: what you can measure gets done). In trying to define the rationality of change processes, development is often back cast when focus is on these characteristics. Emergencies, however, do not fit into this kind of rationale.

---

[1] Of course, any disaster, whether natural or human, certainly creates opportunities for change. The question always as to what is the power angle of change and who is benetting from it. The power aspect of disasters is very well discussed by Naomi Klein (2007: 558). The book elaborates on the assumption that the radical neoliberal free market policies, in some countries, were pushed through while people were scared by disasters or upheavals. The author implies that lobby groups may have intentionally created some of these man-made crises in order to push through unpopular economic reforms. I mention this book as a reference to the change-agent-character of emergencies although I see it oversimplifying the case. One may not too ready see conspiracies where all-too-human pattern of confusion and helplessness, good intentions and greed may as well give a sound explanation.

Whether or not they contribute to change processes depends on how societies react to emergencies. Any society perceives risks as the realisation of possible future emergencies which should be avoided at all costs. The way societies assess these risks is not a matter of physical patterns alone, nor are they the most important part. Technical and social factors are interwoven. Most important is whether a society sees alternatives – technical and social – to avoid or mitigate these risks. The existence of alternatives is fundamental for the scale and extent to which risks are perceived.

Modern emergencies mean that this relation becomes increasingly complex. Emergencies which result from risks and challenges in modern societies, are beyond those which can be pre-calculated. An example of this type of emergency would be a major nuclear accident. They are without pre-set geographic limits and their materiality may easily develop un-predictable (and most probably unmanageable) features. What is more, geographic distances do not translate into social distances; with the opposite more likely to be the case. All of this is demonstrated by the German case in dealing with the Fukushima meltdown.

Another feature adds to this perception of risks and emergencies. It comes with the era of sustainability as policy concept and refers to the character of information and values. The contemporary context seems to accept social values only if they imply a shared meaning (and risks are perceived as something which bear a collective meaning beyond the actual physical impact). Information, in this respect, is only then socially accepted as information if it is based on open sources. If this is not the case, information is simply data and its relevance to public decision-making is denied or under doubt (in 't Veld 2010). Successful emergency response is deemed to be based on open-source information.[2]

Amidst the wide array of environmental and social problems – both global and regional – emergencies call for immediate action and the ethics of help, remedy and facilitation of new thinking. They do not necessarily call for textbook solutions. Emergency response is designed to manage the unexpected, and if this does not happen, then it has failed. Occasionally, the urgency of the situation at hand requires taking action which may contradict the usual logic of procedures without waiting for decisions to be taken by the regular chain of command. Whether this works out or not, is subject to the situational intelligence of those in command. Whether decision routines, both in the public and private sector, are ready to perform transformative action, is characterised to no small extent by the after-event reaction of the governance routine: will it punish or learn? Will it punish those in charge for any mistakes or inappropriate behaviour which might have caused or influenced the situation? Or will it use mistakes in order to learn how to perform better in times of stress, and how to make better use of the knowledge

---

[2] In Germany, the Fukushima event caused meaningful political decision taking. The report Ethics Committee which was in the process of being established right after the initial nuclear accident in Japan, built its report on the facts elaborated here. Most important was the checking of scientific facts and figures which could only then be operational for building consensus when information is based on open sources. Non disclosed information will fail even if the data turns out to be adequate (Ethik-Kommission Sichere Energieversorgung 2011).

available for disaster control, forecast and precaution action? Will it develop
reflectivity or prolong any command-and-control mechanics which it had in place
before? Would the dominant reaction display a reflexive or a compressed attitude:
readiness to u-turn and change mode as opposed to the fortress attitude of simply
building higher walls when under attack? Will it emphasise openness and reflexivity in a time of knowledge democracy and second modernity, or will it devalue and
discourage change?

6.2 Differentiating Change, Transition and Transformation

The discourse on sustainable development seems to use the terms transition,
transformation and change more or less synonymously. A more differentiated
view however, might help to understand the different characteristics of events
and what transformative governance is (or could be) about.

From the above mentioned terms, change is the least specific. It is used to signal
that there is something happening, that we know about this, and that we should take
action to better understand what is going on. An example of this is climate change
or demographic change. Change can mean anything; this term does not denote
going in a direction which is better or worse. Sometimes, the term change agent is
used to characterise an attitude of people. An example of this would be in the
business sector or in a civil society which empowers other people to act in their
own, unrestricted way, but along the pre-set and collective lines.

Transition, I suggest, should be used for pre-defined processes which are
designed to lead from A to B. The access of countries to the European Union is a
good example. A full acceptance of the Aquis Communautaire, the rules and
regulations the European Union has built up, can only be accessed during a (from
country to country different) certain period of time. This time can be called
transition time. Transition is a regulatory administrative action, driven by targets
and timetables (tartim).

Transformation is a term we may speak of when point A is concretely known,
whereas the goal in point B cannot be described in the same concrete way.
Quantitative targets are used as orientation and benchmarks; the most high profile
objective being to keep the global climate change lower than an additional 2°C in
global mean temperature (WBGU 2011). There is also no clear final end or stage of
transformative action. Action is mostly driven by programs and measures (promes).
Another example is the current world's financial debt crisis combined with
destabilising characteristics of the global financial market which are out of control.
Added to these is the transformative process which the globally leading currencies
find themselves submitted to. Another example is the so called Green Economy.
This is part of the private sector which is deliberatively changing the business case
by taking up sustainability solutions, changing gear and performance, or even
developing completely new business models (World Business Council for Sustainable Development 2010). It is characterised by the fact that the green part of the
economy is increasing, but remains far too small to exert dominating power.

Conversely, the conventional part of the economy is still in command although does not seem to be in a position to deliver ways and means which could respond successfully to the crisis of a non sustainable globalised economy.

In other words, change is what we find ourselves in, more or less constantly, without being formally invited. A transition is something which is set up to technically manage a process in order to get from A to B. Transformation marks a way for society to reflexively monitor opportunities and urges in order to formulate decisions. Emergencies may be conceived as a lens which is clustering all of these processes, bringing about a repertoire of action that takes choices and has the openness to change the logic of action.

6.3 TINA Is Not a Friend

From a governance point of view, the most problematic notion is expressed by the acronym TINA. It means 'There-Is-No-Alternative'. The German debate on whether and how to phase out nuclear energy, in essence, is a debate on the choices the society must make in order to come up with alternative, safe energy supply structures (including the demand side, the grid, and most certainly including the informed debate of conflicts of interest regarding costs, climate emissions, dependencies, and so on).

Political decisions can hardly rely on textbook protocol-type solutions if being taken in times of crisis. Obviously, there is a certain urge to defend decisions by arguing that there is no alternative. The no-alternative narrative nourishes the popular expectation that 'things have to go on', and however disruptive an impact might have been, it will not disturb the way of life. This has recently been used in the course of the financial and economic crisis when decisions regarding bail-out options had to be taken (too-big-to-fail or too-big-to-save?). TINA pretends linear steadiness where there is substantial change. Thus, the notion of 'there is no alternative' reproduces its own precondition, defines everything else as not realistic, and accepts this makeup reality as a limitation on development. Most of the governance features which are in operation today are TINA-related. They are designed for permanence in a non-disruptive development. Even the specific governance elements recently called upon by political strategies towards sustainable development are characterised by linearity in this respect. These elements consist of management by quantitative objectives, verification by indicators, management rules, and involvement of stakeholders. They are characterised by an understanding of 'time' as a steady and linear resource. There are no provisions made for sudden and unexpected breaking up of social structures, or emergencies. While these governance features are both necessary and relevant in order to respond adequately to the systemic pressure of non sustainable trends, they are incomplete because they deny the existence of change clusters.

Besides the continued pressure from long term systemic patterns of non sustainable production and consumption or from the emission of green house gases, other sources of pressure might add to change clusters. In particular, environmental, nuclear, or financial emergencies are clustering change: all of a sudden they are erratic, discontinued, and forcefully sporadic, with turnover capacity. From the European industrial transformation period of the nineteenth century as well as the realm of the environmental agenda since 1972, we have seen many examples of accidents, unforeseen impacts, chemical spills, and contaminations. The images of burning landfills, oil on rivers with fish stock floating belly up, abandoned industrial sites, cut away rain forest, arable land turned into desert and soil loss, carry with them iconographic power. Equally powerful are the metaphors of silent spring, the ozone hole, or the extinction of species.

Major disasters and emergencies underline the fact that nothing is without an alternative. The 2011 nuclear accident in Fukushima reminds us that unexpected events and irregularities can happen even if they are clearly beyond what a 'rational' risk calculation can predict. Nuclear accidents and meltdowns are peak events, but basically, they represent a number of emergencies which high-tech societies find themselves confronted with.

In order to sustain a planet with nine billion people who stress the environment and the natural resources, mankind must invent a number of alternatives. While the pursuit of the old myth of economic growth is deteriorating societies and the ecology, the alternative is not. The greening of society is a license to growth, provided growth strategies are informed and guided by the notion of sustainability. Learning from emergencies should enrich the governance debate on how to achieve sustainability.

6.4 Conventional Versus High End Emergencies

The first industrialisation has brought about many emergencies such as explosions of steam machines, railway disasters, mining catastrophes and technical dysfunctions of all sorts. The 'frontiers' (Osterhammel 2010) have been exploited as an unlimited reservoir of resources, both in the 'new world' and in the old world. The notion that the extent of danger and risk can be calculated and reduced to the minimum, is still true for commonplace accidents such as exploding steam machines or discharges of hazardous substances. During the industrialisation, these conventional types of accidents and emergencies led to an incremental improvement of technologies and to advancing liability schemes and concepts for insurance coverage. Unconventional 'high end' emergencies however, such as adverse effects of climate change and geoengineering, ocean degradation, or nuclear meltdowns, are impacting man and nature. There seem to be no immediate limits on these emergencies which have irreversible impacts and the potential to develop follow up impacts beyond control. These emergencies do not stop at borders, they are not linear, and they are beyond the scope of existing governance.[3]

---

[3] The contingency of emergencies is being neglected by mainstream research into governance. It should be noted that one won't find the term 'emergency' on http://bit.ly/rUTSsY

Table 6.1 Types of emergencies

Nineteenth and twentieth century style emergencies	High end emergencies
R-2-C (Ready to Control), accessible for insurance	S-b-L (Systemic beyond Limits), not pre-calculable
Triggering selective learning (if not wasted to oblivion)	Clustering collective change processes (if not wasted to oblivion)

However, a crucial but absent measure is an analysis of how irregularities as expressed by emergencies relate to the governance of change and transformation. An example of this would be during the time of the first and second industrialisation (Brüggemeier 1996).

Today, we are faced with the increasing probability of environmental emergencies. However, these are different to the emergencies which we have come to expect. The industrial style emergencies of the nineteenth and twentieth centuries were ready-to-control, in the sense that the impact could be contained and controlled. In the following years machinery was improved and procedures adapted.

While conventional emergencies still occur in contemporary times, another type of emergency is characterised by non-controllability. Examples of such emergencies are the impacts of nuclear hazards, climate migration, or food disasters. In addition, the examples of major flooding show that the increased vulnerability of human settlements tends to develop natural disasters into social emergencies. It is debatable whether a financial mega crisis such as the financial meltdown of 2008/2009 can be categorised as a high end emergency. Ulrich Beck analyses the delimited and social character of risks in what he calls the second modernity (Beck 1986, 2009).

The definition of risk as a product of likelihood of occurrence and scale of damage is conventional. There are risks (and emergencies) which are beyond this definition, because they are highly complex, the magnitude of impact is beyond being calculable and/or they are on a global scale. This challenges the conventional concept of risk and the way these risks have been dealt with. Not knowing (in the sense of absence of positive information) as well as deliberate ignorance and denial are forming part of modern risks (Table 6.1).

The appearance of extended risks alone is an important change of course. Adding to this, a growing number of emergencies can be expected in the future stemming from the fact that more and more people live in areas which are subject to extremely vulnerable conditions. Another contributing factor is the deteriorating pressure on food and ecosystem services, a factor which is on the increase. Another reason comes with the embedded runaway risks of accidents in nuclear facilities. As demonstrated in Fukushima, a major nuclear dysfunction may cause response action which is far beyond that which is expected and possibly far beyond what is eco-nomically and ecologically maintainable.

It must be understood that not each and every environmental problem causes an emergency. In legal terms, a situation is called an emergency when it places man and nature at immediate harm or a risk, which is not tolerable. This impact is of such a dimension that it must be immediately addressed regardless of whether it is expectable, foreseeable or otherwise predicted or not predicted.

The term emergency is special because of its legal consequence. As it has thus far been applied, it legitimises and enforces 'illiberal' intervention. An emergency legitimises (and requires) governments to directly and immediately intervene. Emergency response will not worry about vested interests such as property rights or facility permits. There is always a momentum of urgency involved. A good role model is the Emergency Response Action carried out under the US Superfund legislation. There are also emergency response routines in Europe which are mostly carried out on the national level by police or army, and by specialised branches of the fire department.

The term emergency is often synonymously used with the concept of 'danger'. This is relevant in a broader sense. Some environmental damages stipulate a danger for humans; others do not (yet). Globally, the man-earth system and the biocapacity are under stress. Increased stress means that the occurrence of emergencies is more probable. Thus, deteriorated and restricted or even denied access to fresh water, fish stock, or food security, and a depletion of sources, may develop into regional emergencies. This assumption is based on knowledge. There is no point in 'crying wolf' or producing gloom and doom messages, but there is also no point in denying this trend.

6.5   Framing Future

The year 2050 is near. Those who will occupy your positions and assume your functions as leaders in sustainability are sitting in our schools and universities. The 40 years leading up to 2050 will mark 40 years of their lives as active members of society, of their business and family life, and of their life in social and local communities. By 2050, the world will look very different, with nine billion people living on it, all with high consumption standards. This world will be resource-constrained, carbon-constrained, and will exhibit profoundly changed geopolitics. It is abundantly clear that governance will be key. The more this is the case, the more the world will care about how to share the ever-increasing wealth of available scientific knowledge. Knowledge and democracy – along with accountability and transparency – are the building blocks for governance.

The European approaches (EU COM 2011) currently brought into play in preparing the 2012 UNCSD, reflect the political dynamics of the European project. For hundreds of years Europe was a byword for permanent war. The European Union, emanating from lessons learned, is a peace project. It is run on a machinery of hard and soft regulation, and builds administrative institutions in collectively-shared responsibility. Still, it is incomplete, and the project continually struggles with how to free up multilateral action and how to link national and European action. In a sense, the story of Europe can be seen as the story of how to integrate diverse views, habits, drivers and cultures.

Europe has learned that 'integration' does not work because good instruments are in place, but instead works on a 'must do' basis designed to achieve collective goals and objectives. The enlargement of the European Union and the specific processes of accession are a good case in point. The European carbon reduction objectives and the long-term goals of the European Union are further examples. They work through strong administrative arrangements, be them legal or communicative, enforced or implemented, voluntarily or on the basis of persuasive instruments. Having established this, it will likely come as no surprise to learn that the EU and Germany, of course, are strongly supportive of the idea of upgrading UNEP and promoting UNEP to the status of specialised agency. In a wider sense, it seems necessary to realign the performance of the UN system with the agenda of sustainability. An umbrella organisation approach seems to be reasonable. The underlying understanding of this position is one of integration and the role that organisation building can play in this respect.

In the UN system, an upgraded UNEP would have to serve as a core element in order to re-integrate environment. International environmental governance, after Rio 1992, seems to have been running in disintegration mode. Additional tasks have been implemented by adding new organisations to the existing ones. The re-unification of the environmental case resembles a piece of homework that is needed to reach out and improve the integration of the environment in the wider task of sustainability and into the Bretton Woods instruments.

- In this respect, the Green Economy poses particular challenges and opportunities. Environmental policies must and can deliver benchmarks and guidelines for roadmapping the green economy. Roadmaps are required and must be moulded into new governance instruments. They are needed in order to tackle upcoming agenda items such as the launch of a recycling exercise for those materials which today are not recycled at all (e.g. rare earth, industrial metals). Roadmaps provide an opportunity to design solutions beyond one-point-regulations.
- Best practice examples may create new ways of thinking and reach agreed objectives (EEAC 2011). Award schemes are best suited to provide a competitive level playing field that may serve for collective sharing of approaches avoiding window dressing from happening and delivering benchmarks for progress (Deutscher Nachhaltigkeitspreis 2008–2011). Peer Review processes may help to benchmark best practice approaches, and to prevent the green economy from developing into exclusive partiality.
- Enforcing capacity building (sustainability skills) and the involvement of the private sector. Business and civil society already play important roles in the transition process towards sustainable development. There are good examples for changing gear, developing new business models, and re-arranging the supply chains by taking sustainability criteria on board. It is for civil society and politics to draw the line and to make progress and success towards the green transition more tangible. Councils for Sustainable Development can make a difference, as demonstrated by the German example of awarding sustainability performance, ranking efforts, and the dialogue-style elaboration of a German Sustainability Code (Rat fürNachhaltigeEntwicklung 2011).

A more visioning governance debate should also cover the aspect of fiscal sustainability, an aspect which all-too-often is completely neglected. However, without any (near-stable) fiscal sustainability, virtually none of the remaining approaches to sustainability will ever come to fruition. By the same token, implementation of a green economy must prove that it will deliver innovation and decent jobs as well as qualitatively justifiable growth, and that it can alleviate poverty (Bachmann 2010).

Change rarely comes from within organisations. This, at least, is true of the concept of organisations as we have come to know them. In addition, it is undoubtedly true that the UN system has so far, not been able to mainstream the sustainability task.

For this reason, the governance debate should highlight the nexus between national and multilateral action. With regard to making the nexus between national and sub-national levels viable, and working towards promoting sustainable development action, National Sustainable Development Councils in European Member states have proven very meaningful and have enabled a broad set of different procedures in governance, and between the public and private sector. For this, multi stakeholder bodies are a good proxy. 'International Environmental Governance' and 'Institutional Framework on Sustainable Development' should encompass also the private sector. If the notion of Green Economy is to be taken for real the governance debate should reflect this. Judging from the German experience, the corporate community (the most advanced part of it, that is) already displays a number of approaches and governance features designed to mainstream 'Sustainable Development' into corporate performance and to distinguish those efforts from mere window-dressing.

In general, the governance debate may gain momentum when it begins to combine administrative, corporate governance, and the governance of social responsibility, and when it takes trajectories into account that are driven by emergency responses. With the concept of the green economy, this step seems compulsory.

6.6 Emergency and Emergency Response

6.6.1 A Knowledge Case

The notion of an environmental emergency associates a knowledge base with the legal right (and contingency) which allows for an enlarged set of interventions, a so called emergency response. The conceptual framework of environmental policies is deeply rooted in dealing with and learning from emergencies, although systematic descriptions of environmentalism tend to dismiss and replace event-enforced learnings by more theory grounded cases of environmental policies (Speth 2005; WBCSD 2010; Radkau 2011). Historically, emergency response action was one of the prime 'sources' of environmentalism, some of those emergencies have been of national significance and required extensive coordination among government agencies in order to prevent, prepare for, and respond to emergencies.

Mostly, responding to emergencies requires immediate action such as shutting down ongoing operations, on site access to facilities and (mostly) emissions discharges and any other law enforcement activities. Emergency response allows for the most effective action providing it is bound to the goal of taking control of an otherwise harmful situation. These actions may deliberately not be constrained by any previously issued emission right or discharge permission. In an extremely dangerous situation there must be no legal limit to site access.

What qualifies as an environmental issue, to be handled as an emergency, is subject to intense debate and profound scientific research. In the context of human toxicology for example, the research focuses on dangerous substances, hazardous substances, exposure pathways, dose–response systems, interpolation from test data and field evidence, and linkages with human disposition.

The notion of emergencies stipulates a burden of proof. It must be proven that a dangerous situation can lead to concrete harm for people or the environment. Furthermore, the dangerous situation may not be 'only' some kind of general (abstract) event which may or may not put people and the environment at unacceptable risk. There are different metrics being used to prove this, all of which are linked to high-end scientific measurement and verification of the evidence:

- Direct measurements (if direct measurement is ethically acceptable and technically possible, however this is seldom the case);
- Epidemiological proof (given the population at risk is large enough in number to calculate the statistics [and to separate a concrete add-on harm from what is perceived as circumstantial or normal risk given the way of life, the terms of operational security, or the ubiquitous background situation]);
- Extrapolation from appropriate field experience given that there is an analogy in the first place;
- Circumstantial evidence such as open burning of hazardous material, evidence of uncontrolled explosives, dead fish stock.

The use of a single metric may not be ruled out. However, it is safe to assume that a combination of these metrics is often used (Bhopal, Love Canal, cases of dioxin spills, children's blood lead levels associated with urban outdoor activities, major cases of groundwater pollution in the US and in Germany, the dangerous exposure to toxics in residential areas that have been built right on top of hazardous waste dumpsites). Combining metrics is a clear choice whenever uncertainties are great, predicted costs for remediation are high, and more people are directly affected (health, mortality).

6.6.2   The Metrics of Adverse Effects and Danger as Scientific Challenge

Environmental governance as expressed by, for example, international regulations and discourse on the International Environmental Governance, has not yet profoundly touched on the case of emergencies. Emergency response action is left to national governments. Following this subsidiary approach, the otherwise well designed research into climate forcing and environmental depletion is not yet linked to the work profile of emergency response and its operational engineering expertise where they exist.

- The most important indicator characterising the state of climate change is given as global mean temperature. Being highly aggregated, it cannot stipulate or even trigger any emergency response measure which responds to the regional impacts of dangerous climate change. This is not coherent.
- What is globally modeled and predicted (the global mean temperature, the global biocapacity, food supply) is not connected to emergencies clusters which require evidence, measurement, reporting, and verification.
- The term 'damage' as in damage thresholds, has been defined in many respects as public health policies[4] and environmental protection. The Commission on the Measurement of Economic Performance and Social Progress, the so-called Stiglitz-Sen-Fitoussi Commission, prominently supports this point (Stiglitz et al. 2009; Rat fürNachhaltigeEntwicklung, Geschäftsstelle 2010). The Commission believes that it is necessary to have indicators that designate thresholds which, when exceeded, give rise to concerns that harmful environmental damage will occur. In particular the Commission emphasises a need for a clear indicator pointing out dangerous levels of environmental damage.

Fundamental questions of knowledge, certainty and burden of proof arise in this context. They demand scientific research into the effects of environmental pollution as well as on the political and conceptual approach to evidence-based decisions.

The definition of a danger correlates to damage. Damage is not a given minor impairment, disruption or inconvenience, but is a serious and unacceptable impairment or burden which is currently happening or for which there is sufficient probability that it will occur. Sufficient probability does not in itself denote a certainty that damage will directly occur; then again, the mere abstract possibility of damage occurring does not warrant the fundamental assumption of danger. Instead, there must be a well-founded concern that the danger will materialise, for example by virtue of a dangerous situation arising if existing trends are allowed to continue unhampered.

---

[4] For example in assessing dangers to public health or human working conditions the critical end point is seen as being an adverse effect that can be traced back to the exposure to a specific contaminant. The damage threshold is largely uniformly defined at the international level. It is determined by impact-related body doses that indicate either no-effect levels, no-observed adverse-effect levels, the lowest observed adverse effect level or any other (barely) tolerable, reabsorbed doses of pollutants. In terms of their definition, methods for derivation and interpolation and the level of protection associated with one of the levels those reference levels are largely stipulated by the World Health Organisation (WHO) or other organisations such as the Environmental Protection Agency in the USA. With regard to carcinogenic effects, statistical probabilities of occurrence are generally considered to be the threshold values (Eikmann et al. 2010).

Warding off danger is linked to the idea of preventing emergencies from happening. This is the purpose of the precautionary principle.[5] As the guiding principle of international declarations such as the 1987 Montreal Protocol, the third North Sea Conference of 1990, as well as the Rio Earth Summit (UNCED) of 1992 the principle is also entrenched in the EU's legal basic documents and various European action programmes. The aim of the precautionary principle is to conserve natural resources and livelihoods in order to preserve their value, efficiency and functions in the long term. A key characteristic of precautionary measures is that often, neither the probability of occurrence nor the extent of damage is specifically known or quantifiable. The IPCC introduces the terminology 'robust findings' (IPCC 2007) and 'key uncertainties', in order to ascertain how much secured knowledge (certainty) is available on the impact of the damage. In other words, how reliably a detrimental impact threshold in a protected property is indicated (forecast).

6.6.3   The Case of Climate Emergency

In 2011 the Security Council finally issued the long debated statement on the possible security implications of climate change.[6] The Security Council notes that in matters relating to the maintenance of international peace and security under its consideration, conflict analysis and contextual information on, *inter alia*, possible security implications of climate change is important, when such issues are drivers of conflict, represent a challenge to the implementation of Council mandates or endanger the process of consolidation of peace. The Security Council expresses its

---

[5] In German environmental policy, this concept has been broached by introducing a soil protection law (*Bodenschutz-Recht*) (Cf. German Government 2000).

[6] At the 6587th meeting of the Security Council, held on 20 July 2011, in connection with the Council's consideration of impact of climate change under the item entitled 'Maintenance of international peace and security', Ambassador Peter Wittig as acting President of the Security Council made a statement on behalf of the Council on the substance of security implications of climate change. Beforehand, the Security Council debate had not reached a consensus. In April 2007, the British government initiated a Security Council debate on climate change as a security risk. If climate change was a threat to international security, intervention by the United Nations should be legitimate. While the British initiative was supported by the European Union and the majority of its Member States, the U.S., G77, China and Russia opposed it. Especially India expressed its worries fearing that acknowledging climate change as a security risk would open the floodgates for industrialised countries to circumvent the sovereignty principle of the UN Charta. In particular they opposed any attempt to widening the interpretation of the application of Chap. 7 of the Charta. The definition of emergency seems to be a crucial argument anyway. The representative of Russia to the UN appealed to 'avoid panicking and overdramatising the situation' and the representatives of Brazil, Pakistan, and China emphasised that climate change is foremost a sustainable development issues. Thus, sudden emergencies due to climate change are meant to be kept outside of the political debate.

It is debatable whether an aggregated indicator such as a global mean temperature can possibly be used as a trigger for emergency response measures.[12] These
impacts, however, react to significantly different vulnerabilities. Large-scale, irreversible and systemic changes in geophysical systems may vary in kind, geographic
dimension, and time.

Key security risks are pointing to indirect reinforcing effects of climate change,
namely border disputes and migration due to incrementally changing landmasses,
energy supply due to increases in competition over scarce energy resources, increased
shortages of other resources such as freshwater supply, and societal stress through an
aggravation of poverty and inequalities through climate change or even unforeseen
impacts of deliberate climate engineering. One key security risk relates to humanitarian crisis in the case of extreme weather events and sudden disruptive climate
change. The Hyogo Framework for Action 2005–2015,[13] provides some guidance
as to the actions which should be taken in response to major natural disasters.

All countries fear that unilaterally defining emergency will undermine their
sovereignty. In general, differentiating ordinary problems and accidents from
dangerous effects and even from emergencies, is an open question. This is true
for climate change as well as for nuclear meltdowns, and the social and infrastruc-
tural impacts which may follow from major natural disasters. What is the degree of
evidence that is substantial enough to legitimate any external intervention? Is there
such a degree anyway? Who would be in the position to clarify different
viewpoints?[14] Who would be legitimised to check data, to detect undisclosed
data, and to ask for missing metrics? How do independency, reliability, and (!)

---

[12] In the Fourth Assessment Report of the IPCC this is especially dealt with in Chap. 19 of WG II
where it says: 'A significant category of key vulnerabilities is associated with large-scale,
irreversible and systemic changes in geophysical systems. [...] central to nearly all the
assessments of key vulnerabilities is the need to improve knowledge of climate sensitivity –
particularly in the context of risk management [...] where the greatest potential for key impacts
lies' (IPCC AR4, WGII, Chap. 19: 804).

[13] The Hyogo Framework for Action is the main document resulting from the World Conference
on Disaster Reduction in January 2005.

[14] Detached from the debate in the Security Council, there exists the idea to widen the interpretation of Chap. 7 of the UN Charta by establishing the 'responsibility to protect' (R2P) as legal norm
in international law. While the concept has been on the agenda for some time now, an initiative by
the former Secretary General of the UN, Kofi Annan, to establish the R2P was not entirely
successful. The International Commission on Intervention and State Sovereignty (ICISS) argued
from the perspective of the concerned population. They recommended establishing the following
criteria to legitimate intervention: 'right intention', 'last resort', 'proportional means', and 'reasonable prospects'. However, the World Summit in 2005 including the high-level preparatory
panel took up a different perspective. To them, the main purpose of R2P was to strengthen the
international security systems (instead of taking the perspective of the concerned population). In
the final document, R2P was taken up. But it was not attached to any criteria. Especially the US
wanted to keep a leeway for wide interpretation of intervention. However, intervention was limited
by the definition of four cases of application, namely genocide, war crime, ethnic cleansing, and
crime against humanity. Crisis due to natural disasters have deliberately been excluded. While
'responsibility to protect' has been confirmed in the sense it was already established in international law, no 'right to protect' has been established.

effectiveness in handling data build up credibility? Is the process of verification part
of the solution or part of the problem?

6.7 Transformation and Governance

The term transformation often suggests big machinery and rightly so. The term
associates itself with big challenges and 'thinking-big-solutions'. Its agenda is
bigger than life. It forces the followers to be part of something bigger. People
generally like this as it does not contradict with what they would do anyway.

The term is used in singular form since it has been used by Polanyi (1978, 1944).
This evokes the idea of a simple solution (a pass-partout thing, a one-way option).
The term somewhat excludes those who are actually carrying out transformation.
It invokes passiveness. A rhetoric example: If there is a revolution, there are
revolutioners. This is not the case with transformation. Who would actually do
the transformation? Transformationers? A Transformateur?

The term suggests, in a way, some kind of 'Big – Bang-ism' where development
constantly needs and provides a sense of scale (something is scaling up). The era of
sustainability proposes proportionality rather than scale as major references. Transformation, probably, has no scaling mechanism at all. Learning from historial
analyses of the first and second industrial revolution one may extract some features
of transformation that might give an idea of what the world is running into with
thriving towards nine billion people with increased life support systems
(Osterhammel 2010; IASS 2011). Discontinuity, purposelessness, locality seem to
be such kind of patterns. There are specific elements of transformation which we
must know about and must accept: how to enhance credibility? There is no way to
force other stakeholders (those responsible) to do something without the next step
being performed by the original self (there is no free meal). How to build trust into
'green economy' when there is no trust in economy? How to ensure the reversibility
of the good action?

6.8 Transgovernance

Ulrich Beck's concept of second modernity expects the old institutions, enterprises
and players to remain in place while the new happens. Change, in this sense, is not
sequential but rather happens through parallel channels and competitive structures.

There is little doubt that the occurrence of emergencies will increase. Indeed
with up to nine billion people living on a planet with carbon constraints and
restricted resources, in 2050 the human settlements will be more vulnerable. As a
runaway problem with a 'fat tail' the climate change will cause a number of
emergency situations. Nuclear facilities are also a potential threat. The recent
nuclear meltdown in Japan has prompted profound and renewed thinking about

the ethics of how much risk a society can bear and whether the idea of risk is still adequate if no society is in the position to simply absorb social distortion by evacuating densely populated areas, dealing with contaminations and, last but not least, with an un-clearly lurking emergency situation arising from not controlling the nuclear power plants.

This think piece is not setting out to emphasise emergencies *per se* as some kind of change making mechanism. Emergencies may catalyse change as they can lead to stand still behaviour. Their political impact is open ended, and we can see examples which have stirred a renaissance style follow up, when the way an entity (society, enterprise, organisation) 'digests' an emergency is trying to rest orate behaviour. Change is not symmetric in time or thoroughness. Rather, it is asymmetric, and this is why there is the case for advanced studies to better understand change and the change as clustered by and in emergency situations. This depends on the responsiveness of democracy and how democracies digest knowledge, in general and under the concrete contingency.

A preventive democracy, in fear of populist and the public debate demagogically destroying forces, may choose to hide itself behind a shade of rules and 'ever existing' procedures denying open political access and fighting change clusters as irregularities which might pass by anyway if not given attention. It would choose to rely on elitist groups of experts legitimised by function and routines. It will hardly accept the perspective of transgovernance.

A flat democracy allows and invites social media and networks of all kinds to directly influence decision-making schemes. It is amorphous and will refrain from taking sides. 'Flat' means that, technically, access is granted to everyone. It may not be media-controlled in the sense of the private sector owning newspapers and tv-channels. It may rather invite market players and especially consumers to act as a crowd and to use demand side power in order to enforce sustainability features in production, product and consumption. Key words and concepts are 'responsible consumption, lifestyle-of-heath-and-sustainability, political consumption, carrot mobs, green procurement'. While these elements may enlarge and improve democracy, a flat democracy is likely to create the notion that those in charge are increasingly alienated from those who run the action on the ground, and maybe this is really the case (Friedman 2008). Seen from a governance perspective, a flat democracy may tend to let governance structure fade away. It replaces procedures by presence. Procedures with checks and balances would then be replaced by the direct influence of leaders who may have no legitimate voice other than through the web-crowd. A flat democracy may even choose to deliberately discard legitimised representative procedures (and their legal derivates, the sitting and permitting procedures) by allowing and enforcing social networks and populist 'leadership' appearances. A transgovernmental perspective will probably be seen as something that is alienated from the flat democracy.

A representative parliamentary democracy that would increase its responsive and participatory lay out options could be called 'transdemocracy'. Building on both procedures and preferences it would count on the democratic lifestyle and social responsibility of people and institutions. It would enlarge legal procedures not by consuming even more time and resources, but by making legal access easier and legal procedures faster. It would clearly not be fail-proof, but by not wasting mistakes for the incremental improvement of the governance approach it will build fire walls against the democratic fatigue and as far as transformation is concerned, against disappointment as well. This concept of democracy will most likely use transgovernmental concepts to better deal with the unpredicted.

In this respect, governance issues, and transgovernance in particular, should also cover the private sector. The corporate community displays different governance approaches for the implementation of sustainability management schemes and addressing social responsibility. Indeed this is a long standing agenda which the private sector and the civil society including the nongovernmental organisations, have in common.[15]

Acknowledgements This paper benefitted from the review done by Prof. Hideyuki Mori, President, Institute for Global Environmental Strategies, IGES, Japan, and Prof. Dr. Carlo Jaeger, Potsdam Institut für Klimafolgenforschung, PIK, Germany. Many thanks to my colleague Sibyl Steuwer for her support. The full responsibility for remaining inconsistencies and possible shortcomings stays with me as author.

References

Bachmann G (2010) Vision: Wissen, Angst, Wagnis. Impuls-statement anlässlich der 14. Benediktbeurer Gespräche der Allianz Umweltstiftung zum Thema 'Früher war selbst die Zukunft besser' (Karl Valentin) – Wirtschaft, Politik und Wissenschaft präsentieren Visionen für eine nachhaltige Zukunft.' am 29 und 30 Apr 2010. http://bit.ly/n8cK1E. Accessed 17Aug 2011
Beck U (1986) Risikogesellschaft. Suhrkamp, Frankfurt/M
Beck U (2009) World at risk. Polity Press, Cambridge
Brüggemeier FJ (1996) Das unendliche Meer der Lüfte. Luftverschmutzung, Industrialisierung und Risikodebatten im 19. Jahrhundert. Klartext Verlag, Tübingen
Deutscher Nachhaltigkeitspreis (2008–2011) http://bit.ly/niB67q. Accessed 17 Aug 2011
EEAC (2011) Challenging, encouraging, innovative: addressing the Green Economy agenda in the context of SD. Learning from long-standing and diverse experience: institutional framework for SD at national level. Statement by the Working Group on Sustainable Development of the European Environmental and Sustainable Development Advisory Councils Network (EEAC): Wroclaw and Brussels, online document to be found on http://www.eeac-net.org/. Accessed 18 Sep 2011
Eikmann T, Heinrich U, Heinzow B, Konietzka R (eds) (2010) Gefährdungsabschätzung von Umweltschadstoffen. Loseblattsammlung. Erich Schmidt Verlag, Berlin
Ethik-Kommission Sichere Energieversorgung (2011) Deutschlands Energiewende. Ein Gemeinschaftswerk für die Zukunft. Berlin. Erstellt im Auftrag der Bundesregierung und vorgelegt am 30 Mai 2011. http://bit.ly/oPO53l. Accessed 17 Aug 2011

---

[15] This is especially true on the UN level of decision taking. So far, there is a lack of effective formats for the private sector to discuss and develop governance strategies in relation to multilateral governance issues. The World Economy Forum and other project level approaches such as Peer Reviews for the purpose of sustainable strategies might be good examples.

EU COM (2011) 363 final, communication from the commission to the European parliament, the council, the European economic and social council and the committee of the regions: Rio + 20: towards the green economy and better governance. Brussels, 20 June 2011. http://bit.ly/r7k3vo. Accessed 17 Aug 2011

Friedman TL (2008) Hot, flat and crowded. Why we need a green revolution and how it can renew America. Farrar, Straus and Giroux, New York

German Government (2000) Gutachten des wissenschaftlichen Beirats Bodenschutz beim BMU: Wege zum vorsorgenden Bodenschutz. Fachliche Grundlagen und konzeptionelle Schritte, Deutscher Bundestag, printed matter 14/2834

IASS (in prep) (2011) Learning from historian analyses of the industrial revolution for the great transformation as we see it today. Workshop document. IASS, Potsdam

In 't Veld RJ (2010) Knowledge democracy. Consequences for science, politics and media. Springer, Heidelberg

IPCC (2001) Climate change 2001: synthesis report online document. http://bit.ly/oUIPT5. Accessed 17 Aug 2011

IPCC (2007) Summary for policymakers. In: Solomon S, Qin D, Manning M, Chen Z, Marquis M, Averyt KB, Tignor M, Miller HL (eds), Climate change 2007: the physical science basis. Contribution of working group i to the fourth assessment report of the intergovernmental panel on climate change. Cambridge University Press, Cambridge/New York. Online document. http://bit.ly/pU2K4e. Accessed 17 Aug 2011

Klein N (2007) The shock doctrine: the rise of disaster capitalism. Metropolitan Books/Henry Holt, New York

Luhmann HJ (2010) Auf welche Wissenschaft beruft sich die Politik beim Zwei-Grad-Ziel? Gaia, München: Oekom Interdisciplinary Journal 19 (2010), 3, see: http://bit.ly/p5zm5J. Accessed 17 Aug 2011

Müller-Jung, J (2009) Warum sollten maximal zwei Grad die Welt retten? Online article of Frankfurter Allgemeine Zeitung. http://bit.ly/oyXbWK. Accessed 17 Aug 2011

Rat für Nachhaltige Entwicklung (2011) Der Deutsche Nachhaltigkeitskodex. Online documentation on http://bit.ly/gn0M0A. Accessed 17 Aug 2011

Rat für Nachhaltige Entwicklung, Geschäftsstelle (2010) Nachhaltigkeits-Indikatoren zur Messung der gesamtwirtschaftlichen Entwicklung. Gutachten an den Sachverständigenrat zur Begutachtung der gesamtwirtschaftlichen Entwicklung zum Bericht der Stiglitz-Sen-Fitoussi-Kommission, online document. http://bit.ly/o8Tlpm. Accessed June 26 2012

Osterhammel J (2010) Die Verwandlung der Welt. Eine Geschichte des 19. Jahrhunderts. C.H. Beck, München

Polanyi K (1978,1944) The great transformation. Politische und ökonomische Ursprünge von Gesellschaften und Wirtschaftssystemen. Deutsche Übersetzung der Originalausgabe von 1944. Suhrkamp Taschenbuch Wissenschaft, Frankfurt

Radkau J (2011) Die Ära der Ökologie. C.H. Beck Verlag, München

Schneider SH, Semenov S, Patwardhan A (2007) Assessing key vulnerabilities and the risk from climate change. In: Parry ML, Canziani OF, Palutikof JP, van der Linden PJ, Hanson CE (eds) Contribution of working group ii to the fourth assessment report of the intergovernmental panel on climate change. Cambridge University Press, Cambridge/New York, pp 779–881

Speth GJ (2005) Red sky at morning: America and the crisis of the global environment, 2nd edn. Yale University Press, New Haven

Stiglitz JE, Sen A, Fitoussi J-P (2009) Measurement of economic performance and social progress. Online document http://bit.ly/JTwmG. Accessed June 26 2012

WBCSD (2010) Vision 2050. The new agenda for business. World Business Council for Sustainable Development. http://bit.ly/q6Qb5U. Accessed 17 Aug 2011

WBGU (2011) Welt im Wandel. Gesellschaftsvertrag für eine Große Transformation. WBGU, Berlin

Weitzmann M (2009) Additive damages, fat-tailed climate dynamics and uncertain discounting. Economics 3(39)

Chapter 7
Taking Boundary Work Seriously: Towards a Systemic Approach to the Analysis of Interactions Between Knowledge Production and Decision-Making on Sustainable Development

Stefan Jungcurt

Abstract The concept of boundary work has been put forward as an analytical approach towards the study of interactions between science and policy. While the concept has been useful as a case-study approach, there are several weaknesses and constraints when using the concept in a more systemic analysis of the interactions between knowledge production and sustainable development decision-making at the international level, such as its inability to capture the diversity of institutions involved in such boundary work. Another inability involves a lack of conceptualisation of the impacts of the specific conditions of intergovernmental decision-making, such as rules for representation and the mode of negotiation. This chapter suggests complementing the concept of boundary work with a configuration approach based on a two-dimensional conceptualisation of the boundary space in international decision-making that allows the positioning of institutions with regard to their degree of politicisation and their position in terms of national and regional representation. Such an approach could be a useful guide in the further conceptualisation and application of the boundary concept.

7.1 Introduction

In the study of interactions between science and policy in sustainable development decision-making, the concept of boundary work has recently emerged as a promising approach which focuses on the social processes at the boundary between the production of scientific and other types of knowledge as well as decision-making processes. The concept goes against earlier representations of the science-policy interface which are based on science and policy as distinct and separate worlds depicting science as the world of neutral and independent facts and policy making as the world of values.

S. Jungcurt (✉)
International Institute for Sustainable Development (IISD) Reporting Services, Ottawa, Canada
e-mail: sjungcurt@gmail.com

L. Meuleman (ed.), *Transgovernance*,
DOI 10.1007/978-3-642-28009-2_7, © The Author(s) 2013

Instead of questioning how scientific knowledge is best transferred into the policy domain in order to 'inform' politics, boundary work focuses on the various types of interactions which take place in the sphere between science and politics. This opens the 'black box' of the science policy interface as a simple line across which knowledge must be transferred. It also questions the implicit assumption that such transfers can take place in a manner that is value neutral and without modification of knowledge content. Furthermore, boundary work assumes bidirectional exchanges and discourse between knowledge production and policy making, leading to the notion that science and social order are co-produced in mutually interdependent processes rather than independent social domains.

While the concept is originally developed in the context of studying science-policy interactions in national decision-making processes, it has also been effectively applied in the context of international decision-making on sustainable development. In their evaluation of the effectiveness of scientific assessments in influencing international decision-making on environmental issues, Mitchell et al. note that assessments

> have influence to the extent that they involve long-term dialogue and interactions in which potential users of an assessment educate scientists about their concerns, values priorities, resources and knowledge of the problem, while scientists educate potential users about the nature causes, consequences and alternatives for resolution of the problem at hand as well as the ways such knowledge is arrived at. Co-production implies that assessments are influential to the extent they are bidirectional, with science shaping politics, but also politics shaping science. (Mitchell et al. 2006: 324)

In their earlier review of knowledge systems for sustainable development, Cash et al. (2003) find that

> those systems that made a serious commitment to managing boundaries between expertise and decision-making more effectively linked knowledge to action than those that did not. Such systems invested in communication translation and/or mediation and, thereby more effectively balanced salience, credibility and legitimacy in the information they produced. (Cash et al. 2003: 8089)

Both of these conclusions are based on the analysis of large numbers of case studies, many of which have used the concept of boundary work as a heuristic guide or as an analytical lens through which to evaluate the effectiveness of specific assessments and other knowledge producing processes in influencing international decision-making on environmental issues.

This contribution will explore the application of the boundary work concept in a broader sense to describe the work of the various types of institutions, actors and processes which populate the space between science and policy on the international level. It will also examine their contributions to managing the boundary, and the interrelations among them. The following section reviews the origins and key features of the concept. Section 7.3 discusses its application to interactions between science and policy making both in general terms, as well as with a view to adopting a more systemic perspective which captures the diversity of institutions and actors involved in boundary work at the international level. Section 7.4 proposes complementing boundary work with a configuration approach which captures this diversity and directs attention to the interactions between the multiple processes involved.

7.2 Boundary Work: The Concept and Its Origins

Thomas Gieryn (1983) describes the phenomenon of boundary work as

> ideological efforts by scientists to distinguish their work and its products from non-scientific activities [. . .by attributing. . .] selected characteristics to the institution of science [(. . .)] for purposes of constructing a social boundary that distinguishes some intellectual activities as non-science. (Gieryn 1983: 782)

Based on an analysis of such demarcation efforts of scientists against religion, 'pseudo-sciences' defend the autonomy of science against efforts to restrict its activities in the name of national security. Indeed, Gieryn shows that boundary work is a rhetorical style which can be used by 'ideologists of a profession or occupation' to: expand authority or expertise into domains claimed by other professions or occupations; monopolise professional authority and resources in order to exclude rivals; protect autonomy over professional activities by excluding members for consequences of their work; and may even be used within science to demarcate boundaries between different disciplines (ibid.: 791). He further concludes that boundaries of science are ambiguous because: the characteristics attributed to science are sometimes inconsistent; boundaries are contested by scientists with different professional ambitions; and boundaries result from the simultaneous pursuit of separate professional goals requiring boundaries that are built in different ways (ibid.: 792).

In short, Gieryn's work shows that the original concept of boundary work is seen as a rhetorical tool applied by scientists primarily to further the interest of their profession rather than establishing unambiguous, scientifically grounded definitions of what constitutes science or how science is defined in a certain discipline. Despite the fact that they are applied by scientists, the establishment of boundaries cannot necessarily be considered a scientific exercise in itself.

Sheila Jasanoff (1987, 1990) applies the concept of boundary work to investigate interactions between scientists and policy makers. She starts from the observation that science has been able to maintain its status as 'provider of truths' even though it is widely recognised that knowledge is indeterminate and can be interpreted in many ways, because of the adherence to shared 'Mertonian'[1] norms 'that foster

---

[1] Introduced by Robert K, Merton, the Mertonian norms are a set of institutional priciples that describe the 'ethos of modern science: Communalism (results of scientific research are common property of the scientific community); Universalism (all scientists can contribute to science regardless of race, nationality and gender); disinterestedness (scientists should not mix personal beliefs or activism with the presentation of their research results); originality (scientific claims must contribute new knowledge); and scepticism (validation through critical scrutiny). See: Merton (1973).

cohesiveness in science, even though its practitioners come from divergent geo-graphic, cultural or linguistic backgrounds' (Jasanoff 1987: 196). The authority derived from these norms is reinforced by a number of rules which govern the practice of science such as high standards for entry into the scientific professions, rules for quality control exercised by 'invisible colleges', 'research circles' or other informal networks that control the diffusion of scientific knowledge' (ibid.). This cognitive authority of science comes under pressure when scientists are called upon by policy makers to provide advice in areas which are at the frontiers of science, and where knowledge is particularly uncertain and indeterminate resulting in a most fragile consensus among scientists (ibid.: 197).

Earlier models depict science advice as a unidirectional process of scientists delivering facts or 'truth' to decision makers as basis for informed decisions about issues affecting, or affected by the physical laws of nature. In direct contrast to this, Jasanoff develops a model in which

> 'truth' emerges from an open and ritualized clash of conflicting opinions, rather than from the delicate and informal negotiations that characterize fact-finding in science. (ibid.)

According to her model, legitimacy in decision-making is achieved through the 'public reconstruction of the scientific basis for regulation'. The process gives rise to competing claims of authority between science and policy making with regard to the interpretation of scientific findings, which in turn challenges the disinterested-ness and certainty of science. The result is a 'partial removal of cognitive authority', which renders explicit the assumptions and uncertainties embodied in scientific research and thereby allows policy makers to show that 'the interpretation of indeterminate facts reflects the public values embodied in legislation as well as the norms of the scientific community' (Jasanoff 1987: 198).

While the process of science's public deconstruction followed by reconstruction of the rationale for decision-making in the policy arena increases the legitimacy of policy making, it challenges the self-image of science as a disinterested search for truth. Furthermore, the public demonstration of uncertainty and disunity among scientists may damage the public image of scientists and may lead to questions about whether or not they truly merit the status as well as the symbolic and material rewards which they enjoy in society. To protect themselves from such negative impacts, scientists have to establish and continuously reinforce the boundaries between science and policy. The boundary is thus a contested space around which scientists and policy makers compete for cognitive authority over the interpretation of indeterminate facts. In essence, the contested boundary arises out of different views over how much decision-making power should be granted to scientists in areas where scientific knowledge is insufficient for decision-making – either because of lack of data and uncertainty, or because of the indeterminacy of knowledge. This gives rise to competing claims which make it impossible to take 'legitimate decisions'.

In her 1987 paper, Jasanoff investigates three 'contested boundaries' (or more precisely three strategies to establish the boundary between science and politics, and thus the distribution of decision-making authority): trans-science, risk assessment and peer-review. Trans-science addresses the grey zone between science and policy, which is characterised by questions such as 'which can be asked of science and yet which cannot be answered by science' (Jasanoff 1987: 201; citing Weinberg 1972). Scientists argue that the cognitive indeterminacy revealed by the policy making process lies outside of 'real' science in the realm of trans-science. This separation is used to argue that, while policy makers may claim authority over issues of trans-science, science itself should remain the undisputed preserve of scientists. Therefore deconstructionist techniques should only be regarded appropriate for issues of trans-science, not genuine science. Jasanoff shows that Weinberg's main objective is to 'shield science against the taints of subjectivity, bias and disharmony that it acquires in the policy environment'. In her conclusion, Jasanoff states that this approach ignores the key procedural concerns of policy making, most importantly the question of who should decide on issues which fall within the boundaries of science and policy, that is, where science is unable to provide unambiguous answers to the questions that policy makers have to address.

Similarly, Jasanoff argues that risk assessment and peer review are used to advance particular views about the extent to which scientists should control decision-making at the frontiers of knowledge. Peer review of suggested regulation by scientific experts, for example is often demanded by the industry in order to shift the balance of decision-making power away from regulatory agencies. Indeed, from the perspective of the industry, these agencies may well be biased towards excessive or overly strict regulation. Because of their impact on the distribution of decision-making power, boundary strategies can be instrumentalised by those who have stakes in the regulatory decisions at stake.

These considerations give rise to further research on the activities which take place at the boundary between science and policy making and the actors and organisations involved in such work. In her 1990 book 'the fifth branch' Jasanoff explores the work of science advisers in policy making (Jasanoff 1990). Here, Jasanoff makes the case for a more detailed analysis of the processes which determine decision-making and the role that science has within these processes. She depicts two schools of thought with regards to the role of science in decision-making. The first school of thought is the technocratic view, according to which, bureaucrats are technically incapable of distinguishing 'good' from 'bad' science and therefore call for a greater involvement and influence from scientists in the decision-making process. The second school of thought is the democratic view, which holds that decision makers fail to incorporate a sufficient range of values in their decision-making, favouring the inclusion of a broader set of viewpoints in the decision-making process beyond narrowly technical viewpoints. In line with the idea that contending views over the role of science in decision-making represent a struggle of different interests over discretion in decision-making, Jasanoff observes that commercial and industrial interests favour the technocratic view, while interest groups such as environmental, labour and consumer movements support the democratic view (Jasanoff 1990: 15–16).

Based on a review of the work carried out by scientific advisory committees in US regulatory agencies, Jasanoff derives a number of conclusions with regard to the characteristics of successful boundary processes. First she highlights that what is considered to be 'good' science in decision-making is the result of negotiations, since 'when stakes are high, no committee of experts, however credentialed, can muster enough authority to end the dispute on scientific grounds' (Jasanoff 1990: 234). Negotiation is at the heart of the construction of regulatory science, which highlights the role of advisory committees as forums where scientific and political conflicts can be negotiated simultaneously. The role of the scientific expert is then to stabilise the results of negotiation against further attempts of deconstruction through his or her ability to validate research, certify scientific methods, define standards of adequacy of scientific evidence and approve inferences made from scientific studies and experiments (ibid.: 237). The conduct of scientific advice as negotiation provides legitimacy to the outcome.

In order to be successful, boundary work should be non-adversarial to avoid an unproductive deconstruction of science and fostering of appearance of capture (ibid.: 246). Committee membership should reflect disciplinary breadth, which may be challenging for small committees. The advisers populating committees also need to be more than mere technical experts to be able to transcend disciplinary boundaries, synthesise knowledge from several fields, and to understand the limits of regulatory science and the policy issues confronting the agency (ibid.: 243). These requirements may make it difficult to find a sufficient number of policy advisers, which may lead to conflicts of interests resulting from long-term and encrusted relationships between agencies and a small group of skilled advisors. Finally, Jasanoff notes that the advisory process must recognise that scientific knowledge is in perpetual flux and demands constant renegotiations, which in turn calls for allowing more flexibility in the rules and norms which govern the work of advisory committees than those of administrative decision-making. The problem of advisory processes then, is not so much to protect decision-making from capture by scientific experts who are influenced by technocratic interests, but to 'harness the collective expertise of the scientific community, so as to advance the public interest' (Jasanoff 1990: 250).

In short, the work of Gieryn, Jasanoff and their colleagues directs attention towards the processes of negotiating the boundary between science and policy as well as the rules and organisations which structure such processes, including the rules for selecting participants of advisory committees, structuring the discourse within these committees, and for the type of outputs expected from them.

Further research concentrates on the role of boundary work and how it stabilises the boundary between science and politics. This is achieved through investigating the role of boundary organisations and their outputs, known as boundary objects or standardised packages. The rationale for conducting such research is, on the one hand, the concern that constructivist arguments about the contingency of these boundaries could lead to a dangerous erosion 'of the cognitive authority of science by legitimizing relativism', and a fear about a decay of the mutually productive relationship between science and liberal democracy (Guston 1999: 89). On the other hand, scholars believe that by clearly portraying science as it is practiced, constructivist accounts can help to improve the position of science in society and 'recover the human face beneath science's rationalist mask' (ibid.).

Star and Griesemer (1989) introduce the notion of 'boundary objects' as common products of negotiations at the boundary between science and policy. Boundary objects are knowledge products, such as reports, methodologies or interpretative frameworks which are 'both plastic enough to adapt to local needs and constraints of several parties employing them, yet robust enough to maintain a common identity across sites' (Star and Griesemer 1989). A similar concept is that of 'standardized packages' which is a means to 'define a conceptual and technical work space [. . . by combining] boundary objects with common methods in more restrictive but not entirely definitive ways'. Standardised packages seek to homogenise and facilitate repeated interactions among practitioners from both sides of the boundary between different social worlds while maintaining their integrity within their respective worlds (Guston 1999, citing Fujimura 1992). In this way, they effectively function as interfaces for the translation and transfer of different kinds of knowledge for the purpose of collaborative knowledge development.

David H. Guston further develops the concept of boundary organisations using principle agent theory. He suggests that the relationship between policy makers and scientists can be represented as a contractual relationship, similar to that between other economic agents, in which policy makers 'hire' researchers to deliver expertise on specified issues. The principal is faced with the problems of adverse selection and moral hazard. Adverse selection may lead to the identification of scientists pursuing a specific agenda in relation to the policy problem at hand, whereas moral hazard describes the agent's incentive to cheat or shirk or otherwise exploit the principal's lack of information. To address these problems, the principal will put into place mechanisms for monitoring the agent's behaviour and for verifying the results of his research. Such mechanisms include procedures for accountability, in particular financial accountability, but will also lead to the development of boundary objects and standardised sets. In Guston's case study of innovation and technology transfer originating from the US National Institute of Health, boundary objects include procedures for 'innovation disclosure' which facilitate the collaborative identification of research results with innovative or market potential, as well as incentives and procedures to facilitate the application for patents. The collaboration between governmental research laboratories and non-federal actors such as private firms, is governed by Cooperative Research and Development Agreements (CARDAs). Indeed, Guston identifies these as the key standardised set of the boundary institutions in his case study. Based on these observations Guston identifies the following shared characteristics of boundary organisations (Guston 1999: 93):

- 'They provide a space that legitimizes the creation and use of boundary objects and standardized packages;
- They involve the participation of both principals and agents, as well as specialised (or professionalised) mediators; and
- They exist on the frontier of two relatively distinct social worlds with definite lines of responsibility and accountability to each'.

The specialised mediators in his case study are technology transfer experts who oversee the collaborative process of technology transfer and report to the governmental agent. The government creates incentives for them which directly depend on the effectiveness of technology transfer, thus establishing a new intermediary agent who is himself in a principal-agent relationship with the researcher.

To summarise, the emergence of the boundary concept has shifted the way in which science and knowledge are perceived in decision-making. They are no longer viewed in terms of the 'pipeline' and 'information deficit' models which presume that knowledge is produced by and delivered to decision makers very much like a commodity or resource towards a model of co-creation or joint fact finding, in which knowledge holders and decision makers work together to develop common understandings of problems and available pathways of action as a basis for legitimate and socially robust decision-making. The boundary model directs attention to the procedural aspects of knowledge creation and use for decision-making. Rules for participation and balance of influence emerge as important factors for successful boundary work next to the mere quality and appropriateness of the knowledge at hand.

7.3 Boundary Work in International Decision-Making in Sustainable Development

Most of the research on boundary work has thus far been carried out in the national context, with the majority of studies analysing science-policy interactions in US decision-making processes. This raises the question of whether the concept can be usefully applied to boundary work on the international level. A number of differences come to mind with regards to both the representation of 'policy' and 'science' as well as in the institutions which frame the interactions. These differences may not be in line with the explicit and implicit assumptions of the boundary work.

Miller (2001) discusses three weak assumptions of the boundary concept which influence its applicability to the international context. First, the concept ignores the diversity of institutions and practices which exist within both science and politics. Scientific practices and discourses vary with disciplines, institutions and networks, and scientists within disciplines frequently disagree about the representation of their knowledge and the implications derived from it. Similarly, perceptions, policy styles and forms of interaction vary across institutions and sectors. On the national level, the assumption of uniformity may nevertheless be acceptable, since decision-making on sustainable development takes place within policy domains which have distinct styles of policy politics – a specific combination of cognitive styles and interaction. Over time, this combination generates particular public epistemologies about the validity and use of different types of knowledge within the domain (Hoppe 2010: 181). On the international level, however, boundary work must

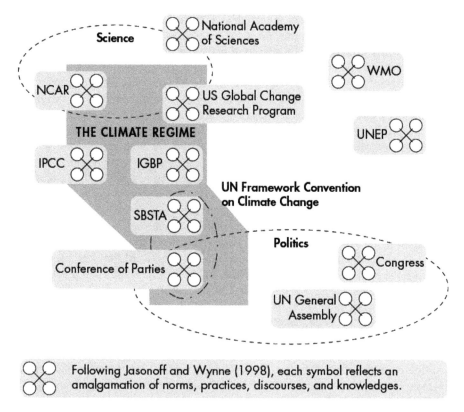

Fig. 7.1 US and international Organisations involved in boundary work on climate change (Source: Jasanoff and Wynne (1998)). Note: NCAR = National Center for Atmospheric Research; WMO = World Meteorological Organisation; UNEP = U.N. Environment Programme; IPCC = U.N. Intergovernmental Panel on Climate Change; IGBP = International Geosphere-Biosphere Programme; SBSTA = U.N. Framework Convention on Climate Change's Subsidiary Body for Scientific and Technological Advice

span the diversity of scientific and political institutions from a large number of countries and policy domains which interact with each other. This leads to confrontations not only between different national policy styles, framings and policy theories, but also between diverse and culturally determined perceptions as well as different ways of identifying and describing problems in different languages (ibid.).

Second, the concept oversimplifies the boundary between science as a 'fine bright line' using inadequate representations of pure science and pure politics. This ignores the diversity of institutions that exist between the two sides which are neither science nor politics 'but combine elements of the two in remarkable different ways'. Miller illustrates this diversity with a map of institutions involved in boundary work on climate change. The map (Fig. 7.1) includes both institutions inside and outside the formal climate change regime, as well as US national institutions. The entire landscape includes 11 institutions, each of which produces its own 'amalgamation of norms, practices, discourses and knowledges' on climate change (Miller 2001: 485, citing Jasanoff and Wynne 1998).

The third weakness identified by Miller is that the boundary concept presents an overly static view of science and politics. The last two decades have seen the emergence of a vast array of new institutions involved in boundary work on the international level which has led to a constant rearrangement of institutions and how they relate to each other. At the same time 'definitions and standards for expertise are deeply contested across cultural and geopolitical divides, as are notions of appropriate political institutions for carrying out public sector management for the planet as a whole' (Miller 2001: 485). This means that institutional arrangements are constantly in flux and there is rarely a negotiation process which ends with the same constellation of institutions involved with which it started.

One may add a fourth weakness here, which becomes relevant if one looks at boundary in an intergovernmental context rather than from the perspective of domestic engagement in international decision-making, as Miller does. Boundary work on the international level takes place within the constraints and practices of intergovernmental decision-making. This means that the discourses and processes of boundary work will always be affected by the rules for representation and decision-making which characterise political processes on the international level. Since states are the main actors of multilateral decision-making, any form of boundary work has to provide for adequate codes of representation in order to be considered legitimate. This has a number of implications for the conduct of boundary work in an international context. First, the criterion of representation of states competes with the criterion of representation of relevant knowledge and scientific expertise. Subsidiary bodies and smaller expert panels in particular suffer from difficult debates about balance in representation either on a country or regional level (Kohler et al. 2011). In most cases, the concern about representation trumps the concern for diversity and relevance of expertise of the individuals that who will be invited as experts. The need for representation limits both the number of experts who can participate from a given country or region, as well as the individuals chosen by countries. The more politicised an issue is, the more countries will tend to send diplomats rather than experts.

Assessment processes attempt to circumvent this problem by establishing criteria for the scientists and experts to be nominated by countries. However many countries will select their participants in a way which ensures that the contribution from those experts is not against their political positions in the negotiation process at hand. Any institution or forum involved in boundary work on the international level will in one way or another be affected by the need to ensure national representation as well as representation with regard to different types of expertise and knowledge. In many cases the intergovernmental negotiation setting will act as a bias which will give primacy to the national requirement.

The second constraint arises from the mode of decision-making in international fora. The great majority of intergovernmental decision-making processes require unanimity by all member states to take decisions. Rules of procedure which allow for majority voting are the exception and are only established within a framework that clearly identifies which decisions can be taken by voting. An example can be seen with repetitive operational decisions such as subjecting new species to the trade restrictions under CITES or adding new chemicals to the list of substances to be monitored by the Rotterdam Convention on Hazardous Materials. In theory, the unanimity rule makes it possible for a single country to block consensus, even if an acceptable amalgamation of values and knowledge has been achieved among all other participants. In reality, the pressure to achieve an outcome often leads to a race to the bottom in terms of the substantive content of an agreement as the majority accepts to water down elements of a decision in order to accommodate minority concerns. Scholars in international relations have identified the pattern that international agreements tend to be either 'broad and shallow', meaning that many states participate in an agreement with limited impacts; or 'narrow and deep', meaning that a small group of states participates in an agreement which yields large benefits from cooperation (Barrett 1999: 525). If one can consider the breadth and depth of an agreement as preliminary measures of the success of boundary work during the negotiation phase of an agreement, then one can expect that the logic of negotiation under the unanimity condition creates an additional hurdle for boundary work on those issues which are most difficult to agree upon.

In decision-making bodies which operate under the one-country-one-vote and unanimity principles, the main participants are country delegates who are bound by the instructions of their capitals. The instructions themselves are the result of processes of policy development and decision-making that may have included boundary work to varying extents, depending on the practices, cognitive styles and modes of interaction of the policy domains involved. Delegates have thus limited flexibility to accommodate the concerns of others both in terms of bargaining as well as with regard to their ability to embrace new concepts and boundary objects that may be developed in the course of the negotiation or presented by other participants, such as civil society actors. On the other hand, delegates' instructions usually do include some flexibility for making concessions in order to be able to strike mutually agreeable deals with their opponents. Whether these flexibilities can be used for the creation of new boundary objects again depends on the political culture and practices in different countries. Some countries give their delegates a lot of autonomy to decide how they will represent the interests of their countries, for instance by providing instructions that are formulated in terms of general objectives. In contrast, other delegates must work with narrowly formulated options for operational text. Delegates from some countries have to ask permission from their capitals for even minor changes, while other countries select their delegations such that the relevant policy domains and fields of expertise are represented at the meeting to allow for the delegation to react to new proposals which could not be anticipated. A typical phenomenon at the final stages of negotiations are delegates who make hectic last minute phone calls to get permission to agree to the final deal, which often involves explanations of a new compromise formula.

The situation is further complicated by the fact that many countries negotiate in coalitions or regional blocks. This involves another level of decision-making at which boundary work may or may not occur. The amalgamation of individual positions inevitably involves further discussions on facts and values within coalitions and regional groups. However, similar to the international level negotiations, it depends on the mode of decision-making and the flexibility of the delegates' instructions as to whether the result will lead to a further increase in the robustness and acceptance of the common position or a watering down of the agreement towards a lowest common denominator.

In assessment processes, the practice of 'negotiating scientific consensus' can lead to oversimplification and inadequate reduction of the complexity of both the science and the values that are behind the effort-reduction to the lowest common denominator replaces amalgamation. While assessment processes publish comprehensive reviews of the state of the art in science, including a consideration of different viewpoints and in some cases even contradictory findings, what gains traction in the policy making process are the severely reduced summaries for policy makers – sometimes even only parts thereof. Only these can be considered as outcomes of completed boundary work, since only these parts become the basis of decision-making. On the other hand, the knowledge produced by assessment processes becomes the basis for boundary work in numerous other institutions and forums that act as additional channels through which they can have an indirect impact. IPCC assessment reports, for instance, are the most important reference for making the case for action against climate change through advocacy groups or policy think tanks. These actors engage themselves in boundary work at different levels which has an influence on national positions as well as the course of the negotiations in international decision-making forums. Many of their outputs should thus be seen as intermediary boundary objects which enable boundary work in other channels.

The third constraint of boundary work emerges from the negotiation mode which prevails in the majority of international decision-making forums. Any outcome of international negotiations is either designed as international law, or will be interpreted in the context of existing international law and obligations. Soft law instruments, such as declarations or non-legally binding treaties and decisions for implementation have proven to exert substantial influence on policy making in many countries and, in many common law countries they can have a direct impact on court decisions. Therefore, many countries treat any negotiations as if the outcome would be legally binding, even if that is not provided for by the mandate, or the decision on the legal nature of an instrument will only be decided at the very end of a negotiation. This means that in the final stages, and often throughout the entire process, negotiations are led by legal, rather than scientific experts. Legal experts however will focus on legal issues, such as consistency with existing international laws and obligations, compatibility with national legal systems and legal clarity. This is often at the cost of scientific adequacy and relevance. Once negotiations have entered into the legal 'codification' mode, they tend to become less receptive to new knowledge and ideas, at least as long as this knowledge is communicated by non-legal actors. Furthermore, the final stages of a process are often marked by a decrease in trust among participants as confusion over legal concepts may lead negotiators to accuse others of trying to reverse previous agreements or of using existing decisions and other legal arguments strategically to their own advantage. The erosion of trust is further aided by the fact that countries become increasingly aware of the costs and benefits of proposed agreements and therefore switch to strategies of distributive bargaining: ensuring fairness in the distribution of costs and benefits takes primacy over the common objective of solving global problems.

The final stages of a negotiation are therefore carried out under the shadow of both existing law and the anticipated legal impact of the text under negotiation. Under certain conditions, this shadow can extend far into the early stages of a negotiation, thus leading participants to engage in distributive legal bargaining at a point in time when there has not yet been enough boundary work done to provide a basis for a successful completion of the negotiation. In other words, the amalgamation of facts and values is not yet sufficiently mature to withhold the erosion of trust in the process of legal bargaining.

This extension of the shadow of the law can occur for several reasons. One is the informal rule prevailing in many negotiating forums not to reopen text for discussion which has been previously agreed. Despite the formal rule that agreements are adopted as a package and thus 'nothing is agreed until everything is agreed', the request to make changes in agreed language is often interpreted as bad faith by other delegations. While this rule is a necessary convention to prevent legal negotiations from backlash and endless circles of revisiting the same issues, it can also prevent delegates from testing different framings and approaches during a negotiation to select the most suitable approach. What is more, when trust is low, some delegates will categorically disagree with any text on the table in order to ensure that they keep their options open until the very end. In extreme cases this practice can neutralise previous informal agreements, including any boundary objects which may have been embodied or referred to in the initial text.

Another factor is that boundary work and legal negotiations are often carried out under the auspices of the same institution, notably subsidiary bodies that provide advice to a governing body or conference of the parties of the same process. Without a mandate which clearly identifies the nature of the subsidiary body's work and delimits it from the actual negotiation process, such bodies tend to transform into preparatory meetings for the actual negotiation process. If countries expect that the outcome of a subsidiary body will be a draft decision which may be difficult to reopen for further discussion, they will send legal experts rather than scientists to represent them in these processes. The longer the shadow of the law, the more reluctant countries will be to let non-legal experts speak and engage in an open form of discourse, for the fear that their proposals will become fixed into legal concepts that may be interpreted against their own interest or original intention.

To summarise, the intergovernmental setting and the shadow of the legal negotiations of international sustainable development decision-making have fundamental impacts on the way in which boundary work is conducted at the international level. Much more than in national decision-making processes, it must be recognised that boundary work is not a simple bridging process between science and policy that is carried out in a single locus, but that it is composed of many processes in a complex web of loci which deliver partial amalgamations of facts and values from different perspectives. These partial amalgamations are often complementary, but in many cases they compete, since they represent different configurations of values brought forward by different sets of stakeholders and scientists who have participated in the process.

Secondly, it must be recognised that the negotiation process is an integral part of boundary work on the international level. Similar to the inseparability of discussions on facts and values, international boundary work is inseparable from the multilateral negotiation process. Delegates as well as scientists will always be influenced by, or even pressured to, represent the positions of their countries. The degree to which representation influences the outcome of boundary work can be depicted as the distance from the actual negotiation process. The more influence a process of boundary work can be expected to have on a negotiation process, the stronger the participants' bias towards their countries' positions in the negotiation processes itself. The closer a boundary organisation is located to decision-making, the more politicised its deliberations will be.

7.4   From Boundary Work to Boundary Configurations

In order to account for these factors, it is useful to conceive of the boundary between knowledge production and decision-making in an intergovernmental setting as a two-dimensional space defined by the axes of science and policy as well as national and international processes (Fig. 7.2). The further to the right a process or institution is located, the more politicised it can be expected to be. The higher up it is situated, the stronger will be the constraints of representation in the conduct of boundary work. An exception may be the intermediary organisations which are depicted here in the middle of both axes. This group itself represents a large diversity of institutions which may or may not be internally organised according to representative principals. Such organisations may participate as experts in assessment processes, as observers or representatives of civil society or major groups in subsidiary bodies and negotiations. These organisations often provide different types of knowledge to boundary work in other institutions or processes, or are themselves loci of boundary work.

The main assumption underlying this representation is that the boundary space is populated by different institutions and organisations which produce partial amalgamations of facts and values that are influenced by their position within the space as well as other factors such as membership, interests or ideological conceptions. This representation should also allow for the location of different

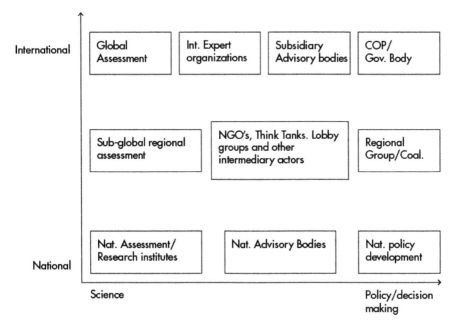

Fig. 7.2 Institutions and processes involved in boundary work on international sustainable development decision-making (compiled by the author)

institutions and processes involved in boundary work with regard to their position in terms of politicisation and constraints through presentation and negotiation in intergovernmental decision-making processes. This enables the development of a more differentiated conceptualisation of the types of processes and discourses taking place within these organisations and how these are influenced by their position. In addition, this also develops their relationships to one another. For instance, it may be possible to develop conjectures about the conditions under which institutions or actors occupying similar or overlapping positions within the boundary space will behave competitively or collaboratively. Similarly, one may ask under which conditions institutions positioned on opposite ends will complement each other. This is done either by delivering 'compatible' partial amalgamations of facts and values which can be consolidated into inclusive and robust decisions, or by further developing boundary objects provided by other actors in the space.

Figure 7.3 presents the configuration of institutions and processes involved in boundary work from a US perspective, based on the work of Jasanoff and Wynne (1998) presented in Fig. 7.2. It should be noted that the location of the different institutions is for illustration only. The exact locations would need to be determined based on extensive empirical research including a methodology for comparing the degree of politicisation in each organisation and the extent to which the work is influenced by the mode of representation. Nonetheless, some interesting questions can be asked based on this representation. The first is that of potential divergences

These examples illustrate the potential of a configuration approach to a more systemic study of boundary work in international sustainable development decision-making. Such an approach would combine a number of theories and methods to expand and complement the concept of boundary work.

7.5 Conclusion

This chapter has explored the challenges and constraints of applying the concept of boundary work to interactions between knowledge production and international sustainable development decision-making. The analysis finds that, while the concept of boundary work has proven useful as a case-study approach, it must be complemented in order to gain a more systemic view of international science-policy interactions. Several conceptual weaknesses must also be addressed, including its inability to capture the diversity of institutions involved in boundary work at the international level and the implications and constraints of the modes of representation and negotiation present in boundary work in the context of intergovernmental decision-making. The chapter suggests the development of a configuration approach which allows the positioning of institutions involved in boundary work with regard to their degree of politicisation and mode of representation. Such an approach would yield a more systemic understanding of boundary work for international sustainable development decision-making. In addition it could guide the development of theories and specific hypotheses on how the positioning of institutions influences the processes of boundary work taking place within them, as well as their behaviour towards other boundary institutions and organisations.

Acknowledgements This chapter greatly benefited from numerous discussions with my colleagues from the Earth Negotiations Bulletin in particular the comments by Pia Kohler, Alexandra Conliffe and Eugenia Recio. I am also indebteded to my colleagues at IASS for their critical comments and support. All remaining errors and inconsistencies are of course my fault alone.

References

Barrett SA (1999) A theory of full international cooperation. J Theor Polit 11(4):519–541
Cash DW, Clark WC, Alcock F, Dickson NM, Eckley N, Guston DH et al (2003) Knowledge systems for sustainable development. Proc Natl Acad Sci U S A 100(14):8086–8091
Fujimura JH (1992) Crafting science: standardized packages, boundary objects, and "translation". In: Pickering A (ed) Science as practice and culture. Chicago University Press, Chicago, pp 168–211

Gieryn TF (1983) Boundary-work and the demarcation of science from non-science: strains and interests in professional ideologies of scientists. Am Sociol Rev 48(6):781–795

Guston DH (1999) Stabilizing the boundary between us politics and science. Soc Stud Sci 29(1):87–111

Hoppe R (2010) From "knowledge use" towards "boundary work": sketch of an emerging new agenda for inquiry into science-policy interaction. In: In 't Veld RJ (ed) Knowledge democracy: consequences for science, politics and media. Springer, Heidelberg, pp 169–185

Jasanoff S (1987) Contested boundaries in policy-relevant science. Soc Stud Sci 17(2):195–230

Jasanoff S (1990) The fifth branch: science advisers as policy makers. Harvard University Press, Cambridge, MA

Jasanoff S, Wynne B (1998) Science and decisionmaking. In: Rayner S, Malone EL (eds) Human choice and climate change. Batelle Press, Columbus, pp 1–87

Kohler PM, Conliffe A, Jungcurt S, Gutierrez M, Yamineva Y (2011, in press) Informing policy: science and knowledge in global environmental agreements. In: Chasek PS, Wagner LM (eds) The roads from Rio: lessons learned from twenty years of multilateral environmental negotiations. RFF Press, an imprint of Taylor and Francis, New York

Merton RK (1973) The normative structure of science. In: Merton RK, King R (eds) The sociology of science: theoretical and empirical investigations. University Press, Chicago

Miller C (2001) Hybrid management: boundary organizations, science policy, and environmental governance in the climate regime. Sci Technol Hum Val 26(4):478–500

Mitchell RB, Clark WC, Cash DW, Dickson NM (2006) Global environmental assessments: information and influence. MIT Press, Cambridge

Star SL, Griesemer JR (1989) Institutional ecology, 'translations' and boundary objects: amateurs and professionals in Berkeley's museum of vertebrate zoology, 1907–39. Soc Stud Sci 19(3):387–420

Weinberg A (1972) Science and trans-science. Minerva 10(2):209–222

Chapter 8
Transgovernance: The Quest for Governance of Sustainable Development

Roeland Jaap in 't Veld

Abstract In this chapter, the Summary and Recommendations are included of the first report of the TransGov project of IASS, Potsdam, authored by Roeland J. in 't Veld. For this report the contributions to this volume were used as source of inspiration.[1]

8.1 Summary: Rethinking Sustainability Governance

8.1.1 Points of Departure

This report aims for innovation by adopting and amalgamating advanced insights in order to add value to the debate on the governance of sustainable development. We adapt a specific view on the present patterns of evolution of the world using the term *knowledge democracy* (in 't Veld 2010a). We interpret the recently developed theories on transitions and transformations with respect to governance, and accept thinking on *second modernity* (Beck 1992) as a background idea. Moreover, we concentrate on dynamics, because the term development necessitates a dynamic view, and because each societal phenomenon or system is simultaneously influenced by endogenous and exogenous dynamics. Furthermore, we add ideas from *reflexivity theory*, *configuration theory* and *governance theory*. We will argue that the proposed combination of these advanced concepts leads to a new approach of sustainability governance which we call *transgovernance* (Fig. 8.1).

---

[1] The full final report can be downloaded as open source publication at http://www.iass-potsdam.de/fileadmin/user_upload/documents/Transgovernance_-_The_Quest_-_Nov_2011.pdf.

R.J. in 't Veld
Waterbieskreek 40 2353 JH Leiderdorp, Netherlands
e-mail: roelintveld@hotmail.com

L. Meuleman (ed.), *Transgovernance*,
DOI 10.1007/978-3-642-28009-2_8, © The Author(s) 2013

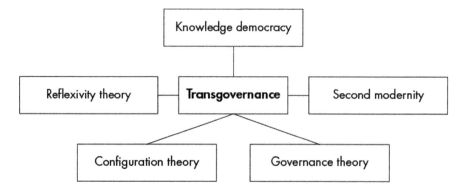

Fig. 8.1 Combination of theories and concepts leading to transgovernance

8.1.1.1 Knowledge Democracy

We refer to the evolutionary pattern of democracy as knowledge democracy because the interactions between politics, media and science have adapted a new shape with far reaching consequences, in many nations, regions and localities and on a global level. Representative democracy, as the dominant concept, appears to be in decay. Its ability to govern the present complex problems is met with wide spread scepticism. The mediatisation of both politics and science has changed the character of both, but also their interaction. As a consequence, the problem-solving potential of societies is affected.

The Curse of Success?

During the last decade, an influential debate has been conducted on the 'knowledge-based economy'. This concept has even become the main policy objective of the European Union, the Lisbon Strategy. However, there are signs that the strength of the argument for the knowledge-based economy is weakening rapidly.

The current worldwide economic crisis leads to new, very challenging questions. These questions refer mainly to the institutional frameworks of today's societies. It is therefore time for a transition to a new concept which concentrates on institutional and functional innovation. As the industrial economy has been combined with mass democracy through universal suffrage and later by the rise of mass media, one might suggest that the logical successor of knowledge economy is a new type of governance context, which has been called *knowledge democracy* (in 't Veld 2010) (Fig. 8.2). Knowledge democracy is an emerging concept with political, ideological and persuasive meaning. The relations between politics, science and media in the twentieth century, the corners in the triangle, are prone to profound change, indicated in second-order relationships (Fig. 8.3):

- The bottom-up media do not only supplement the classical media, but also compete with them.

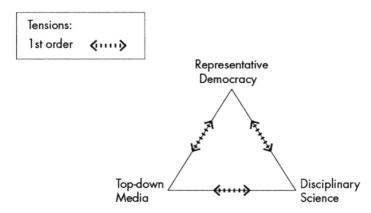

Fig. 8.2 Twentieth century relationships between politics, science and media

- Participatory democracy is complementary to representative democracy but is also considered as a threat to the latter.
- Transdisciplinary design or research is not only a bridge between classical science and the real world but also produces deviant knowledge and insights.

As a consequence we are confronted with tensions, threats and opportunities which are indicated in third-order relationships, also shown in Fig. 8.3. The tensions are those we find in second modernity. Society is enriched by the extensions of the corners of the triangles but it has to cope with the tensions. The first- and second-order tensions do not disappear in a knowledge democracy but do change character in the presence of third-order tensions. With regards to empirical research on this matter, comprehensive studies have not yet been conducted.

As we may observe, the outer points of the extended triangle also strengthen and stimulate each other. Transdisciplinarity nears participatory democracy, and social media play crucial roles in large scale communication processes. With this, the tensions relate mainly to the inside-outside relations in the triangle while the stimuli relate to the outer point of the corners. Moreover, we might observe relations between each inner and each outer corner (Fig. 8.4).

This has far reaching consequences for the governance of sustainable development in knowledge democracies. We can combine other insights here. The concept of change from within (*intraventions*, see Sect. 8.1.4 [in this chapter]) is brought into practice both in transdisciplinarity and in participatory democracy. Social change is designed or brought about here bottom-up, out of deliberations between individuals who are concerned.

The fruitful development of relationships between science and policy making has been characterised by co-evolution, but as we shall see the conditions for that are not always met. Indeed, even less than before, the so-called wicked problems which require a 'dealing with' approach rather than an approach which defines simple solutions, dominate political and corporate agendas. Knowledge democracy marks the transition of representative democracy to a more mixed political system in which more direct participation in decision-making by citizens and societal

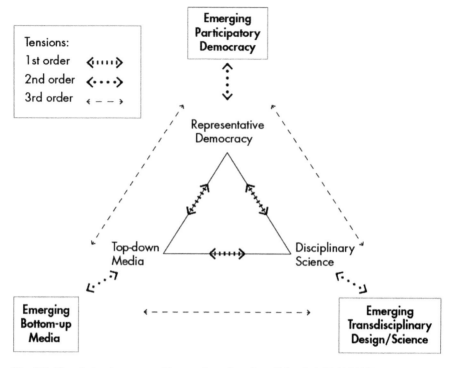

Fig. 8.3 Knowledge democracy: Three orders of tensions (After in 't Veld 2010)

groups is introduced. It also sees the appearance of social media as an alternative to the classical media, and the rise of transdisciplinarity to accompany the predominant disciplinary character of science. For the corporate community, knowledge democracy marks the transition of mere business cases (the business of business is business) to a responsible 'green economy' business case. This involves stakeholders, and public reporting, with a vision towards the future roadmaps of producing and consuming, and a sustainable corporate performance.

These developments cause new societal relationships between old and new institutional arrangements, which are full of tensions. They should neither be ignored nor can they be solved: they have to be dealt with and if possible made productive.

> I think it is the direction in which we all have to go. Whether you call it green economy or sustainable development, basically it is aimed at finding production and consumption patterns that are more in line with the natural limitations of the planet. They are unavoidable. They are a must. We are coming up to relatively short term turnaround points; we must take a U-turn in the next five decades. (Karl Falkenberg)[2]

---

[2] This is the first of a series of quotations taken from interviews with influential decision makers or experts, held for the TransGov project in May/June 2011.

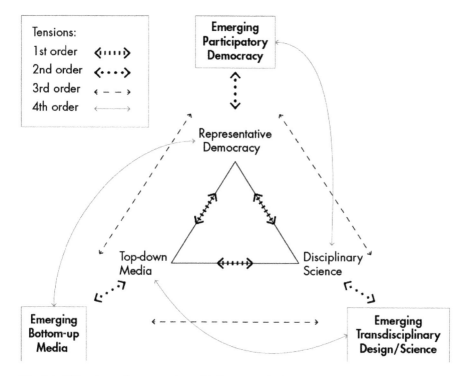

Fig. 8.4 Old and new forms co-exist and influence each other

8.1.1.2 Second Modernity: 'And' Instead of 'Or'

The second concept we embrace is the *second modernity* viewpoint (Beck 1992).[3] This notion states that today's societal evolution is characterised by the emergence of tense relationships between contradictory phenomena, by 'and' instead of 'or'. We accept the viewpoint of Ulrich Beck and others, that the specific character of the era we live in is no longer determined by the substitution of the former institution by a new one, but by the emerging tense coexistence of both. They need each other although there are controversies, and continuous tense relationships. Rosenau's (2005) definition of *fragmegration*, identifying sustainability both as fragmentation and integration, is a typical example of that character. Another instance of this is globalization, which on the one hand describes the simultaneous enlargement of scales of economics, of institutional arrangements and of thinking, whilst also arguing for local identities and intimacy. In order to properly understand the meaning of this observation we must digress on globalisation. This phenomenon, made possible by technological innovations, has led to unknown potentials to

---

[3] Beck's research focus is 'reflexive modernization' (1992), which explores the complexities and uncertainties of the process of transformation from 'first' to 'second' modernity.

influence economic and other developments elsewhere in a massive manner within a split-second by transactions on capital markets and others.

Knowledge democracy also has second modernity characteristics: representative democracy does not disappear because of the rise of participatory democracy. The classical media stay alive while social media grow, and disciplinary science goes on, while transdisciplinarity begins to flourish. The relationships however are full of tensions, and governance in the context of sustainable development will either be effective or ineffective depending on its ability to handle such tensions.

8.1.1.3 Techno-social Systems: Reflexivity

We have organised our worlds in order to master technologies, to produce goods and services according to human preferences, to enable people to pursue happiness, and to avoid as well as fight disagreeable actions and events. The patterns of organisation are immensely varied and interconnected.

People have organised themselves in stable social systems like tribes, villages, cities, regions and states, but can be observed also as flows of fugitives, masses, publics, crowds and other temporary shapes. Moreover, people live in a technological manner, that is, they are surrounded by applications of technologies in nearly every aspect of their activities, and themselves are increasingly becoming parts of technological systems. Moreover, people are (parts of) ecological-biological systems, or at least are surrounded by such systems.

All systems are due to change over time, but they evolve in very different ways. Some seem to change according to an S-curve, while others show tipping points. We may be able to analyse the change of ecological-biological systems with the support of natural sciences which lean heavily on regularities, often formulated as causalities. These regularities shape bodies of knowledge. This type of knowledge is accumulative in nature: our knowledge about stars nowadays is better than it was a century ago. Indeed, it can be utilised to forecast, to steer, and to develop.

Social systems however are functioning according to the way in which reflexivity, as we refer to it, operates. This concept is concerned with human competence to learn, and to adapt. This competence enables people to learn from any source, experience, practice, information, knowledge, theory, and so on, and to re-orientate behaviour subsequently. The inner logic of this learning process is unknown to any outside observer. As a consequence, the future behaviour of a social system in general cannot be forecast properly. It is doubtful whether knowledge regarding social systems can be characterised as accumulative: social systems will learn from any knowledge known to them. As a consequence, the knowledge may lose its validity. Knowledge on social systems is volatile in principle.

These considerations about the reflexive nature of social systems and interactions shed more light on one point addressed further (Sect. 8.4 [in this chapter]) under the rubric of configurations theory. Systems can often be influenced from outside. We call a purposeful attempt to influence a system from outside an *intervention* (or steering action). We call an attempt to influence a system from inside an *intravention*. The volatility of knowledge concerning social systems provides a major hindrance in attempts to formulate adequate outside policies for interventions pointing at change, because the knowledge base is not trustworthy as far as the functions and characteristics of social systems are concerned. Reflexivity, or in Giddens' (1991) terminology reflexive monitoring, leads to intraventions.

8.1.1.4 Configuration Theory and Intraventions

In order to grasp the way in which actions of a certain actor may influence other actors, we can build on configuration theory (e.g. Van Twist and Termeer 1991). This theory offers a profound insight into the essential aspects of organising and the specific approach of organisations. It helps us to develop a more satisfactory vision on multi-level governance. Organising, according to this theory, takes place via reflexive processes of argumentation and communication. These processes are taking place repeatedly and intensely between the members of a group. They gradually shape a common understanding, a common sense, a common frame, a common view on reality, and moreover a common idea of meaning within the group. We call the result of such processes a configuration. A configuration develops along two dimensions, the social and the cognitive dimension and thus truth claims emerge with regards to both substance and social relations.

As argumentation and communication decrease in intensity because of the internal consensus found, fixation begins. The configuration has grown up, but the danger of a standstill starts to grow. The disappearance of reflection creates stability but learning stops. Innovation becomes problematic. Inclusion and exclusion go hand in hand.

How can grown-up configurations still then innovate? Not by steering from outside, but also not primarily by impulses from the leader, the centre, because the centre is the centre due to social fixation – firm beliefs, vision, leadership, and so on. The centre, to a certain degree, could even be called the least plausible source of innovation.

People however live in different configurations: the peer group, the firm, the church, and so on. They are *multiply included* in several configurations. Multiple inclusion may be a 'burden', however, it also enables the multiply included actor to introduce ideas existing in configuration A and also in configuration B. He or she will be more credible in this role as he or she is engaged in both worlds and hence in a position to 'transfer' meaning. The fact that such an actor may be more often than not a marginal actor in both configurations may rather contribute to his or her capacity to bridge divides rather than hindering them. Configuration theory teaches us to abstain from naïve classical planning, steering or instructing, because the overwhelming majority of configurations live in the phase of fixation.

> We have to reform the existing institutions from within. That is a slow and gradual approach which requires leadership – and at the moment there is no leadership – but that is what we need to do. [...] The pressure to reform and strengthen existing international institutions is necessary, and needs to come from civil society too, with a call for reform

many cases try to find a satisfactory relationship with the surrounding nature. Their visible world is not abstract or systemic but specific and concrete. Entrepreneurs make attempts to design and apply more sustainable technologies. These are also specific.

Therefore, major discrepancies may exist between views on the systemic world on one hand and the daily life world on the other. In governance concepts both views are legitimate, and both should be taken care of. Transgovernance, in the context of sustainable development and transformations (plural), must also embrace the human view and must not restrict itself to the systemic view. Restricting governance notions to the latter might prohibit people and other societal actors from utilising their competences in order to change the path of development.

We are more aware of what sustainable development is than what it is not. We feel more comfortable with judgements on improvements of unsustainable technologies than with notions of optimal sustainability. In some theories on social integration, the core of social integration is understood as shared unvalues, more than values. Sharing unvalues, give recommendations as to what should not been done, and leave more space for variety than the necessity of consensus on necessary action. The analogy is clear: getting rid of unsustainable technologies leaves room for varied roads (and roadmaps) towards sustainability.

8.1.2.2 Values

Values are social and psychological concepts. They are rooted in cognition and emotion, and they can be informed by various sources, including insights. They concern the beautiful, the good, the true, and the trustworthy. Values urge for reflection, interventions and intraventions. Socialised values lead to norms that regulate human behaviour. People live values. Values that are lived, albeit in the shape of explicit norms, constitute culture. The specific culture of a certain social system is its identity. Cultures and identities may change over time. This change however takes place in a reflexive manner. Developments in accordance with values make sense.

Well-understood self-interest might lead to collective action which respects ecosystem services and social welfare, and may even produce collective goods. Egocentricity and free-rider behaviour however demand violence monopoly over a group in order to ensure sufficient collective goods production.

8.1.2.3 Cultural Diversity

Views on sustainable development vary with cultural backgrounds. How should we deal with cultural diversity in relation to sustainability, and in particular to the precautionary principle?

Culture is the production of meaning, and meaning relates to values. Without values there is no meaning, and no culture. Humankind has brought forward many varied cultures. In a certain normative orientation we experience cultural variety as richness. However, our basic attitude to cultural diversity is more critical than our attitude towards biodiversity. A society needs a certain cohesion, which is produced as a moral order, based on consensus on some fundamental values and norms. Indeed, culture within a society is also sharing some common substantial and relational values. A society consists of configurations. A configuration possesses a specific culture but as observed earlier, this leads to outside walls and thus tensions arise. In particular, the tensions between emerging identities on one side, accompanied necessarily by outer walls, and the need for cohesion and collective action on the other will never disappear. Shaping governance therefore, is walking a high wire.

We may conclude that biodiversity and cultural diversity are both components of sustainability. We may mourn the loss of a language somewhere on this planet as much as we may about the loss of a species. However, this does not represent our general insight. We do not believe that each culture is intrinsically good. On the contrary, some cultures are horrifying to many. As sustainability also implies the economic and social dimension, we realise that 'diversity always is a bedfellow of inequality' (Van Londen and De Ruijter 2011: 14). Inequality might be a threat to sustainable development and thus our attitude towards cultural diversity is ambiguous.

> I think that what is missing is a clear regional and culturally rooted process of development management. It is not the same to do something for the Arctic people as for people in El Salvador. Both have the same problems but have very different outcomes. (...) At the local level one of the key issues is to involve women, especially as they are directly related to survival, and especially in the very poor countries. The World Bank has understood that in the micro credit system they have a better return rate if they do it with women than with men. (Úrsula Oswald Spring)

According to *second modernity* it is probable that from the tense relations between emerging opposites, variety further increases. Striving for sustainable development urges us to take these tensions fully into account when dealing with governance. Governance is a relational concept. Hierarchy needs dependent subjects, network governance requires interdependency between partners, and market governance necessitates independent relationships.

Hence, it is fair to assume that different governance styles also reveal how people consider other people's values. Complex metagovernance combines the different archetypes, so that different patterns of relational values are also assembled. In system theory it is held that diversity promotes resilience, while uniformity breeds fragility. This may also be the case regarding cultural diversity. Diversity alone leads to chaos; what is probably needed is institutional redundancy, similar to redundancy in ecosystems.

Reflexivity is the strongest engine of social dynamics. It also relates to governance. The interaction of the general laws of diminishing effectiveness and of subsequent policy accumulation as indicated above, lead to crises which enable a phoenix to arise from the ashes, and to invent new governance arrangements. We are aware of the inevitability that government as a major component of governance will consciously destroy variety according to predominant substantial values, but also profoundly influence social relations and relational values. How the latter evaluate is due to reflexivity. We may better observe, with the support of the foregoing schemes, how these evolutions emerge. We will realise in shaping governance that tensions are not going to disappear but tend to intensify as governance solidifies. We understand that the precautionary principle sometimes demands the destruction of cultural variety. We know that biodiversity and cultural diversity have similarities but also major differences.

Governance of sustainable development is extremely complex as it must deal with all the tensions described above and their dynamics, while at the same time it is itself subject to reflexivity. Aiming at compatibility instead of assimilation appears to be a useful recipe.

> Putting all your eggs in one basket and relying on government seems dangerous, I think you have to find other ways to do this. Maybe social media will help here – I think the private sector can also be very helpful here, although they can also cause a backlash. So you have to try all of these things in the absence of strong government and of institutions that aren't that effective – you need a multidimensional, multi-track approach. (Eileen Claussen)

8.1.2.4 Planetary Boundaries

Recently a powerful new concept about global developments has been published: the idea about planetary boundaries. How to deal with the governance implications of this concept? The major difficulties that the concept causes are the following (Schmidt 2012):

- The boundaries are solely formulated in one of the three dimensions.
- The aggregate level of the truth claims seems to necessitate central decision-making.
- It remains unclear how to disaggregate the boundaries in order to create a frame of reference for other, de-central decision-makers.

Regarding the first cause, it is worthwhile, or maybe even necessary, to identify planetary boundaries in the other dimensions of sustainability, in order to restore equilibrium again. In economics for instance, the concept of a 'positional good' resembles the boundary concept. The core idea here is that the utility of certain goods and services decreases once the supply enables mass consumption. This decrease may be gradual, but the loss of sociability which Hirsch forecasts as a fatal consequence of the expansion of the relative share of positional goods in total consumption, might bear a tipping point character.

When dealing with cultural diversity we have already concluded that a minimum of social cohesion within a society is needed in order to produce the worthwhile public goods. This cohesion may be protected by the existence of a democratic nation-state, but the minimum condition is valid in other regimes too. With this in mind, loss of social cohesion as it is described in the literature on social capital, also leads to the awareness that we trespass a critical boundary if we lose too much cohesion, for instance either by intense individualisation or by the predominance of greed in economic affairs.

The third cause should be seen as challenging scientific excellence: The concept of co-evolution between decision-making and science must be focussed on this cause. Further research is required as well as think pieces which dig deep into the question of whether and how global boundaries would be derived from local and regional boundaries. Transgovernance (as a concept, a method, as a dialogue-style policy) is again the key here. Geopolitical stratification (the world of a nine billion population with emerging economies, and new alliances, a multipolar power system) will be in desperate need for this kind of – as we suggest calling it in line with our transgovernance concept – mosaic-style way of putting planetary boundaries together and making them useful for policies.

8.1.2.5 Dealing with Emergencies

Uncertainty prevails in long term decisions. The consciousness of threats or emergencies creates the sense of urgency which is often necessary to take decisions at all. As Bachmann (2012) points out, historically emergency response action has been one of the prime 'sources' of environmentalism. However, here the distinction between the two categories of long term problems is also decisive for the kind of action to be taken. If the objectives of actions to meet threats are formulated too roughly, like greening the economy or a change of less than two degrees in mean global temperature, it remains unclear which measures should be taken, and whether one should aim at resilience or at persistent interventions.

Adoption of the resilience approach might lead to delay of decision as the best approach, because in the case of a long lead time between action and effect we may delay as long as we respect the lead time.

The whole domain of sustainable development is filled with dangers, threats, risks, emergencies, and related phenomena, but also with options, opportunities, chances, beginnings and stories of success and progress. Often, environmental emergencies may serve in a lens-like way to clarify options and problems. In conventional governance systems – due to their focus on institutions and regulations – the 'sudden chance' and the unforeseen impact are frequently excluded.

In addition, here we should examine both sides of the coin: on the one hand these phenomena produce a sense of urgency, a momentum for action. This may be important and precious because many political systems in general are rather lethargic as the transaction costs of action appear high or are deliberately perceived as high even when, in fact, they are not higher than the costs of non-action.

On the other hand, hypes, momentum, and the like, are volatile: 'they do not keep longer than fish'. Additionally, the transaction costs of regaining momentum are often considerably higher. Indeed, unless the emergency is gradually converted in more fundamental components of value patterns and competences in knowledge and responsible action, the net result of an emergency as far as sustainable development is concerned might still be negative. This, again, is a field for transgovernance concepts which bring knowledge and action, responsibility and awareness, engagement and reasoning together. Letting options for transforming pass by unused is the worst result of a crisis or an emergency.

8.1.2.6 Transformations

Sustainable development is often described as a great transformation in Polanyi's (1944) terminology. Our insights into the nature of profound change are deepened by recognising the insights produced by the advanced transition/transformation theory – as developed, for example, by Grin et al. (2010). It deals with the multi-level and multi-scale evolution of technical and social systems utilising a multi-level approach along the distinction landscape-regime-niche. What happens in the niches is not altogether separated from regime changes, but the relationships are loose and complex.

We suggest using the term transformation in its plural form. In a world of high complexity and multifactor drivers of development it seems reasonable not to single down transformation into a one-size-fits-all approach. The notion of 'wicked problems' supports concepts for transformations that always include a variety of pathways and features. Furthermore, by using the singular, a large-scale perspective is often applied or suggested. Yet many if not most of transformative changes are taking place at a very small-scale level ranging from technological innovations in niche-markets to adjustments in individual behavioural patterns leading to pro-found changes if aggregated. Transgovernance is rather about finding and nurturing such small-scale transformative changes instead of neglecting them for the sake of large-scale systemic interventions.

8.1.2.7 Towards Transgovernance: Beyond Conventional Governance

How does sustainability governance look when we recognise the concepts of knowledge democracy and second modernity? The best answer might be that we do not need a new paradigm, a new orthodoxy, but should develop the sensitivity to look beyond governance conventions. This implies an approach beyond traditional forms of governance, beyond disciplinary scientific research, towards more transdisciplinarity; beyond borders formed by states and other institutions, towards trans-border approaches; beyond conventional means to measuring progress, towards new and more interactive measuring methods; beyond linear forms of innovation, towards open innovation; beyond cultural integration or assimilation, towards looking for compatibility. In other words, governance for sustainable transformations requires thinking beyond standardised governance recipes, towards a culturally sensitive metagovernance for sustainable development. The combination of these steps beyond familiar sustainability governance, we call *transgovernance*.

Transgovernance is an approach rather than a recipe. Using this approach, solutions may differ. We have suggested a number of these possible solutions, such as global innovation networks of governments and corporations, innovation tournaments for small and medium enterprises, nation states in a new role as process architect, and a new diplomacy for international agreements.

The challenges for sustainability governance leadership go beyond designing solutions. It is essential to have a long-term orientation, in order to understand the complexity of our time and to understand the lesson that changes of real-world configurations often come from inside (*intra*ventions). Leadership needs sustainability skills. The conventional hard skill/soft skill approach is being challenged.

We see today that individuals play a big role. There are a few leaders in their countries making a difference. I also think it cannot be just individuals. We need to make sure that all the things we talked about there is proper information, we organise structures, discussions we collectively set frameworks that behaviour is moving in a more knowledgeable, knowledge-based direction. We do need leaders. Leaders dependent on polling results are not what we need for the fundamental change (Karl Falkenberg).

8.2 Recommendations

Our Summary introduces several concepts which are crucial for rethinking sustainability governance: knowledge democracy, cultural diversity, planetary boundaries and reflexivity, as well as structural changes through emergencies. Below, examples are provided of possible consequences of using and linking these conceptual cornerstones. These insights are formulated as recommendations and are presented on ten sustainability governance themes:

- Developing societal networks that trespass the traditional boundaries of governance arrangements, involving private and public actors: 'co-decentral' arrangements.
- Conditions for better long-term decisions.
- A new diplomacy for international agreements.
- Conditions for a more transdisciplinary science system.
- Checks and balances in science communication.
- Upgrading the relevance of city initiatives.
- Nation states in a new role of process architect.
- Crowdsourcing and volatile publics.
- Creating space for new institutions, and allowing for old institutions to be phased out or to be transformed into new ones.
- Measuring progress through metrics which are to be found in dialogue-style search procedures.

8.2.1 New Private-Public Networks: Co-decentral Arrangements for Technological Evolution

Conventional governance respects boundaries between public and private actors. Hierarchy and regulatory power are reserved for public actors. Our insights into reflexivity bring the observation that many conventional arrangements are useless as far as fundamental change is concerned. In order to further this we need new, semi-horizontal relationships. We call these relationships co-decentral. It is possible to design a private-public network, consisting of corporations, citizen groups and scientific bodies, that will further sustainable technologies, while public bodies ensure a level playing field.

Technology and sustainable development have complex and crucial relationships. On one hand, the precautionary principle produces critical attitudes towards technological developments that may bring with them considerable risks and possibly produce irreversible and unfavourable effects. On the other hand, new technologies may enable humankind to take production in a far more sustainable direction. An important example is renewable energy.

The technological development in a number of domains lies mainly in the hands of large enterprises, but in other less mature developments multitudes of very small firms are responsible for innovations.

> Big business has a huge role – the Walmarts of this world – they have a huge possibility of putting demands down the whole demand chain, the whole structure. And by that – in combination with what politicians do, in combination with the right price structure, in combination with civil society and the awareness rising among citizens – they start to just do things differently to what they did only five years back. (Connie Hedegaard)

We design two institutional arrangements which cope with this diversity:

Proposal 1: A Global Sustainable Innovation Network

Most technology driven markets for consumer goods and services are worldwide oligopolies. Because of this a limited number of enterprises are in a leading position. Although they cooperate with universities and other scientific centres, they themselves provide the leadership for the direction in which the technological development moves. In many cases they operate in business to business chains with suppliers and subcontractors. Nowadays they report to the public at large about their general position towards sustainable development.

The employees in the higher ranks within large companies are – more than on the average – sensitive to sustainability issues. Within R&D departments, professionals develop value patterns which are often closely linked to those of important NGOs in the same domain. Therefore employers with a high sustainability profile are very attractive to conscious and competent professionals, and vice versa. Thus such a profile is rewarding in at least two relationships, with clients and with employees.

Public authorities may regulate broadly, in attempts to prohibit unsustainable developments or to further innovations, but they can hardly influence the paths of technological evolution chosen by large companies because governments neither sufficiently understand the most advanced elements of technologies nor the crucial trade-offs which entrepreneurs are confronted with. Moreover, in large parts of the world, public authorities cannot dispose of policy instruments which force entrepreneurs to select a specific critical path for their technological innovation.

> Sustainability is one of the main challenges for the decades ahead and the market will not produce sustainable outcomes – so then there is a major task for international institutions –

for international institutions, for national government, but also for local government to set standards and to issue laws within which and on the basis of which sustainability can advance. The market itself will not produce sustainability to the extent that is necessary. (Jan Pronk)

However, the competitors and subcontractors, and even remote enterprises which utilise either identical or related technology, in general have a far better understanding of these positions.

Generally speaking there are various roads towards more sustainable technologies. Competitors and scientific partners can make reasonable judgements with regards to the direction which a certain company chooses.

Consumers, clients – also being citizens – are increasingly sensitive in the long run to matters of sustainable development. They organise themselves in numerous ways. These consumer organisations could be powerful allies in the combat for sustainable development.

We need a regulatory framework in which individual companies function. We all want market economies, but we all know that they don't work without rules. Environmental collateral damage needs to be taken into account. There are cost-producing damages that society is not capable of shouldering anymore. We have to stop polluting in the way we have so far, and there are only two ways of getting there: (1) regulate what emissions are acceptable, and (2) put a price in order to incentivise innovation, in order to better accommodate the limits of the planet. (Karl Falkenberg)

If we consider the aforementioned chains, networks and other relevant relationships as a potential landscape for the evolution of governance, we might envisage the following scenario, which is of course not a blueprint:

- Public authorities may design a regulatory regime which ensures level playing fields for enterprises that strive for sustainable technological evolution. That means among other things the following: the competitive advantage that is collected by entrepreneurs utilising a less sustainable technology should be considered as false competition. The public market regulators could be enabled to burden these entrepreneurs with fines, or peculiar taxes.
- The 250 largest companies in the world will set up a co-decentral network in order to make judgements regarding the preferable patterns of technological evolution in many different sectors. They will promote the erection of networks within each sector which encourage the empathic cooperation of suppliers, manufacturers and subcontractors in sustainable directions. The (global) network will provide a system of communication that produces possibilities for naming, faming and blaming.
- The existing national and international competition authorities spend the income they collect on fining to fund prizes and rewards for excellent entrepreneurial performances in sustainable solutions.
- The network is connected with communities of clients and NGOs who contribute to dialogues and the collection of information on entrepreneurial practices. Crowd sourcing is not only used in order to detect data on facts, but is also

have reached the end of their policy life cycle. Long-term decision-making may require policy mechanisms that prolong the policy lifecycle of policy issues.
- It is also important to be transparent and realistic about the limitations of decision support systems, and to ensure that ethical and political assumptions in decision support systems are chosen in the political arena.
- The knowledge basis for long-term decisions requires a comprehensive approach. Knowledge production for long-term decision-making should be a combination of future orientation, design and research (F-ODR[4]) bearing many elements of transdisciplinarity. This demands different process requirements than the requirements for 'normal research' and conventional 'future-oriented research'. Participation of actors is one of the key requirements.
- Investing in increasing the long-term oriented values of citizens may make long-term decision-making more politically feasible: it will be less risky in terms of losing support from voters.
- The consequences of using the wrong 'best practices' in long-term decision-making processes may be even more damaging then in short-term decisions. Instead of copying 'best practices' it is better to translate them into a form which works in a specific situation, tradition and culture. The crucial question is: What works where and why?

> Whether we like it or not, we are locked into each other going forward in a way were not in the past. When we look at these partnerships, there is the question of the role of civil society. I see civil society as the supplier of trust for these solutions. Even if we are in agreement in government and business about what should be done, none of us enjoy a high degree of trust. So we need cooperations with civil society to provide trust for the solutions and to gain political acceptance of some of the solutions going forward. (Björn Stigson)

8.2.3   A New Diplomacy for International Agreements

Until recently, international agreements have played a major role in the furthering of sustainable development. It seems, however, that the past years have hardly shown any further progress.

> The speed by which climate agreements are reached at is determined by the slowest player. For that reason I think that measures at the national level also have to take place in parallel to these international agreements for us to make progress. (Bärbel Dieckmann)

Widespread dissatisfaction on the effectiveness of many treaties and other international agreements is one explanation for the stagnation. Our second possible explanation is that the reflexivity on behalf of the younger nation-states as to the predominant approaches, concepts, methods and instruments which are put into practice in international relations has founded the sentiment of being victims of hegemony.

---

[4] See Meuleman and in 't Veld (2009).

There is this discussion if we should, every time we have a new convention, create a new institution around it. For biodiversity, for Montreal, for climate, for whatever...The tricky thing is: if we spend a lot of time fighting over these institutional things, while we really need to get some action done, how do we balance these things? ... I think that what will bring us most is a structure that supports the mainstreaming and [does] not isolate. (Connie Hedegaard)

With this in mind, the call for institutional but also cultural variety in governance is increasing. Indeed, the attempt at agreeing on percentages of reduction of emissions must resemble a postcolonial hegemonic gesture for those former colonies which had earlier experienced a delay in economic development and are only now seeing their economic growth percentages increase. This has produced a lot of resistance to continuation of the routines leading to yet another binding treaty. The second modernity viewpoint does not allow the recommendation that from now on we should abstain from efforts on the global stage to reach agreements, but that they need to be modified considerably in the following directions:

- Because we have to deal with wicked problems, the complexity of solutions should match the complexity of the problems, as Hoogeveen and Verkooijen (2010) rightly argue. This is because such complexity may be better met by a variety of arrangements working towards a common goal rather than a mono-lithic, holistic arrangement which tries to capture every aspect of it itself.
- Each party has to realise that cultural variety does not only relate to the substance of sustainable development but also to the scope, shape and instruments of binding arrangements themselves; also with respect to these components fear of hegemony might cause stagnation.
- If on a global scale the differences are too considerable in order to reach unanimous agreements, it might be wise to concentrate on regional agreements which would unite a number of more homogenous countries. These differences may be between actors, which includes culture variety, differences in their stages of 'development', differences in power, or belongings to powerful sub-groups such as the EU or G77/China.
- Each international agreement must be accompanied by efforts of nation-states to bring about national and sub-national complementary and synergetic additional arrangements.
- A new diplomacy is needed, because the variety of relevant actors has increased, and because the complexity exceeds the competences of traditional diplomats. In addition, here transdisciplinary trajectories are indispensable, leading both to cooperation between policy-makers and scientists, as well as between policy-makers and stakeholders.
- A single treaty, a single instrument is in many cases inferior to a portfolio approach, if the portfolio successfully arranges for a level playing field.
- Under certain conditions, voluntary agreements with a strong moral appeal, accompanied by effective naming, blaming and faming mechanisms, might be at least equivalent to legally binding agreements.

8.2.4 The Organisation of the Scientific System

> One thing that troubles or occupies me greatly is how one can have uncontested knowledge
> and information – and yet not act upon it. (Bärbel Dieckmann)

Has science lost public authority? If so, than the support for action perspectives
based upon knowledge has lost its legitimacy. Maybe it is too easy to argue that
public authority as such has disappeared in any societal domain to a considerable
degree. Some specific explanations are offered here.

8.2.4.1 Science and Media

The first explanation is primarily concerned with the manner in which scientists
often behave while appearing in the mass media. Modern science has developed
mainly evolutionary patterns of specialisation into disciplines. Disciplines deal with
an aspect of the world: economics studies choice under scarcity, astronomy studies
the physical and chemical aspects of the universe, and so on. As a consequence, the
main product of scientific activity, namely knowledge, is formulated in terms of
regularities concerning relations between independent and dependent variables
under the condition *ceteris paribus*.[5]

> All facts have only a value if they can stand the criticism. So you need validation. The
> IPCC, which is a huge validation machine and the fact all these researchers wherever they
> come from talk to each other, and argue, you know it is quite expensive in terms of
> investment but that needs to be done. (Jos Delbeke)

The validity claim is formulated within the specific methodological constraints
agreed upon within the discipline. The methodology serves as an internal tool for
communication, but also as a device in order to immunise against outside criticism.
Contradictory viewpoints may arise, and are even normal, but will be analysed
according to the methodological rules of the game. Among many scientists it is *in
confesso*,[6] that the roots of scientific knowledge are hypothetical in nature.

Scientific disciplines have outer walls. Representatives of different disciplines
may communicate but they will experience language problems. Specific words
have specific meanings within a specific discipline. In the political realm however
societal problems are dealt with. They never bear a monodisciplinary character and
thus monodisciplinary knowledge is never immediately applicable in the solution of
a real world problem. Therefore it has to be amalgamated with other scientific
insights, and moreover with value judgements.

If a scientist responds to the invitation to present scientific insights to a broader
public, he is tempted to leave out all of the complicating remarks about the

---

[5] Latin: 'All other things being equal or held constant'.
[6] Latin: 'Acknowledged'.

methodological constraints under which the insight has been formulated. Journalists do not like such considerations. Moreover it is often assumed that the scientist's viewpoint is immediately relevant in relation to the solution of societal problems. Indeed, the scientist is systematically invited to publically exaggerate the unconditional character of the truth claim of his insights. In the scientific world he would make himself vulnerable or even ridiculous by doing so, but in the media realm this behaviour is a condition for survival as a commentator. Contradictory viewpoints then become conflicting truth claims, and even real world controversies. The scientist has entered the world of politics.

Politics is a power game. In politics all weapons are admissible. One of the popular techniques in politics while dealing with wicked problems is to play two-level-games: the fight on the level of substance is supplemented with an additional fight on the truthfulness of the different knowledge sources. In this manner politicians become interested in blaming the quality of the knowledge producers who support the hostile viewpoint. This of course results in a decrease of the public authority of science.

8.2.4.2 Science and Politics: Transdisciplinarity

The second explanation concerns the way in which the scientific system relates to the other actors in the political realm. As explained above, the satisfactory management of so called wicked problems – that nowadays dominate political agendas – demands transdisciplinary trajectories. Sustainable development is the prime wicked problem on this globe. Orthodox scientists hesitate to participate in these exercises, because they hate to move outside of their comfort zones.

The scientific system is organised in such a way that monodisciplinary products earn the highest prestige. Transdisciplinarity is the trajectory performed by scientists and policymakers together in order to develop robust action perspectives by amalgamating scientific and normative political viewpoints. Transdisciplinarity is seldom punished because the participant in the aforementioned trajectories will easily step on hostile political toes. In addition, politicians decide on the allocation of many resources for science.

In some European nation-states we have even observed recently that many interdisciplinary scientific institutes have disappeared. Moreover, many boundary work organisations which have built bridges between science and politics have been abolished.

According to principles of second modernity, the organisation of the scientific system following distinctions in scientific disciplines should not disappear but be supplemented with constructions – not necessarily permanent ones – that could further transdisciplinarity. With this in mind, reorganising the scientific system in the direction of positive incentives for participation in transdisciplinarity is a necessary condition for better fits between science and politics in relation to sustainable development. A number of splendid examples exist which could be multiplied. Jungcurt (2012) suggests complementing the concept of boundary work

Habitat, 2011. Cities appear to be able to develop private-public partnerships in this
domain easier and quicker than national governments.

Cities tend to learn from each other faster than many other actors. Sustainable
cities are attractive cities and attractive cities are strong cities. Strong cities can be
selective with regards to the access granted to new enterprises. Prioritising sustainable new firms will make accumulative progress possible.

> I would say that for challenges on a global level, the bottom-up is still important and
> needed. The local or city level will agree on policy because it is an easier landscape of
> actors. We see that cities are driving things much more than countries, and countries more
> than international institutions and agreements. In light of the disillusionment with interna-
> tional processes, that local level is what you have to set your hopes on. [...] Activities at
> that level can help us really move towards sustainability – quickly. (Sören Buttkereit)

City democracy adapts more easily than other public bodies to the new potential
of participatory democracy. Moreover cities, when compared to others, may better
recognise the niche players who bring real innovation and try to connect these to
related actors and 'regime' decision makers. Glocalisation is also related to cities.
A strong movement is developing that urges food producers to be nearby. Regional
and local food gain in popularity and moreover metropolitan agriculture is a
winning concept.

It would be a quiet revolution if national governments would be able to redefine
their positions towards cities in such a way that they would feel responsible for the
optimisation of the constraints under which cities could strive for sustainable
development, instead of trying to prescribe to cities how to act. A striking analogy
could be found with the position of nation-states in the domain of fair competition
aiming at the provision of level playing fields.

8.2.7   National Governments in Transition

Although nation-states are embedded in trans-, multi-, inter- and supra-national
networks, they also still possess a considerable amount of power and discretionary
space themselves. They will not disappear as relevant actors, but their functions and
duties are complicating: they can no longer behave as the authorities which simply
decide either to regulate an aspect of life themselves or to contribute in an international global environment the willingness to close binding treaties which will settle
things on a global scale.

The reflexive nation-state will continuously reveal combinations of substantial
and relational values that guide the choices as to the metagovernance of sustainable
development. These choices concern:

- Where to rely on existing/emerging markets;
- Where and how to encourage or regulate private-public partnerships that concern aspects of sustainability;

- How to improve the implementation of existing international environmental treaties, and how to deal with expiring global environmental treaties, as well as where to support new initiatives;
- Where and when to create or close transnational or regional agreements;
- Where and when to stimulate local internal public programmes;
- How to produce a brand of representative and participatory democracy in decision-making;
- How to build transdisciplinary trajectories towards decisions;
- When and where to utilise crowd sourcing and involvement of publics.

The choices are interrelated: once you leave a matter of concern to a private-public partnership you cannot at the same time regulate it one sided in any legal text. With this in mind, the governance arrangements are partially substitutes, but as we will see below they are also complementary, and reinforce each other. The argumentation that should be constructed has at least the following building stones:

- How close will the result of a certain arrangement be to the defined optimum?
- How large is the probability of success in the preparation of a decision?
- How large is the probability of successful implementation of the decision?
- How large are the transaction costs of action and how large are the costs of non-action?
- How synergetic will a certain arrangement function in relation with others?
- Most importantly, who is legitimised to pass judgement on all of this, in particular in transgovernance setups?

Accepting second modernity fully one has to argue that the effectiveness of global institutions is furthered by the simultaneous existence of local and regional institutions. This demands a well thought out division of scarce attention. If agreements between neighbours are generally more effective, the streamlining through a global organisation only would even be harmful.

Indeed, the complexity of the position of nation-states is illustrated by this: reasoning in second modernity terms they will continuously ask themselves how a certain arrangement on a certain level, for instance a global treaty, should be accompanied by arrangements on other levels in order to produce synergies. They will accept the need for complementarities. Although the world has become more polycentric than before, nation-states appear to be the natural process architects in order to both operate in a global landscape and combine the complementary efforts on different levels by a varied collection of actors.

> If you look for what could come out of Rio+20 [...] about sustainable development, in the best case you can have some agreements on a general goal, but the real action has to be done on the ground floor – at the level of states and local governments. And as you said of course it's also all about the individuals' behaviour. If each of us uses electric lights or other electric machines – normally we use them because this is what all people need and do. So changing behaviour will be a big step. Just because we still think that what 'I' do will not really affect much or anything. (Staffan Nilsson)

8.2.8   Crowds/Publics/Social Tipping Points

The world has become connected, flat, spiky and lateral. Traditionally we speak about levels of governance, ordered by hierarchy, but this type of order is in disarray. The vertical order is not disintegrating altogether but lateral arrangements, enabled by the Internet and communication technology, could possibly mean that a local initiative becomes a global hype within a very short time. Our analysis of societies must therefore also take into account new shapes of social organisation with potential influence like crowds and publics.

The wisdom of crowds may prove to be doubtful as universally characteristic (see Barbara Tuchman's The March of Folly, 1984), but crowd sourcing is often effective. Of course it demands a thorough approach to define the objectives of the search, the nature and size of the crowd, and the method used to select the collected information. A crowd is not necessarily a random crowd. Expertise within the crowd is relevant.

> If you look now, we have spring; there are a lot of observations in the nature of birds, of animals, of the flora, of what is happening. And a government can never, never monitor this without the help of engaged people in organizations looking for the birds' life or walking in the forest reporting, to take just an example or two. So it is really in my view a bottom-up approach which is needed, both when we make and when we implement policies. (Staffan Nilsson)

'Publics' are even more difficult to approach. Publics are event related. As Basten (2010) argues, publics may gain political momentum, once there is an institutional void in the respect that the traditional democratic institutions fail to solve problems. However, it is also possible to utilise publics: the supporters of soccer clubs have convinced many local public authorities that it would be proper to subsidise professional soccer.

Each actor who is interested in sustainable development may attempt to activate the existing or emerging publics in that domain. With this, the repertoire of each actor is enriched but also complicated. The choice of the mix of approaches to apply is a matter of primary concern: the classical method of building alliances with the well-established actors like governments on different levels, or designing networks can be supplemented with crowd sourcing and the utilisation of publics. In some instances publics – for instance gathering on a large square – mark a social tipping point, and may gain so much political influence that regimes topple down, as can be seen once more in the spring of 2011. It appears that not only governing bodies but also and maybe in particular NGOs should reflect upon the opportunities offered by the potential meetings with crowds and publics.

8.2.9   New Institutions and Fading Away of Old Ones

> I don't think you have support for new institutions. Not at the moment. I certainly can't see the U.S. subscribing, and it's going to be a struggle to keep up our ability to work within the already existing ones. (Eileen Claussen)

8.2.9.1 Courts and Truth Committees

New institutions belong to the dreams of many structuralists in the dialogue. We have already discussed the continuous plea for a global decision-making body which would enable strong coordination. We have also raised doubts about the question of whether such a body would be able to cope with the existing cultural heterogeneity.

Some have formulated ideas on new institutions for conflict resolution. The erection of an international court is one of them.[7] Indeed, in 2002 a large international group of judges had already concluded that 'an independent judiciary and judicial process is vital for the implementation, development and enforcement of environmental law'. The idea of the Forum is that that the Court could impose sanctions such as declaratory relief, fines and sanctions of restoration and rehabilitation of damaged habitats. Not only states but also NGOs, corporations and citizens would have access to the Court. It appears inevitable however to agree on a treaty that would establish the Court. Every one shares the opinion that it would take quite some time to decide on such a treaty. It is improbable that all nation-states will become Signatory States, which would harm the universal character of the judiciary.

Meanwhile, there is room for other mechanisms of conflict resolution. As the long run future of sustainable development should be characterised by harmony, the installation of truth committees operating according to the South African example would maybe be preferable. The moral authority of such committees would not necessarily be inferior to that of the Courts.

8.2.9.2 Informal Communities

The rapid rise of the social media enables all kinds of new communities. Many of them will be quite volatile, like publics and crowds, but some might become stable and unfold actions, or even programmes. In an earlier paragraph we have designed a private-public network, consisting of corporations, citizen groups and scientific bodies, which will further sustainable technologies, while public bodies ensure a level playing field.

> We need an international level playing field for companies – otherwise they will only compete on the basis of cost reduction and not on the basis of sustainability. (Jan Pronk)

The level playing field is, however, not an undisputed concept. Level playing fields are more or less paradoxical because they define equality in conditions in order to enable market actors to cause inequality.

---

[7] See for instance www.earthsummit2012.org for the Stakeholder Forum published in February 2011: Environmental Institutions for the twenty-first century: An International Court for the Environment.

> There are no level playing fields. It is nice to say but it will never happen. When I was in business, I wanted a playing field that was supportive of what I was trying to do – not what others were trying to do. (Björn Stigson)

Building institutions is a slow process. Attempts at acceleration are dangerous. When we deal with long term problems we have already formulated a number of recipes: depending on the character of the problem either persistent or resilient action is needed. The gradual establishment of institutions demands persistency during a longer period of time. As we argued while dealing with configurations, gradual solidification both in the cognitive and in the social dimension takes place.

Such institutions might avoid the usable market failures, but maybe also the non-market failures which states inevitably reproduce. The existing actors should become aware of the possibly benign functioning of such new institutions and create spaces where initiatives could breed.

The dynamic conservatism and the resilience of unsustainable institutions are matters of concern for many observers. Some argue in favour of a crusade against such anomalies. In our approach we would not prepare for external interventions, but would instead aim at the possibility of intraventions, hollowing out such institutions from the inside. Implosion would be the ultimate success.

8.2.10   Governance Indicators and Assessments

Many people are fond of performance indicators. They clarify the details of the test which must be passed by accountable decision-makers. They create a transparent dialogue. They specify what it is all about. Alas however, the empirical results are often disappointing because:

*   The indicators apparently do not adequately reflect the values of the parties concerned.
*   Behavioural reactions and immunising strategies gradually devastate the meaning of the indicators.
*   The indicators appear insufficiently flexible, and so became obsolete.

The points mentioned above are only a few of the many explanations for failure. In reaction to the observation of failure some policy designers have returned to the world of principles, and have re-introduced principle based accountability as opposed to indicator or rule based accountability and supervision.

In earlier situations the indicators themselves are decided upon by the highest hierarchical actor. In a knowledge democracy the performance indicators (what counts?) would be decided in societal dialogues. Those would bear an iterative character. Learning experiences would be collected continuously. Relevant changes in values would become visible at the earliest possible moment.

To sustain these dialogues, periodical societal 'balance sheets' on aspects of sustainable development would be produced by knowledge brokers such as advisory councils, think tanks and planning bureaus, whereby progress or deterioration would be mentioned. Such balance sheets, sometimes using the metaphor of traffic lights, have already become more popular over the last few years.

Thermometers for the quality of democracy, in particular participatory democracy, could also be designed. Even very specific assessment on the evolution of the green arrows in our knowledge democracy scheme could take place. Timely renewal of all decision support mechanisms would be crucial.

8.2.11 Concluding Remarks

We have concentrated on governance, not on domains. By doing so, we do not suggest that the distinction in domains is irrelevant. Of course the situation with regards to forestry differs from the carbon emissions environment. Of course, a contingent approach is necessary for each domain. However, the interdependencies of all biosphere systems also demand overview and linkages.

We have hardly touched on the myth of urgency, of momentum, and of *opportunita*. Macchiavelli has already said a lot on the latter. It is the genius of leadership, or the collective intuition of communities which will be the decisive factor here.

8.2.12 Who Should Do WHAT and WHEN?

In open societies the reflection upon and creation of governance are a matter for all citizens, and many private and public organisations. In accordance with values and responsibilities each organisation will act in its own way. Firms will accept their responsibilities for fair markets and more sustainable technologies, while public actors will provide level playing fields, collective goods and redistribution in accordance with preferences on distributive justice. Everyone can accept a morally binding obligation, but the monopoly on creation of legally binding arrangements is in the hands of states. Complementary positions demand empathy as relational value all the time.

The complex interactive relationships which characterise transitions necessitate for each actor a high degree of consciousness on possible options for new combinations, and continuous learning capacity. In knowledge democracies, 'mindfulness' marks the competence to operate in cultural diversity, and to aim at compatibility and congruence of values and actions. Action perspectives have to be multi-fold.

Transdisciplinarity and participatory democracy contain the intraventions that enable change, transition, and transformation. As sustainable development should be rooted in adequate value patterns and frameworks of competences, the efforts of many should be directed towards learning processes that further these values. The value of setting up time tables and indicators is well understood if those are used a

In 't Veld RJ (ed) (2000/2009) Willingly and knowingly: the roles of knowledge about nature and the environment in policy processes. RMNO, The Hague

In 't Veld RJ (ed) (2001/2008) The rehabilitation of Cassandra: a methodological discourse on future research for environmental and spatial policy. WRR/RMNO/NRLO, The Hague

In 't Veld RJ (2009) Towards knowledge democracy. Consequences for science, politics and media. Paper for the international conference towards knowledge democracy, Leiden, 25–27 Aug

In 't Veld RJ (ed) (2010a) Towards knowledge democracy. Consequences for science, politics and media. Springer, Heidelberg

In 't Veld RJ (2010b) Kennisdemocratie. SDU, The Hague

In 't Veld RJ, Verhey AJM (2000/2009) Willingly and knowingly: about the relationship between values, knowledge production and use of knowledge in environmental policy. In: In 't Veld RJ (ed) Willingly and knowingly: the roles of knowledge about nature and environment in policy processes. RMNO, The Hague, pp 105–145

In 't Veld RJ, Maassen van den Brink H, Morin P, Van Rij V, Van der Veen H et al (eds) (2007) Horizon scan report 2007: towards a future oriented policy and knowledge agenda. COS, The Hague

Jasanoff S (1990) The fifth branch: advisers as policy makers. Harvard University Press, Cambridge

Jasanoff S (2003) Technologies of humility: citizen participation in governing science. Minerva 41:223–244

Jasanoff S (ed) (2004) States of knowledge: the co-production of science and social order. Routledge, London/New York

Jasanoff S (2005) Designs on nature: science and democracy in Europe and the United States. Princeton University Press, Princeton

Jungcurt S (2012) Taking boundary work seriously: towards a systematic approach to the analysis 3 of interactions between knowledge production and decision-making on sustainable development. Chapter 7 in the volume

Kickert WJM, Koppenjan JFM, Klijn EH (1997) Managing complex networks: strategies for the public sector. Sage, London

Lindblom CE, Cohen DK (1979) Usable knowledge: social science and social problem solving. Yale University Press, New Haven

Meuleman L (2008) Public management and the metagovernance of hierarchies, networks and markets. Springer, Heidelberg

Meuleman L (2010) The cultural dimension of metagovernance: why governance doctrines may fail. Public Organ Rev 10(1):49–70. doi:10.1007/s11115-009-0088-5

Meuleman L (2010b) Metagovernance of climate policies: moving towards more variation. Paper presented at the Unitar/Yale conference. Strengthening institutions to address climate change and advance a green economy, Yale University, New Haven, 17–19 Sept 2010

Meuleman L (2011) Metagoverning governance styles: broadening the public manager's action perspective. In: Torfing J, Triantafillou P (eds) Interactive policy making, metagovernance and democracy. ECPR Press, Colchester, pp 95–110

Meuleman L (2012a) Cultural diversity and sustainability metagovernance. In: Meuleman L (ed) Transgovernance: advancing sustainability governance. Springer, Heidelberg

Meuleman L (ed) (2012b) Transgovernance: advancing sustainability governance. Springer, Heidelberg

Meuleman L, In 't Veld RJ (2009) Sustainable development and the governance of long-term decisions. RMNO/EEAC, The Hague

Napolitano J (2012) Development, sustainability and international politics. In: Meuleman L (ed) Transgovernance: advancing sustainability governance. Springer, Heidelberg

Newman P, Kenworthy J (1999) Sustainability and cities: overcoming automobile dependence. Island Press, Washington, DC

Nowotny H, Scott P, Gibbons M (2002) Re-thinking science: knowledge and the public in an age of uncertainty. Polity Press, Cambridge, UK

Nussbaum MC (2006) Frontiers of justice. Belknap, Cambridge

Perez-Carmona A (2012) Growth: a discussion of the margins of economic and ecological thought. In: Meuleman L (ed) Transgovernance: advancing sustainability governance. Springer, Heidelberg

Petschow U, Rosenau J, Von Weizsaecker EU (2005) Governance and sustainability. Greenleaf, Sheffoeld

Peverelli P, Verduyn K (2010) Understanding the basic dynamics of organizing. Eburon, Delft

Pohl C, Hirsch Hadorn G (2007) Principles for designing transciplinary research, proposed by the Swiss academies of arts and sciences. Oekom, München

Pohl C, Hirsch Hadorn G (2008) Methodological challenges of transdisciplinary research. Natures Scoiences Sociétés 16(1):111–121

Polanyi K (1944) The great transformation: the political and economic origins of our time. Beacon Press, Boston

Pollitt C, Bouckaert G (2000) Public management reform: a comparative analysis. Oxford University Press, Oxford

Putnam RD (2002) Democracies in flux: the evolution of social capital in contemporary society. Oxford University Press, Oxford

Regeer B, Bunders JFG (2009) Knowledge co-creation: interaction between science and society. RMNO, The Hague

Regeer B, Mager S, Van Oorsouw Y (2011) Licence to grow. VU University Press, Amsterdam

Rockström J, Steffen W, Noone K, Persson Å, Chapin FS III, Lambin E et al (2009) Planetary boundaries: exploring the safe operating space for humanity. Ecol Soc 14(2):32

Rosenau J (2005) Stability, stasis, and change: a fragmegrating world. In the global century: globalization and national security, vol I. National Defense University Press, Washington, DC

Schmidt F (2012) Governing planetary boundaries – limiting or enabling conditions for transitions towards sustainability? In: Meuleman L (ed) Transgovernance: advancing sustainability governance. Springer, Heidelberg

Scholz RW (2011) Environmental literacy in science and society: from knowledge to decision. Cambridge University Press, Cambridge

Scholz RW, Stauffacher M (2007) Managing transition in clusters: area development negotiations as a tool for sustaining traditional industries in a Swiss prealpine region. Environ Plann A 39:2518–2539

Schumpeter J (1943) Capitalism, socialism and democracy. Allen and Unwin, Londen

Schwarz M, Elffers J (2011) Sustainism is the new modernism. DAP, New York. See also: www.sustainism.com

Selin H, Najam A (eds) (2011) Beyond Rio + 20: governance for a green economy. Boston University Press, Boston

Sen AK (1999) Development as freedom. Knopf, New York

Stakeholder Forum (2011) Environmental institutions for the 21st century: an international court for the environment. ICE Coalition. Retrieved from http://bit.ly/ooMlzs

Stern N (2006) Stern review on the economics of climate change. HM Treasury, London

Surowiecki J (2004) The wisdom of crowds: why the many are smarter than the few and how collective wisdom shapes business, economics, society and nations. Little Brown, London

Teisman GR, Van Buuren MW, Gerrits L (2009) Managing complex governance systems. Routledge, New York

Thompson M, Ellis RE, Wildavsky A (1990) Cultural theory. Westview Press, Boulder

Tuchman BW (1984) The March of folly: from Troy to Vietnam. Random House of Canada, Ontario

UN-DESA (2011) The transition to a green economy: benefits, challenges and risks from a sustainable development perspective. United Nations, New York

UNDP (2007) Governance for sustainable human development. New York, UNDP

UNEP (2011) Towards a green economy. UNEP, Nairobi

UNESCO (2001), Universal declaration on cultural diversity. Paris, 2 November 2001

United Nations Habitat (2011) Global report on human settlements 2011 – cities and climate change. Nairobi, Kenya

Van Londen S, De Ruijter A (2011) Sustainable diversity. In: Janssens M, Bechtoldt M, De Ruijter A, Pinelli D, Prarolo G, Stenius V (eds) The sustainability of cultural diversity. Edward Elgar, Cheltenham, pp 3–31

Van Twist MJ, Termeer CJ (1991) Introduction to configuration approach: a process theory for societal steering. In: In 't Veld RJ, Termeer CJAM, Schaap L, Van Twist MJW (eds) Autopoiesis and configuration theory: new approaches to societal steering. Kluwer Academic, Dordrecht, pp 19–30

Voß J, Bauknecht D, Kemp R (2006) Reflexive governance for sustainable development. Edward Elgar, Cheltenham

Walker B, Holling CS, Carpenter SR, Kinzig A (2004) Resilience, adaptability and transformability in social-ecological systems. Ecol Soc 9(2):5

Walter A, Helgenberger S, Wiek A, Scholz RW (2007) Measuring societal effects of transdisciplinary research projects: design and application of an evaluation method. Eval Program Plann 30:325–338

WBCSD (2010) Vision 2050: the new agenda for business. WBCSD, Geneva

WBGU (2011) Welt im Wandel: Gesellschaftsvertrag für eine Große Transformation. WBGU, Berlin

Weber L (1979) L'Analyse économique des dépenses publiques. Presses Universitaires de France, Paris

Webler T, Tuler S (2000) Fairness and competence in citizen participation: theoretical reflections from a case study. Adm Soc 32(5):566–595

Weick K (1995) Sensemaking in organisations. Sage, London

Weick KE, Sutcliffe KM (2001) Managing the unexpected: assuring high performance in an age of complexity. Jossey-Bass, San Francisco

Weinberg A (1972) Science and trans-science. Minerva 33:209–222

Weingart P (1999) Scientific expertise and political accountability: paradoxes of science in politics. Sci Public Policy 26:151–161

Wynne B (1991) Knowledges in context. Sci Technol Hum Value 16:111–121

Wynne B (1996) May the sheep safely graze? A reflexive view of the expert-lay knowledge divide. In: Lash S, Szerszynski B, Wynne B (eds) Risk, environment and modernity, towards a new ecology. Sage, London/Thousand Oaks/New Delhi

Wynne B (2006) Public engagement as a means of restoring public trust in science – hitting the notes, but missing the music? Community Genet 9:211–220

Wynne B (2007) Risky delusions: misunderstanding science and misperforming publics in the GE crops issue. In: Taylor I, Barrett K (eds) Genetic engineering: decision making under uncertainty. University of British Columbia Press, Vancouver

Yearley S (2000) Making systematic sense of public discontents with expert knowledge: two analytical approaches and a case study. Public Underst Sci 9:105–122

About the Authors

The TransGov Team

The first research project of the IASS cluster 'Global Contract for Sustainability', TransGov – Science for Sustainable TRANSformations: Towards Effective Governance – began in the summer of 2010. This cluster also works very closely with GeoGovernance collaborating with various partner institutes in the region. Further research projects are currently under development and will be presented on the IASS website www.iass-potsdam.de. The composition of the TransGov team was as follows:

Dr. Günther Bachmann studied landscape planning and ecology, and received his Ph.D. from the Technical University Berlin in 1985 with a thesis about soil functions. He was a researcher in environmental sciences at the Technical University Berlin until 1983, and then became a scientific assistant with the Federal Environmental Agency. With post doc grants he researched hazardous waste issues and the environmental transition policies. In 1992 he became a director and professor with the Federal Environmental Agency and took over responsibility for upcoming soil regulation in Germany. In 2001, Günther Bachmann started his work for the German Council for Sustainable Development, an advisory board to the Federal Government, since 2007 in the position of its General Secretary. He is networking sustainability solutions, both in the international level and through comparative instruments within the private sector. He publishes on environmental policies, energy and climate, and on sustainability issues. He is cooperating with the IASS, and was member of the IASS TransGov steering group.

Contact: guenther.bachmann@nachhaltigkeitsrat.de.

Prof. Dr. Roeland J. in 't Veld is professor at the Open University of the Netherlands and professor of Governance and Sustainability at the University of Tilburg. Moreover he chairs a number of national and international research programmes, and societal organisations. Roel in 't Veld has editorial responsibility

L. Meuleman (ed.), *Transgovernance*,
DOI 10.1007/978-3-642-28009-2, © The Author(s) 2013

for a wide range of publications, including works on process management and the Handbook on 'Corporate Governance' as well as Knowledge Democracy.

During the last 15 years Roel in 't Veld was Chair of the Advisory Council for Research on Spatial Planning, Nature and the Environment in the Hague and has held such positions as Director-General for Higher Education and Scientific Research at the Ministry of Education, Culture and Science, Secretary of State for Education and Science and Chairman of the Supervisory Board of the Dutch Railway Infraprovider, as well as IBM. He fulfilled duties as a professor at seven European universities. He served as an advisor of the World Bank, OECD and the Council of Europe. He was also Dean of the Netherlands School for Public Administration, Rector of SIOO, the Interuniversity Centre for Development in the field of Organisation and Change Management. Roeland in 't Veld was member of the IASS TransGov Steering group. More on www.roelintveld.nl.

Contact: roelintveld@hotmail.com.

Dr. Stefan Jungcurt has a Ph.D. in agricultural sciences from Humboldt University, Berlin. His Ph.D. research focused on institutional interplay in the international regulation of the conservation and use of plant genetic resources for food and agriculture. Stefan works as research associate at the Council of Canadian Academies, a not for profit corporation that provides independent, science-based assessments that inform public policy development in Canada. Stefan is also, writer, team leader and thematic expert for the International Institute for Sustainable Development (IISD) Reporting Services. He is reporting for the Earth Negotiations Bulletin at international negotiations in the areas of biodiversity, biosafety, forests, wetlands and food and agriculture. His work as thematic expert for the biodiversity policy and practice website (www.biodiversity-l.org) focuses on developments in international biotechnology research and policy. Stefan has also worked for IISD as project officer on capacity building for negotiators in the negotiations on reducing emissions from deforestation and forest degradation in developing countries (REDD). He has worked as research assistant and project associate on numerous projects in the areas of sustainable agriculture, genetic resources for food and agriculture, and linkages between international regulation on biodiversity conservation and other issue areas such as trade and climate change. He was research fellow at the IASS TransGov project.

Contact: sjungcurt@gmail.com.

Dr. Louis Meuleman was the director of the TransGov project. He has 30 years of public sector experience, serving as a policy-maker, project manager, head of unit, process manager and project director, on national, regional and international issues, mainly in the fields of environment, sustainable development and spatial (land use) planning. He works currently as seconded national expert in DG Environment of the European Commission. Until January 2010 he was Director of the Dutch Advisory Council for Research on Spatial Planning, Nature and Environment (RMNO) in The Hague. He was until May 2011 Chair of the Netherlands Association for Public Management (VOM), is senior fellow at the Center for Governance and

Sustainability of the University of Massachusetts, Boston, USA, and research fellow at the VU University, Amsterdam. He is the author of a.o. Public Management and the Metagovernance of Hierarchies, Networks and Markets (Ph.D. dissertation; Springer, 2008) and The Pegasus Principle: Reinventing a credible public sector (Lemma, 2003). See also www.louismeuleman.nl.

Contact: louismeuleman@hotmail.com.

Dr. Jamel Napolitano graduated summa cum laude in Sociology from the University of Naples with a dissertation in Development Studies. She completed her Ph.D. with a scholarship from the University of Bologna, undertaking research in the field of International Politics. While a Ph.D. student, she was invited as a visiting scholar at the School of Advanced International Studies at Johns Hopkins University in Washington DC to carry out her doctoral research on the topic of world hegemony. After she received her Ph.D. in Political Science in 2009, she had some teaching experiences at the undergraduate level both in International Relations and in History of Sociology. She has also been translating into Italian a monograph on development and globalisation and a number of chapters from a book on the 2008 American presidential elections. Finally, she has been involved as a scientific consultant in the field of EU-funded projects. Jamel Napolitano was research fellow at the IASS TransGov project.

Contact: jamel.napolitano@gmail.com.

MSc Alexander Perez-Carmona received his Diploma as agronomist at the National University of Colombia in 2000. In 2003 he participated in an investigation about the conditions for the entrance of Poland in the European Union from the perspective of agricultural, environmental and institutional economics. In 2003/2004 he was involved in a project in the Philippines helping to build an ecological network and investigating negative externalities arising from a mining megaproject in the island Palawan. In 2005 he received the master degree 'Sustainable Land Use' from the Humboldt University of Berlin with emphasis in the topics: power, environmental and institutional economics. In 2007 he started his Ph.D. The doctoral investigation addresses different economic and institutional perspectives of the environmental conflict labelled as the not-in-my-backyard phenomenon (NIMBY) arising in Colombia by the siting of landfill facilities. His intellectual interests lie in institutions from the 'old' tradition, and the sub-fields of collective action and game theory from New Institutional Economics; and economics/environment from the perspective of Ecological Economics. He is research fellow at the IASS TransGov project.

Contact: alexandrop@gmx.net.

Dr. Falk Schmidt is the Academic Officer in the Executive Office of IASS, Potsdam. He was one of the research fellows of the IASS TransGov project, too. Before joining IASS, he has been an Academic Officer at the Secretariat of the International Human Dimensions Programme on Global Environmental Change (IHDP) and head of its Science Management Unit from 2006 to 2010. Among other things, he acted as Officer-in-charge of the IHDP in 2009. He has a special expertise in water and governance research and he is involved in several international interdisciplinary research projects and initiatives in the realm of global (environmental) change. Furthermore, science-policy interaction has been a key feature of his work, both as a topic of social science research as well as a participant in international policy processes, for example within global water governance. He holds a M.A. in Practical Philosophy, Business and Law. He defended successfully his doctoral thesis at the Otto-Suhr-Institute of Political Science of the Free University Berlin, currently in the process to be published. In his doctoral thesis he proposes a new way of understanding current global water governance in the light of a renewed regime theory. In doing so, he addresses a significant research gap, i.e. the global level of water governance, related to one of the most important challenges nowadays: the global water crisis. However, his focus on the institutional dimensions present important insights for many issue areas and his methodological considerations aim, more generally, at clarifying the role of the social sciences within global change research.

Contact: falk.schmidt@iass-potsdam.de.

Prof. Dr. Klaus Töpfer is the founding Director and current Executive Director of the Institute for Advanced Sustainability Studies (IASS) based in Potsdam. He is also the former Executive Director of the United Nations Environment Programme (UNEP) based in Nairobi and Under-Secretary-General of the United Nations (1998–2006). He graduated from Mainz, Frankfurt and Munster in 1964 with a degree in Economics. From 1965 to 1971 he was a Research Assistant at the Central Institute for Spatial Research and Planning at the University of Münster, where he graduated in 1968 with a Ph.D. on 'Regional development and location decision.'

From 1971 to 1978 he was Head of Planning and Information in the State of Saarland, as well as a visiting Professor at the Academy of Administrative Sciences in Speyer. During this period he also served as a consultant on development policy on the following countries Egypt, Malawi, Brazil and Jordan. From 1978 to 1979 he was Professor and Director of the Institute for Spatial Research and Planning at the University of Hannover. In 1985 he was appointed by the University of Mainz Economics Faculty as an Honorary Professor. He has since 2007 been a Professor of Environment and Sustainable Development at Tongji University, Shanghai. He is also a visiting Professor at the Frank-Loeb Institute, University of Landau.

Klaus Töpfer is a member of the CDU party in Germany and has been since 1972. He is the Former minister for Environment and Health, Rheinland-Pfalz (1985–1987). He was Federal Minister for the Environment, Nature Conservation and Nuclear Safety from 1987 to 1994 and Federal Minister for Regional Planning, Housing and Urban Development from 1994 to 1998. He was also a member of the German Bundestag during the period 1990 to 1998. He has received numerous awards and honours, including in 1986, the Federal Cross of Merit and in 2008 the German Sustainability Award for his lifetime achievement in the field of sustainability. Klaus Töpfer was chairman of the IASS TransGov steering group.

Contact: klaus.toepfer@iass-potsdam.de.

Author Index

L. Meuleman (ed.), *Transgovernance*,
DOI 10.1007/978-3-642-28009-2, © The Author(s) 2013

Printed by Printforce, the Netherlands